MODELING AND COMPUTATION IN ENGINEERING II

PROCEEDINGS OF THE 2nd SREE CONFERENCE ON MODELING AND COMPUTATION IN ENGINEERING (CMCE 2013), HONG KONG, 22–23 JUNE 2013

Modeling and Computation in Engineering II

Editor

Liquan Xie

Department of Hydraulic Engineering, Tongji University, Shanghai, China

CRC Press
Taylor & Francis Group
Boca Raton London New York Leiden

CRC Press is an imprint of the
Taylor & Francis Group, an **informa** business

A BALKEMA BOOK

CRC Press/Balkema is an imprint of the Taylor & Francis Group, an informa business

Typeset by V Publishing Solutions Pvt Ltd., Chennai, India
Printed and bound in Great Britain by CPI Group (UK) Ltd, Croydon, CR0 4YY

Published by: CRC Press/Balkema
P.O. Box 11320, 2301 EH Leiden, The Netherlands
e-mail: Pub.NL@taylorandfrancis.com
www.crcpress.com – www.taylorandfrancis.com

ISBN: 978-1-138-00058-2 (Hbk)
ISBN: 978-1-315-88491-2 (eBook)

Modeling and Computation in Engineering II – Xie (ed)
© 2013 Taylor & Francis Group, London, ISBN 978-1-138-00058-2

Table of contents

Modeling and Computation in Engineering II – Xie (ed)
© 2013 Taylor & Francis Group, London, ISBN 978-1-138-00058-2

Preface

The 2nd SREE Conference on Modeling and Computation in Engineering (CMCE 2013) is held during 22–23 June 2013 in Hong Kong and hopes to offer researchers an occasion to exchange their experiences of modeling and simulation technology, which include modeling technology, simulation technology & tools, computation methods and their engineering applications. The 3rd SREE Workshop on Applied Mechanics and Civil Engineering (AMCE 2013) was held in CMCE 2013. This workshop is a forum for enhancing mutual understanding between scientists, engineers and experts in the civil, municipal and hydraulic engineering field, and rich contributions have been demonstrated from many experts to understand challenging mechanics in engineering.

50 technical papers are published in the proceedings. Each of the papers has been peer reviewed by recognized specialists and revised prior to acceptance for publication. This book review recent advances in several areas that are important for applied mechanics & civil engineering and modeling & simulation in engineering. The papers related to modeling and computation in engineering mainly address on advanced technology of numerical simulations to improve accuracy and engineering applications of modeling and computation. The papers related to applied mechanics and civil engineering mainly focus on design theories, construction science and advanced material applications in building structures, underground structures, bridge structures, hydraulic engineering, municipal engineering, port and coastal engineering, road and transportation engineering.

Last but not least, we would like to express our deep gratitude to all authors, reviewers for their excellent work, and Léon Bijnsdorp, Lukas Goosen and other editors from Taylor & Francis Group for their wonderful work.

Modeling and Computation in Engineering II – Xie (ed)
© 2013 Taylor & Francis Group, London, ISBN 978-1-138-00058-2

Sponsor

Sponsored by *Society for Resources, Environment and Engineering*

Modeling and Computation in Engineering II – Xie (ed)
© 2013 Taylor & Francis Group, London, ISBN 978-1-138-00058-2

Comparison of empirical calculations for venting of gas explosion with the experiment

M. Mynarz & M. Bárta
VSB—Technical University of Ostrava, Czech Republic

ABSTRACT: Empirical calculation methods leading to determination of blast load on building structures are verified in this paper. For determination of overpressure during explosion of gas or vapour of combustible liquids, various computational approaches exist but their corresponding to real values is not very clear and their utilization in practice is not demonstrated. The contribution presents the results of evaluation of 8 equations, pointing out the most universal formula, together with the most suitable formula for given geometry and type of flammable gas.

1 INTRODUCTION

Explosion venting belongs to the group of construction arrangements for explosion protection. This arrangement by itself could not prevent the explosion but instead danger effects are limited to reasonable size. If it is not possible to avoid the creation of explosive atmosphere and to eliminate explosion hazard with the help of active explosion prevention (or if these precautions are not suitable), then the objects should be designed in such a way that explosion effects are lowered to the safety margin. The arrangements ensure that damage of structure and exposure of people would be minimised.

In practice, most internal explosions are solved as venting explosions. At pressure increase in the object or room, so-called vent areas (e.g., windows, doors, partition walls etc.) are relieved at certain value. In the case the explosion is vented, final generated pressure is much lower. Many authors have been focused on the determination of this reduced pressure and lots of calculation formulas have been published. Chosen formulas are presented in this contribution.

2 TERMS AND DEFINITIONS

Table 1 shows used terms and definition. It also specifics single variables and constants used further in this text.

3 EQUATIONS ASSUMPTIONS

The measurements were executed at stoichiometric concentration of homogeneous methane-air mixture in cubic equipment with volume of 0.250 m^3. Table 2 introduces the constants used for calculations of reduced pressures.

Testing equipment was built as a reduced physical model of the object. In the envelope of the model, vent opening was installed in the middle of one side. Area of square-shaped opening was 0.040 m^2 and the diaphragm with average value of static activation pressure of

Table 1. Used terms and definitions.

Term	Definition
p_{red}	Reduced pressure
π_{red}	Dimensionless reduced pressure
p_{stat}	Static activation pressure
p_0	Initial pressure
A_s	Internal area of enclosure
A_v	Vent area
V	Volume of enclosure
E_0	Expansion ratio
c_0	Sound speed
χ	Turbulence factor
π_v	Dimensionless air-release parameter
π_0	Ludolf number
γ_b	Adiabatic coefficient of combustion
γ_u	Adiabatic compression coefficient
S_{u0}	Normal burning velocity
S_0	Laminar burning velocity
W	Weight of m^2 of vent area
S_{fl}	Burning of flame velocity
C_d	Coefficient of discharge (resistance)
K	Ventilation coefficient
A_{min}	Minimal size of envelope areas
A_{max}	Maximum size of envelope areas
d, f, g, h	Constants
S_0^-	Dimensionless parameter
A^-	Dimensionless parameter
B_r, B_{rt}	Bradley number
χ/μ	Number expressing relation between deflagration and turbulent flow
D	Diameter
C	Coefficient
A_{sv}	Area of surface with vent area

Table 2. The constants used for calculations of reduced pressures (Razus 2001).

Gas mixture	S_{u0} (m/s)	C_d (m/s)	c_0 (m/s)	E_0 (–)	γ_u (–)	γ_b (–)
Methane 9.5%	0.44	0.6	353	7.48	1.38	1.18

0.226 bar was used as a filling of the opening. Area of internal surface of the model was 2.418 m^2, area of one side with the opening was 0.393 m^2, weight of 1 m^2 of vent area was 0.092 kg, minimum size of envelope areas was 0.605 m and maximum size of envelope areas was 0.650 m. In the experiment enclosure, any obstructers for increase of the turbulence did not occur. Ignition of the mixture was supposed to be in the centre of the model (Mynarz 2010, 2012).

4 EQUATIONS USED FOR OVERPRESSURES CALCULATION

This chapter describes particular computational procedures for calculation of reduced explosion pressure, or more precisely blast load of the structure, by various sources and authors. For particular procedures, important notes or limitations that should be followed to obtain correct results are introduced.

Table 3. Computational procedures for calculation of reduced explosion pressure.

Procedure	Formula	Limitations
Runess (Runess 1972)	$p_{red} = 1{,}804 \cdot 10^{-4} \cdot \left[D^2 \cdot S_{fl} \cdot (E_0 - 1)\right]^2 \cdot A_V^{-2}$	
Bradley (formula No. 1) (Bradley 1978), (Bartknecht 1981)	$p_{red} = 2{,}43 \cdot \left(\frac{\overline{A}}{\overline{S_0}}\right)^{-0{,}6993} \quad for \quad p_{red} > 1\, bar\, g$ $p_{red} = 12{,}46 \cdot \left(\frac{\overline{A}}{\overline{S_0}}\right)^{-2} \quad for \quad p_{red} < 1\, bar\, g$ Calculation of parameters $\overline{A}$ and $\overline{S_0}$: $\overline{A} = \frac{C_d \cdot A_V}{A_S}$ and $\overline{S_0} = \frac{S_{u0}}{c_0} \cdot (E_0 - 1)$	
Bradley (formula No. 2) (Bradley 1978), (Simpson 1986)	$p_{red} = 4{,}82 \cdot p_{stat}^{0{,}375} \cdot \left(\frac{\overline{A}}{\overline{S_0}}\right)^{-1}$ For calculation of $\overline{A}$ and $\overline{S_0}$ see Bradley (formula No. 1)	
Cubbage and Simmonds (formula No. 1) (Cubbage 1955), (NFPA 68 1998)	$p_{red} = 0{,}365 \cdot \left(\frac{\overline{A}}{\overline{S_0}}\right)^{-1}$ For calculation of $\overline{A}$ and $\overline{S_0}$ see Bradley (formula No. 1)	
Rasbash (Rasbash 1969), (Bangash 2009)	$p_{red} = 0{,}15 \cdot p_{stat} \cdot 0{,}365 \cdot \left(\frac{\overline{A}}{\overline{S_0}}\right)^{-1}$ For calculation of $\overline{A}$ and $\overline{S_0}$ see Bradley (formula No. 1)	$0{,}2 < A_v/A_{sv} < 1$, $p_{stat} < 0{,}48$ bar g
Yao (Yao 1974)	$p_{red} = \left[\frac{0{,}375 \cdot \chi^{0{,}675} \cdot E_0^{7/6}}{E_0 - 1}\right]^2 \cdot \left(\frac{\overline{A}}{\overline{S_0}}\right)^{-2}$ For calculation of $\overline{A}$ and $\overline{S_0}$ see Bradley (formula No. 1)	
Molkov (formula No. 1) (Molkov 1998)	$\pi_{red} = \frac{p_{red} - p_0}{p_0} = 9{,}8 \cdot \left[\frac{B_r \cdot (E_0 - 1)}{(36 \cdot \pi_0)^{1/3} \cdot \sqrt{\gamma_u}} \cdot \frac{\mu}{\chi}\right]^{-2{,}4}$ Calculation of parameter B_r (Bradley number): $Br = \frac{A_V}{V^{2/3}} \cdot \frac{C_0}{S_{u0} \cdot \left(E_0 - \frac{1 - \frac{1}{\gamma_b}}{1 - \frac{1}{\gamma_u}}\right)}$	

(*Continued*)

Table 3. *Continued*

Procedure	Formula	Limitations
Molkov (formula No. 2) (Molkov 1999)	Calculation of parameter χ/μ: $$\frac{\chi}{\mu}=0{,}9\cdot\left[\frac{\left(1+10\cdot V^{1/3}\right)\cdot(1+0{,}5\cdot Br)}{1+\pi_v}\right]^{0{,}37}$$ Calculation of parameter π_v: $$\pi_v=\frac{p_{stat}}{p_0}$$ $$\pi_{red}=Br_t^{-2{,}4}\quad for\quad \pi_{red}<1$$ $$\pi_{red}=7-6\cdot Br_t^{0{,}5}\quad for\quad \pi_{red}>1$$ Calculation of parameter B_{rt}: $$Br_t=\frac{\sqrt{\frac{E_0}{\gamma_{u0}}}}{\sqrt[3]{36\cdot\pi_0}}\cdot Br\cdot\frac{\mu}{\chi}$$ For calculation of B_r, χ/μ and π_v see Molkov (formula No. 1)	

5 CALCULATED RESULTS

Only eight of presented twelve equations could be used for given conditions. Remaining equations did not satisfy mentioned limitations of particular formulas. Calculated values are shown in table 4.

Experiments were executed on testing equipment at atmospheric pressure of 0.995 bar and internal temperature of 12°C. For each of the concentration of methane-air mixture in the range of explosion limits LEL (lower explosion limit) and UEL (upper explosion limit), four experiments were realized. At stoichiometric concentration of the mixture, average value of reduced explosion pressure was 1.046 bar (Serafín 2012, 2013). This value was compared to values calculated according to formulas mentioned above.

Value of relative deviation δ between calculated and measured reduced pressures was determined as follows:

$$\delta=100\cdot\frac{\left(p_{calc}-p_{exp}\right)}{p_{exp}} \tag{1}$$

Figure 1 illustrates values of relative deviations of calculated and measured reduced pressures.

6 CONCLUSIONS

This contribution introduces calculation of reduced explosion pressure according to several methods and following comparison of calculated and experimentally found values. Result of comparison shows that calculation of reduced pressures according to the formulas is rather approximate. In the case of the second formula according to Bradley, deviation from experimentally measured value was almost 104%. The most precise result was determined according to Runes. Calculated value differed from measured value by less than 32%.

Table 4. Calculated values of reduced pressures.

Formula	Runes	Bradley 1	Bradley 2	Cubbage and simmonds 1
Reduced pressure (bar)	0.715	2.104	2.133	0.297
Formula	Rasbash	Yao	Molkov 1	Molkov 2
Reduced pressure (bar)	0.331	1.577	0.410	0.332

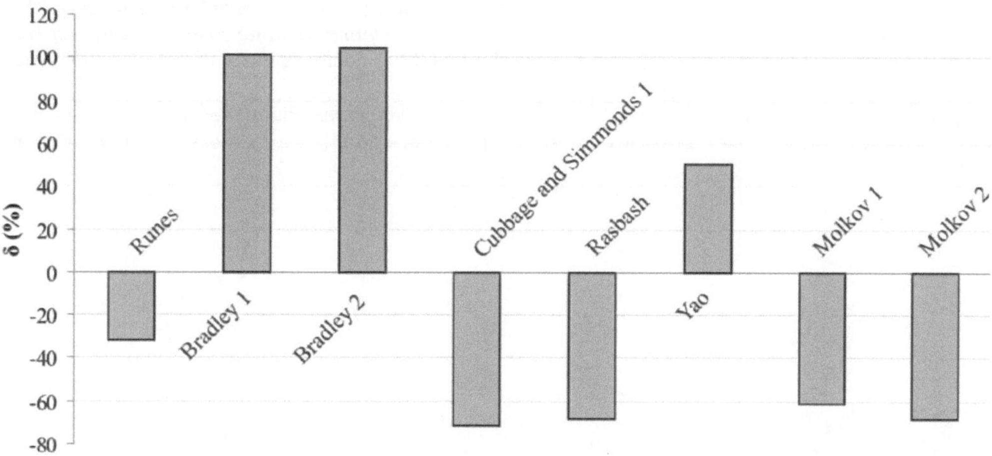

Figure 1. Values of relative deviations of calculated and measured reduced pressures.

It should be pointed that results of formula used in this paper can differ significantly. Venting explosion pressure depends on many factors that's why it could not be uniquely determined which of the formulas is the most precise. It can only be assessed which formula is the most precise for concrete model, on the basis of used combustible mixture, volume and geometry of equipment and size, shape, way and conditions of venting.

REFERENCES

Bangash, M.Y.H. 2009. *Shock, Impact and Explosions-Structural Analysis and Design.* Springer-Verlag Berlin and Heidelberg GmbH & Co. KG, p. 1410.

Bartknecht, W. 1981. *Explosions: course, prevention, protection.* New York: Springer.

Bradley, D. & Mitcheson, A. 1978. The venting of gaseous explosions in spherical vessels, I—Theory. *Combust Flame*, Nr. 32, pp. 221–236.

Cubbage, P.A. & Simmonds, W.A. 1955. An investigation of explosion reliefs for industrial drying ovens—I Top reliefs in box ovens. Trans. Inst. Gas Eng.

Molkov, V.V. 1999. Explosion safety engineering: NFPA 68 and improved vent sizing technology. *Proceedings of 8th International Conference on INTERFLAM'99, vol. 2, Edinburgh*, pp.1129–34.

Molkov, V.V., Dobashi, R., Suzuki, M. & Hirano, T. 1998. Venting of deflagrations: hydrocarbon-air and hydrogen-air systems. *In proceedings of the Colloquium on Gas, Vapor, Hybrid and Fuel-Air Explosions, The International Symposium on Hazards, Prevention and Mitigation of Industrial Explosion, Schaumburg. Ill. USA*, pp. 297–314.

Mynarz, M., Lepík, P. & Serafín, J. 2012. Experimental determination of deflagration explosion characteristics of methan-air mixture and their verification by advanced numerical simulation, *Twelfth international conference on Structures under Shock and Impact, Kos, Greece, WIT Transactions on The Built Environment,* Vol. 126, s. 169–178, ISBN: 978-1-84564-612-7, ISSN: 1746-4498 (print).

Mynarz, M. & Šimoník, Z. 2010. Assessment Methods for Explosion Load Parameters of Chosen Building Structure (in Czech). *Spektrum.* Ostrava, SPBI, pp. 215–217. ISBN: 978-80-7385-087-6.

NFPA 68. 1998. Guide for Venting of Deflagrations. National Fire Protection Association, Quincy, MA, USA.

Rasbash, D. 1969. The relief of gas and vapour explosions in domestic structures. Fire research note No. 759.

Razus, D.M. & Krause, U. 2001. Comparison of empirical and semi-empirical calculation methods for venting of gas explosion. *Fire safety journal.* vol. 36, issue 1, pp. 1–23. ISSN 0379-7112.

Runess, E. 1972. Explosion venting; *Loss Prevention,* Nr. 6, pp. 63–67.

Serafín, J., Bebčák, A., Bernatík, A., Lepík, P., Mynarz, M. & Pitt, M. 2013. The influence of air flow on maximum explosion characteristics of dust-air mixtures, *Journal of Loss Prevention in the Process Industries,* Volume 26, Issue 1, p. 209–214, ISSN: 0950-4230.

Serafin, J., Mynarz, M., Konderla, I., Bebčák, A. & Lepík, P. 2012. Study of Influence of Combustible Gas on Explosion Parameters of Black Coal Dust. In: *Chemical Engineering Transactions: PRES 2012, 15th Conference on Process Integration, Modelling and Optimisation for Energy Saving and Pollution Reduction.* First-edition 2012. Czech Republic: ČSCHI in colaboration with AIDIC, s. 955–960. Vol. 29. ISBN 978-88-95608-20-4 ISSN 1974-9791.

Simpson, L.L. 1986. *Equations for the VDI and Bartknecht Nomograms*, Plant/Oper Prog 5, pp. 49–51.

Yao, C. 1974. Explosion venting of low-strength equipment and structures, *Loss Prevention,* Nr. 8, pp. 1–9.

Modeling and Computation in Engineering II – Xie (ed)
© 2013 Taylor & Francis Group, London, ISBN 978-1-138-00058-2

Mechanical behavior of steel anchor beams on pylons of cable-stayed bridge under construction

Lu-Bing Song & Chong Wu
Tongji University, Shanghai, China

ABSTRACT: Steel anchor beam is a main kind of cable-pylon anchorage for cable-stayed bridge, and three different construction schemes are considered for it: Scheme 1, keep steel anchor beams simple supported on the brackets; Scheme 2, keep steel anchor beams simple supported on the brackets until initial tension is finished; Scheme 3, keep steel anchor beams fixed with brackets all the time. In this paper, the spatial finite element model of New Yalu River Bridge, a cable-stayed bridge, was established. Different construction schemes of steel anchor beams were proposed and the stress, as well as force distribution of anchorage zone under different construction schemes had been analyzed.

1 INTRODUCTION

As shown in Figure 1, the New Yalu River Bridge between China and the Democratic People's Republic of Korea is a cable-stayed bridge with span length of (82.7 + 229 + 636 + 229 + 82.7) m. Both of two 193.5 m high rectangular box pylons are made of concrete. The section of pylon at cable-pylon anchorage zone is 5.0 × 7.0 (m). The wall thickness of the pylon is 1.2 m and 1.0 m in longitudinal and transverse direction respectively (Figure 2). Layout of pre-stressed tendon is shown in Figure 4 and Figure 5, pre-force is 1041 kN for N1, and 465 kN for N2.

Steel anchor beams are installed on pylon for cable staying. And three construction schemes are considered for it: Scheme 1, keep steel anchor beams simple supported on the brackets; Scheme 2, keep steel anchor beams simple supported on the brackets until initial tension is finished, after that, fix steel anchor beams with brackets; Scheme 3, keep steel anchor beams fixed with brackets all the time.

Steel anchor beams are supported on steel brackets which are connected to inner face of concrete pylon by shear studs. Anchorage zone lie above upper beam of pylon, from 3# cable to 19# cable, 17 steel anchor beams are installed at each limb of pylon. From bottom to top, there are four kinds of vertical distance for steel anchor beam, 2.8 m, 2.7 m, 2.6 m, and 2.5 m. In this paper, GML18 was taken as the typical segment to analyze the mechanical responses of steel anchor beams.

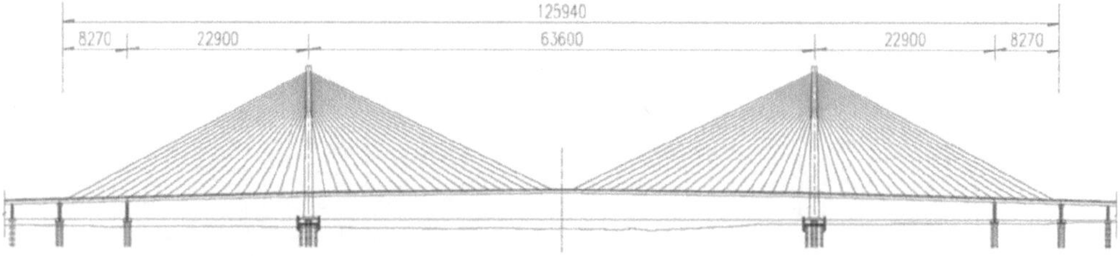

Figure 1. Layout of New Yalu River Bridge (unit: mm).

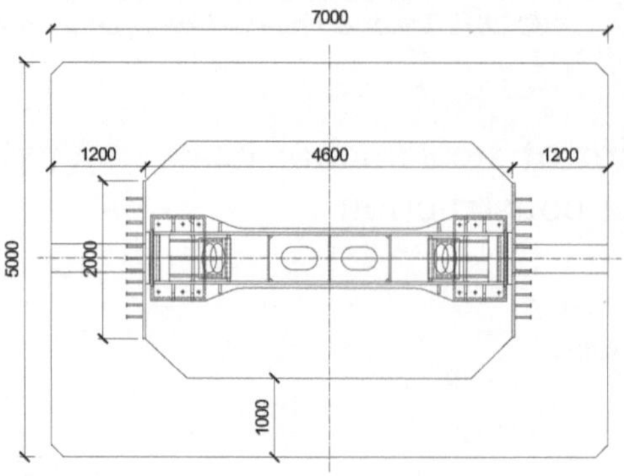

Figure 2. Section of pylon (unit: mm).

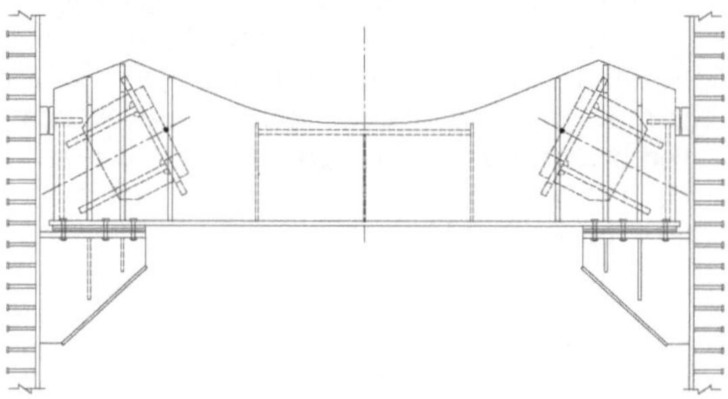

Figure 3. Elevation layout of steel anchor beam.

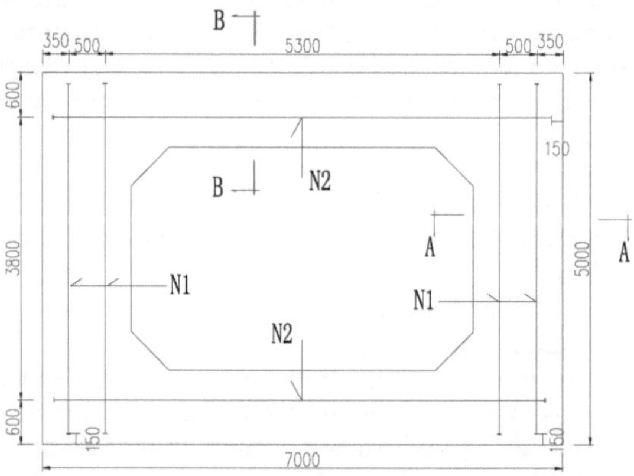

Figure 4. Horizontal layout of pre-stressed tendon (unit: mm).

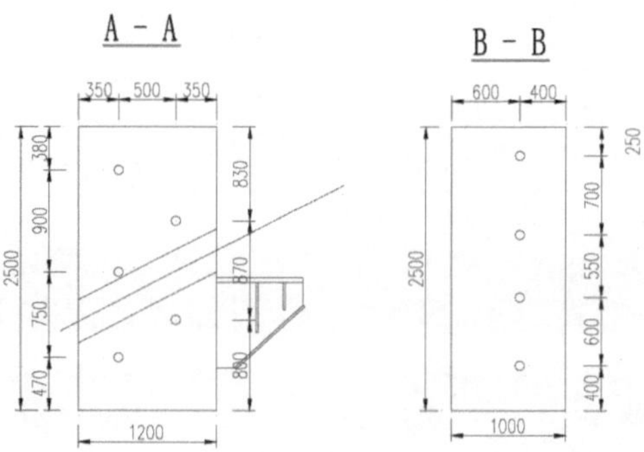

Figure 5. Elevation layout of pre-stressed tendon (unit: mm).

2 FEM MODEL

General finite element analysis software ANSYS was employed to investigate the mechanical behaviors of the steel anchor beams in New Yalu River Bridge. To reasonably simulate boundary conditions, the spatial model contained the top five segments of steel anchor beams and the top part of pylon, as shown in Fig. 4. In the FEM model, the sheet of both steel anchor beams and brackets were simulated with shell element SHELL43, concrete pylon were simulated with solid element SOLID65, reinforcing bars in the concrete pylon were taken into consideration by real constants of SOLID65, and the prestressing tendons were simulated with truss element LINK8. Nonlinear spring element COMBIN39 was employed to study the mechanical behavior of the shear studs, and three spring elements were arranged in the point of the stud to simulate the behavior of the studs in all directions. The shear stiffness of the studs in longitudinal and transverse direction was 3.76 × 105 kN/m, and the tensile stiffness of the studs was 3.99 × 105 kN/m. Surface-to-surface contact element CONTACT173 and TARGET170 were employed to simulate the touching behavior between steel anchor beams and brackets, as well as steel wall and concrete pylon. Node coupling were employed to simulate the screw joints.

The design loads considered in the finite element analysis included self-weight, prestress and cable force. Loss of prestress was calculated in accordance with the Chinese standard "Code for design of highway reinforced concrete and prestressed concrete bridges and culverts" (JTG D62-2004), and the cable force was listed in the table below.

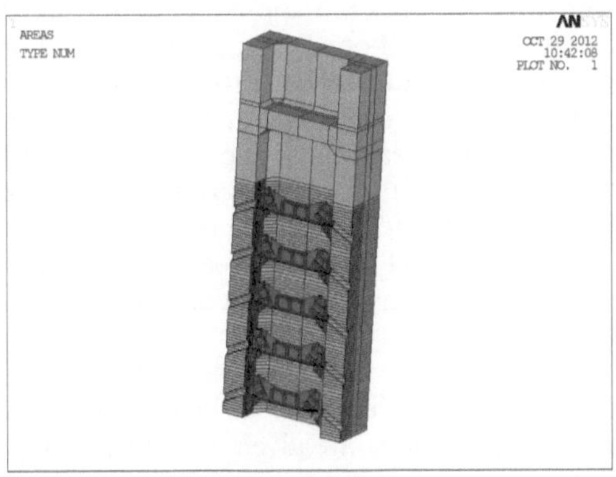

Figure 6. Half of partial model.

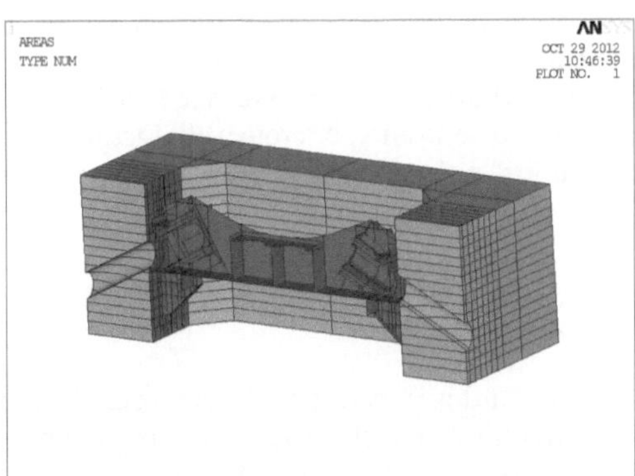

Figure 7. Half of GML18 segment.

Table 1. Cable force on the spatial model.

Cable no.	Dip angle of cables °	Cable force for initial tension kN	Maximum cable force kN
A19	27.732	3748	5510
A18	28.574	3583	5300
A17	29.497	3410	5270
A16	30.515	3181	4929
A15	31.642	3007	4829

3 RESULTS AND DISCUSSION

Stress levels of steel anchor beam under different construction schemes are approximate, but stress distributions are pretty different, and differences are mainly shown on side plates. As shown in Figure 8, range of stress is 2.29 MPa~200 MPa. For scheme 1, as steel anchor beams are almost free along the bridge, tension stress at the middle part of side plates is nearly uniform. For Scheme 3, as steel anchor beams are fixed with brackets all the time, deformation of bottom part of side plates are restricted, as a result, stress distribution in the side plates present arch-type shape. Scheme 2 is a compromise scheme, and stress distribution is between that of scheme 1 and scheme 3.

The force of shear studs is important for cable-pylon anchorage zome, detail date both shear force and axial force of shear studs is listed in talbe 2. Shear forces and force distributions of shear studs under three schemes are approximate. But axial forces of shear studs under three schemes are quite different. For scheme 1, as steel anchor beams are nearly free along the bridge, the direction of load transforming to brackets is mainly vertical. As a result, the shear studs that near brackets bear shear force and large bending moment. Due to the large bending moment, axial force of shear studs is no uniform. For scheme 3, as steel anchor beams are fixed with brackets, the load transforming to brackets contain horizontal load, as well as vertical load. The horizontal load reduce the bending moment that beared by shear studs on a large scale. As a result, axial force of shear studs is more uniform that that of scheme 1. Scheme 2 is a compromise scheme, and axial force of shear studs is between that of scheme 1 and scheme 3.

As pre-stressing force is employed, the overall stress levels of concrete pylon are approximate. The most obvious difference can be found at inner face that close to steel brackets, and the mechanical behaviors of this area are given in Figure 9, in which range of stress is –3 MPa~3 MPa. Besides pre-stressing force, the force that concret wall bear is mainly transmited from shear studs. So the stress distribution of concrete wall is similar with force distribution of shear studs, and the reason for difference of stress distribution shown in Figure 9 is similar with that of shear studs.

While the steel anchor beams are free along the bridge, most of horizontal force is taken by steel anchor beams, concrete pylon wall only take tiny part of horizontal force that transmitted by friction. While the steel anchor beams are fixed with brackets, horizontal force is distributed between steel anchor beams and concrete wall. Detail data is given in Table 3. For GML18 segment, Horizontal force of cables is 4654.5 kN, the percentage that steel anchor beam take are 97%, 90% and 66%, from scheme 1 to scheme 3.

4 CONCLUSIONS

Based on the analysis presented in this paper, the following conclusions can be drawn.

Firstly, the calculation results show that stress levels of steel anchor beams under different construction schemes are approximate, but stress distributions are pretty different. When steel anchor beams are fixed with brackets, the stress distribution present arch-type shape in the side plates.

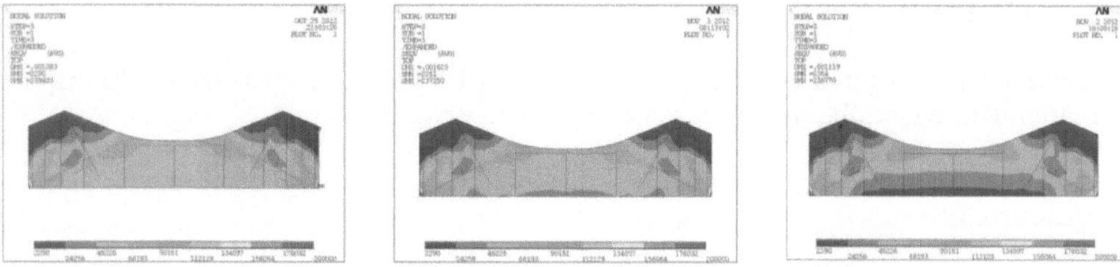

(a) Scheme 1 (b) Scheme 2 (c) Scheme 3

Figure 8. Stress distribution of side plates (unit: kPa).

Table 2. Shear force and axial force of shear studs (kN).

Construction schemes	Max of shear force	Max of axial force
Scheme 1	24.4	61.9
Scheme 2	24.7	47.0
Scheme 3	25.1	14.6

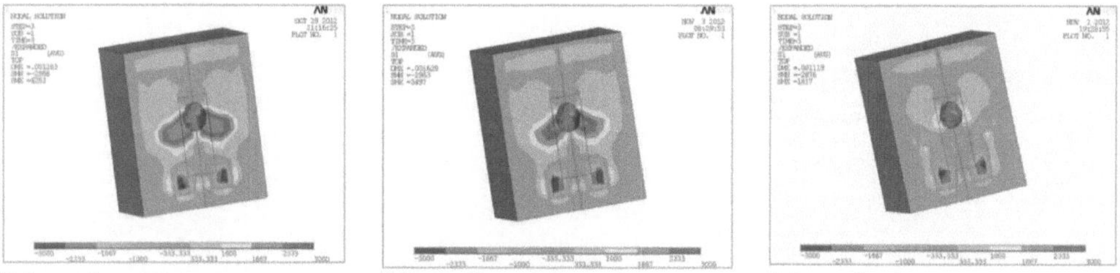

Scheme 1 Scheme 2 Scheme 3

Figure 9. Stress distribution of inner side of concrete pylon (unit: kPa).

Table 3. Distribution of horizontal force between steel anchor beam and concrete wall (kN).

Construction scheme	Horizontal force of cables	Horizontal force of steel anchor beam	Horizontal force concrete wall
Scheme 1	4654.5	4524.2	130.3
Scheme 2	4654.5	4188.9	465.6
Scheme 3	4654.5	3088.5	1566.0

Secondly, the construction scheme that keeping steel anchor beams fixed with brackets is favorable to axial force of shear studs. Because horizontal force can reduce bending moment at the section that near the concrete wall on a large scale, as a result, the distribution of axial force is more uniform.

Thirdly, as distribution of axial force is more uniform under scheme 3, the stress distribution of concrete wall (inner face) is more uniform.

Last, distribution ratio of horizontal force between steel anchor beams and concrete pylon are pretty different under different construction schemes. When steel anchor beams are fixed with brackets, only 66% horizontal force is taken by steel anchor beams. Proper time to fix can get perfect distribution ratio of horizontal force.

ACKNOWLEDGEMENTS

This research is sponsored by Science and Technology Program of Department of Transport of Liaoning Province, the support is gratefully acknowledged.

REFERENCES

He, Z.L., Zhong, W.P. & Zhang, W.J. et al. 2012. Mechanical Behavior of the Shear Studs on Composite Deck of Steel Arch Bridge. *Advanced Materials Research. Vols. 374–377*: 2408–2483.

Ministry of Communication of China. 2004. General code for design of highway bridges and culverts. JTG D60-2004. *China Communications Press*. Beijing.

Ministry of Communication of China. 2004. Code for design of highway reinforced concrete and prestressed concrete bridges and culverts. JTG D62-2004. *China Communications Press*. Beijing.

Su, Q.T., Yang, G.T. & Wu, C. et al. 2012. Investigation on the horizontal mechanical behavior of steel-concrete composite cable-pylon anchorage. *Journal of Constructional Steel Research. Volume 72*: 267–275.

Modeling and Computation in Engineering II – Xie (ed)
 ISBN 978-1-138-00058-2

Cyclic loading tests on RC beam-column subassemblages

H.J. Jiang, Y.K.Wang & X.F. Liu
Research Institute of Structural Engineering and Disaster Reduction, College of Civil Engineering, Tongji University, Shanghai, P.R. China

ABSTRACT: As one of the principal deformation indices, the inter-story drift ratio is generally applied to seismic performance assessment of building structures. As for moment-resisting frame structures, the inter-story drift (global deformation) is the integration of local deformation of constituent components, such as the beam, column and joint. In this study, nine half-scale Reinforced Concrete (RC) beam-column subassemblages extracted from a moment-resisting frame structure were tested under cyclic loading to study the relationship between the global deformation and the local deformation. The main design parameters of the tests are the amplification coefficient of column moment, the axial compression ratio, and the shear-span ratio of beam. The individual contribution to the inter-story drift from the beam, column, and the joint was analyzed from the test data. The results obtained in this study can be incorporated into displacement-based seismic design of RC frame structures.

1 INTRODUCTION

Lots of the structural and nonstructural damage in a building is induced by deformation during an earthquake. Deformation rather than force has been recognized as the most suitable and direct performance or damage indicator. To some extent, performance-based design and displacement-based design have been used interchangeably. A lot of efforts have been made on displacement-based seismic design (Kowalsky 2002). The inter-story drift ratio, closely as well as directly related with structural and nonstructural damage, is the primary deformation index most popularly applied in seismic design codes around the world to check or control the seismic performance of building structures. Researchers as well as practitioners are quite familiar with this deformation index.

The inter-story drift is the integration of deformation of individual structural component. The limitation on inter-story drift should be realized by controlling the deformation of the related structural component. It is necessary to study the relationship between the deformation of structural component and the resulted inter-story drift. However, few researchers have carried out such kind of research (Ghobarah and El-Amoury 2005, Jiang and Lu 2009). In this study, nine half-scale RC beam-column subassemblages extracted from a moment-resisting frame structure were tested under cyclic loading to study the contribution of the deformation of beam, column, and joint to the inter-story drift.

2 EXPERIMENTAL PROGRAM

2.1 *Description of test specimens*

The beam-column subassemblage was extracted from a typical moment-resisting frame structure. For the specimens, the end of the column is the mid-height point of the prototype structure, and the end of the beam is the mid-span point of the prototype structure. The mid-height point of the column and the mid-span point of the beam are considered as the approximate contra flexural points of the members. Nine 1/2-scale specimens which were classified into three groups were designed and constructed. The variables of design parameters are the amplification coefficient of column moment, defined as the ratio of flexural moment resistant capacity between the column and the beam, the axial compression ratio, and the shear-span ratio of beam. For all specimens, the dimensions are identical. The cross-section of beam is 150 mm × 250 mm rectangle, and that of column is 200 mm × 250 mm rectangle. The main design parameters of the specimens are shown in Table 1. KJJ1a, KJJ1b,

Table 1. Main design parameters of the specimens.

Group	Group 1			Group 2			Group 3		
Number	KJJ1a	KJJ1b	KJJ2	KJJ3	KJJ4	KJJ5	KJJ6	KJJ7	KJJ8
Axial compression ratio	0.21	0.31	0.42	0.63	0.21	0.21	0.21	0.21	0.21
Length of beam (mm)	1150	1150	1150	1150	650	1790	1150	1150	1150
Beam longitudinal steel	2Φ14 +1Φ12	2Φ14 +1Φ12	2Φ14 +1Φ12	2Φ14 +1Φ12	2Φ14 +1Φ12	2Φ14 +1Φ12	2Φ14 +1Φ12	2Φ14 +1Φ12	2Φ14 +1Φ12
Beam stirrups	Φ6@ 60/45	6@ 60/45	Φ6@ 60/45	Φ6@ 60/45	Φ6@ 60/45	Φ6@ 60/45	Φ6@ 60/45	Φ6@ 60/45	Φ6@ 60/45
Column longitudinal steel	8Φ12	8Φ12	8Φ10	4Φ14 + 4Φ12	8Φ12	8Φ12	8Φ8	8Φ10	4Φ14 +4Φ12
Column stirrups	Φ6@ 60/45	Φ6@ 60/45	Φ6@ 60/45	Φ6@ 60/45	Φ6@ 60/45	Φ6@ 60/45	Φ6@ 60/45	Φ6@ 60/45	Φ6@ 60/45
Strength coefficient of the joint core area	1.18 (1.17)*	1.18 (1.17)	1.21 (1.20)	1.21 (1.20)	1.18 (1.17)	1.18 (1.17)	1.23 (0.92)	1.18 (1.17)	1.18 (1.17)
Amplification coefficient of column moment	1.30 (1.34)	1.32 (1.36)	1.30 (1.34)	1.34 (1.34)	1.30 (1.34)	1.30 (1.34)	0.85 (0.91)	1.08 (1.05)	1.50 (1.50)

* The figures in the parentheses are the actual value calculated from the actual strength of the material.

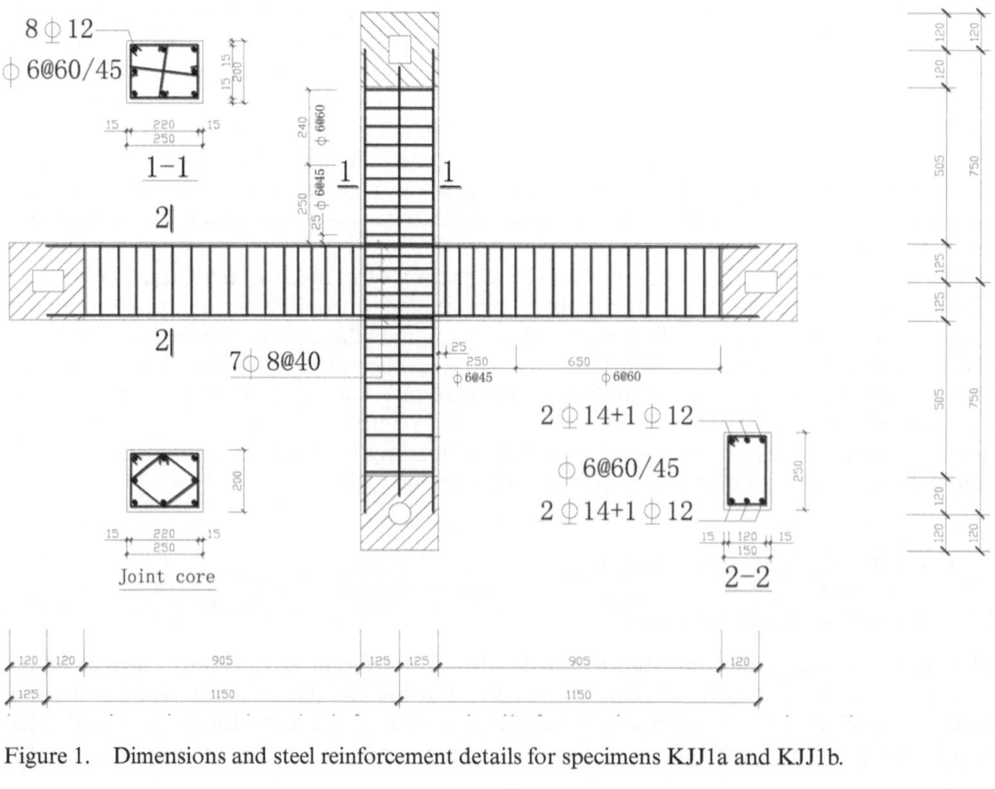

Figure 1. Dimensions and steel reinforcement details for specimens KJJ1a and KJJ1b.

KJJ2, and KJJ3 belong to Group 1. KJJ4 and KJJ5 belong to Group 2. KJJ6, KJJ7, and KJJ8 belong to Group 3. The dimensions and steel reinforcement details of the specimens KJJ1a and KJJ1b are shown in Figure 1.

2.2 *Instrumentation and test procedures*

The test setup and arrangement of linear variable displacement transducers (LVDT) are illustrated in Figures 2 and 3. Sixteen rolling bearings were installed in the four ends of the specimen so as to reduce friction during the test. The vertical load was exerted on specimens by hydraulic jacks. The lateral load was applied by the SCHENCK servo actuator. The specimens were extensively instrumented to monitor global responses (e.g., applied lateral load and displacement) as well as local ones (e.g., strains).

Firstly the vertical load was exerted and kept constant. Then cyclic lateral load was applied until the failure of the specimen. Prior to the lateral load arriving at the yielding force of the specimen, the lateral loading was controlled by the force. After the yielding of the specimen, the lateral load was applied by displacement control, with displacement amplitude increment of specimen yielding displacement. During the whole test process, the displacement, force, and the steel strains were recorded electronically. The maximum crack width and maximum residual crack width were measured at the peak horizontal force or displacement points and the zero-force points within each loading and unloading cycle, respectively.

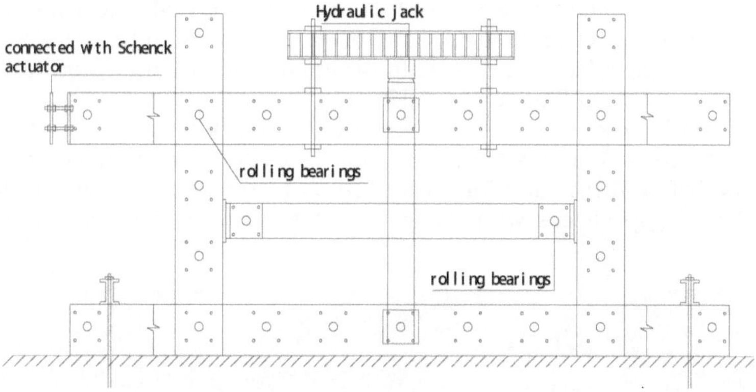

Figure 2. Test setup.

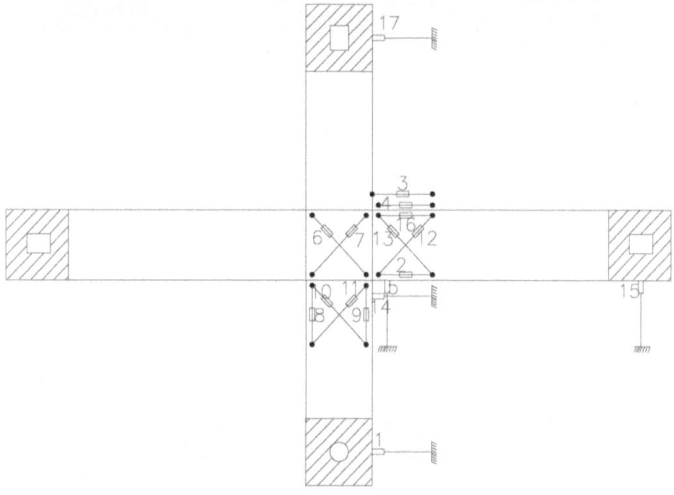

Figure 3. Arrangement of LVDT.

3 EXPERIMENTAL RESULTS

3.1 *Failure pattern*

Three types of failure pattern were found in the tests, as shown in Figure 4. KJJ1a, KJJ1b, KJJ2, KJJ3, KJJ5 and KJJ8 failed with the forming of plastic hinges at beam ends. Slight damage occurred in columns and joints. As KJJ4 and KJJ7 are concerned, shear failure occurred in the joint after the plastic hinges formed in the beam ends. For KJJ6, shear failure occurred in the joint after the plastic hinges formed in the column ends.

3.2 *Deformation contribution of each component*

The lateral displacement of the subassemblage, i.e., the inter-story drift, is generated from the deformation of the beam, the column, and the joint. According to the test data, the global inter-story drift of the subassembladge and the corresponding local deformation of the beam, the column and the joint can be obtained. The test data shows that the stirrup of beams and columns did not yield during the tests. In addition, the shear deformations of beams and columns are very small. Therefore, the contribution of the shear deformations of beams and columns was not considered in this study. According to the test phenomenon, the plastic hinges and the slippage of the longitudinal steel bars occurred in the beam ends of all specimens except KJJ6. The relationship between the plastic rotation angle of the beam end and the inter-story drift ratio, and the relationship between the beam slip rotation angle and the inter-story drift ratio are shown in Figures 5 and 6 for all specimens except KJJ6. Similarly, the relationship between the plastic rotation angle of the column end and the inter-story drift ratio, and the relationship between the column slip rotation angle and the inter-story drift ratio are shown in Figure 7 for KJJ6.

The relationship between the shear deformation of the joint and the inter-story drift ratio is shown in Figure 8. The contribution ratio of each deformation component to the total inter-story drift and the variation with ductility factor for KJJ1a and KJJ6 is shown in Figure 9. For the specimens with the same failure pattern, the characteristics of the

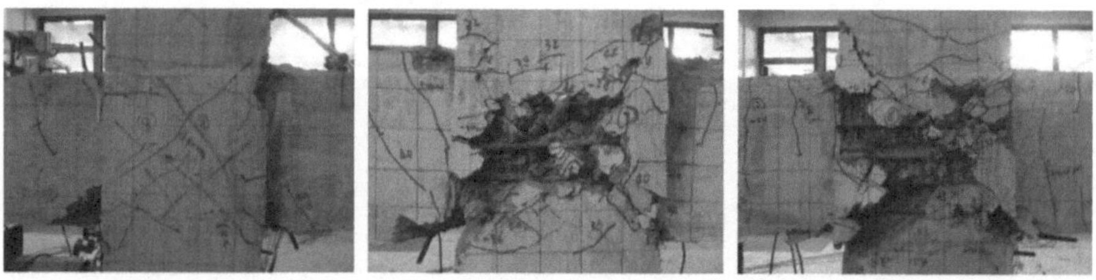

(a) KJJ1a (b) KJJ4 (c) KJJ6

Figure 4. Failure pattern.

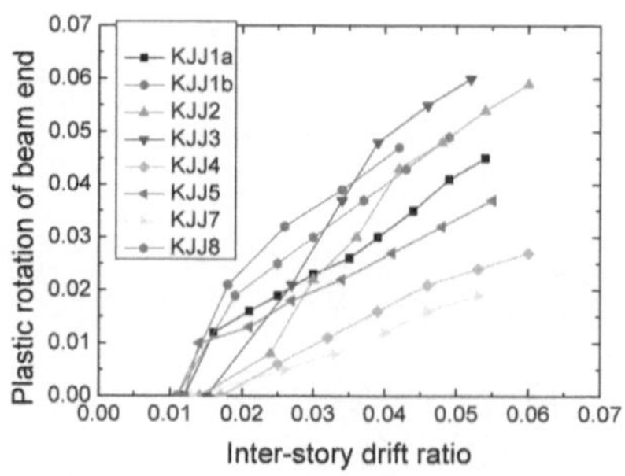

Figure 5. Relationship between beam plastic rotation angle and inter-story drift ratio.

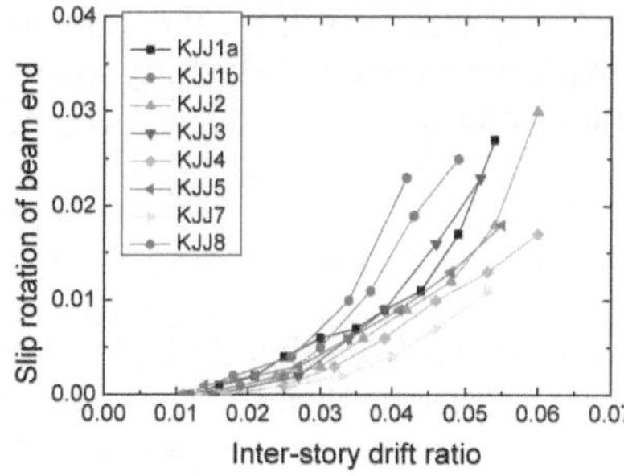

Figure 6. Relationship between beam slip rotation angle and inter-story drift ratio.

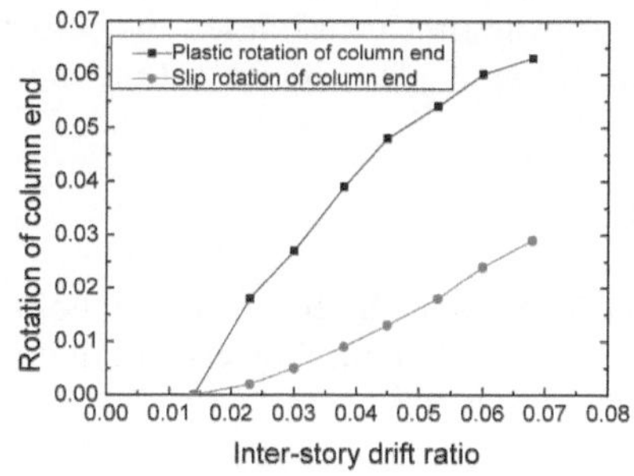

Figure 7. Relationship between column deformation and inter-story drift ratio.

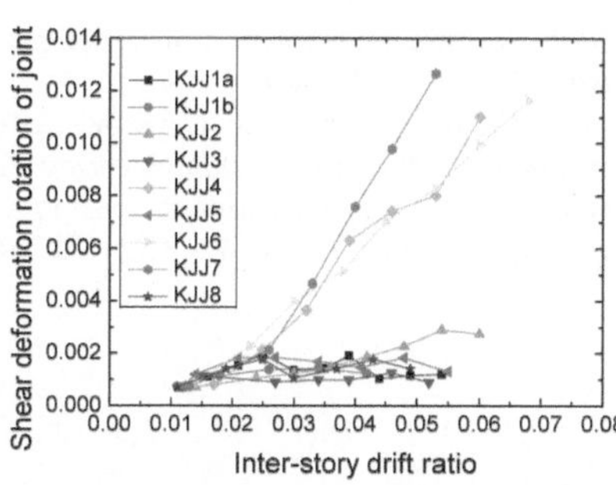

Figure 8. Relationship between shear deformation of joint and inter-story drift ratio.

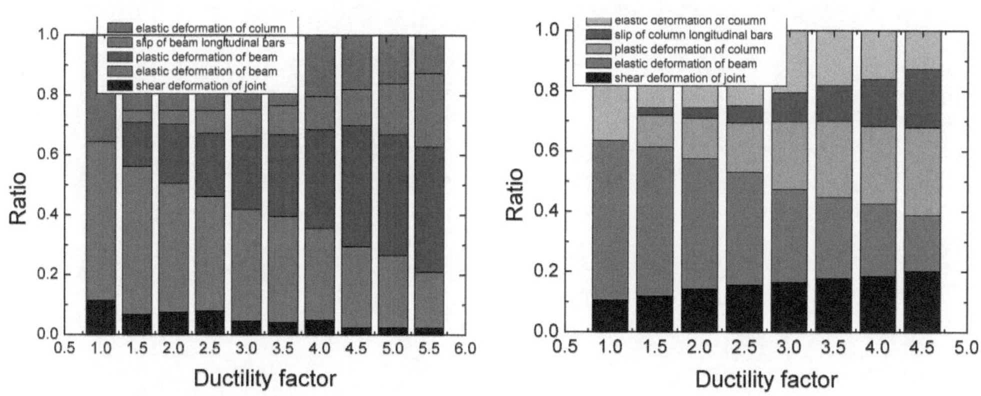

(a) KJJ1a

(b) KJJ6

Figure 9. Contribution ratio of each component to inter-story drift.

contribution ratio are similar. Compared to KJJ1a, the contribution ratio of column and joint is much larger for KJJ6. With the increase of ductility factor, the contribution of elastic deformation of beam or column decrease.

4 CONCLUSIONS

The following conclusions can be drawn from this study: (1) Three types of failure pattern were found for RC beam-column subassemblages, which depends on the relative strength relationship between the column and the beam, and the shear strength of the joint. (2) For the specimens without the shear failure of joint, the contribution of joint deformation to the total inter-story drift is very low and could be omitted. On the contrary, it would be very high for the specimens with the shear failure of joint. (3) With the increase of the displacement amplitude as well as the degree of nonlinear responses, the contribution of elastic deformation decreases while the contribution of plastic deformation and steel bar slip increases.

ACKNOWLEDGEMENTS

The financial support from the Ministry of Science and Technology of China through Grant No. SLDRCE09-B-10 and the National Natural Science Foundation of China under Grant No. 51078272 are gratefully acknowledged.

REFERENCES

Ghobarah. A. & El-Amoury. T. 2005. Seismic rehabilitation of deficient exterior concrete frame joints. *Journal of Composites for Construction*. 9(5): 408–416.

Jiang, H.J. & Lu, X.L. 2009. Study on the relationship between story drift and element deformation for reinforced concrete frames. *Earthquake Engineering and Engineering Vibration*. 29(2): 66–72.

Kowalsky, M.J. 2002. A displacement-based approach for the seismic design of continuous concrete bridges. *Earthquake Engineering and Structural Dynamics*, 31(3): 719–747.

Modeling and Computation in Engineering II – Xie (ed)
© 2013 Taylor & Francis Group, London, ISBN 978-1-138-00058-2

Practice on improvement of urban thermal environment with evaporation combined CFD simulation

Wenbin Hu & Xiaoqin Liao
Architecture Design and Research Institute, South China University of Technology, Guangzhou, China

ABSTRACT: Thermal environment analysis plays an important role in urban planning. For complex heat exchange between airflow and urban underlying, simplified evaporation model combined with CFD simulation was proposed to describe surface evaporation cooling effect in hot-humid regions. Based on improved CFD simulation, three approaches of outdoor natural ventilation promotion, improvement of thermo-physical characteristics of underlying surface and outdoor ecological compensation were analyzed, and designing factors involved in the approaches were elucidated in detail. Finally, based on the analysis of urban thermal environment of a hot-humid county in South China, the conclusion was arrived that approaches were effective in the improvement of outdoor thermal environment in hot and humid regions.

1 INTRODUCTION

With the development of urbanization in China, the problem about urban thermal environment, such as urban heat island effect, is more and more outstanding. In the process of urban construction, the negative effects on outdoor ventilation due to unreasonable planning of large-volume and high-density buildings, worsening of urban heat island due to large area of hardened and impermeable underlying surface have a serious influence on the quality of outdoor thermal environment, which makes building energy consumption grow further. A serious fact is that poor urban thermal environment leads to building energy consumption growth, which deteriorates urban thermal environment in return. Therefore, improvement of urban thermal environment based on CFD simulation attracts worldwide attention. However, for complex heat exchange between airflow and underlying surface, especially with evaporation combined CFD analysis, the troublesome whole process makes poor contribution in practice, as Matthias (2007) stated. Based on most used commercial CFD software, simplified evaporation model with effective coefficient of solar absorption was proposed to describe surface evaporation cooling effect, which is critical in modification of thermo-physical properties of underlying surface by vegetation and permeable pavement in hot-humid regions.

2 EVAPORATION MODEL OF PERMEABLE PAVEMENT

In following analysis, a layer of permeable medium with depth d is for consideration. When heat exchange happens between upper surface of the medium and airflow, moisture will diffuse from the inside to the surface and evaporate there, which can produce cooling effect.

According to the theory of heat and mass diffusion, the equations governing the diffusion are following, as written by Luikow (1980):

$$\frac{\partial u}{\partial \tau}=\frac{\partial}{\partial z}\left(D_m\frac{\partial u}{\partial z}\right)+\frac{\partial}{\partial z}\left(D_t\frac{\partial t}{\partial z}\right) \tag{1}$$

$$\frac{\partial t}{\partial \tau}=\frac{\partial}{\partial z}\left(a\frac{\partial t}{\partial z}\right) \tag{2}$$

For heat balance of upper surface of the medium:

$$q_0+q_{ar}-q_{sr}=q_e+q_c+q_d \tag{3}$$

where

$$q_0=\alpha_s I;\ q_{ar}=\sigma[t_a(\tau)+273.16]^4[0.802+0.004t_d(\tau)];\ q_{sr}=\varepsilon\sigma[t_s(\tau)+273.16]^4$$

$$t_d=8.22+12.4\ln(P_{ab})+1.9[\ln(P_{ab})]^2;\ q_c=h_c[t_s(\tau)-t_a(\tau)];\ q_e=E_s(\tau)L$$

Therefore, the heat fluxing into the medium can be expressed as:

$$q_d=q_{0,e}+q_{ar}-q_{sr}-q_c \tag{4}$$

where

$$q_{0,e}=q_0-q_e=\alpha_{s,e}I \tag{5}$$

In which, $\alpha_{s,e}$ is called as effective coefficient of solar absorption. As for the calculation of evaporation rate, the method proposed by Van Bavel et al. (1976) to calculate the evaporation rate on the surface of soil is used in the calculation.

$$E_s(\tau)=\frac{d_s-d_a}{\xi_c} \tag{6}$$

Evaporation resistance can be calculation by following formula:

$$\xi_c=\xi_a S_t \tag{7}$$

where

$$\xi_a=\frac{[\ln(z/z_0)]^2}{0.16V\gamma_a};\ S_t=\frac{1}{1-10R_t};\ R_t=\frac{9.81(z-z_0)(t_a-t_s)}{(t_a+273.16)V^2}$$

To calculate the evaporation rate on the upper surface of the medium, the modification item for the resistance is following, as proposed by Qigao Chen (1991):

$$\xi_s=-1523.59+55.89(u_{sb}-u_s) \tag{8}$$

Then, the evaporation rate can be written as:

$$E_s(\tau)=\frac{d_s-d_a}{\xi_c-\xi_s} \tag{9}$$

Based on above models, four cases with different initial temperature and humidity were selected to calculate effective coefficient of solar absorption for permeable pavement. The

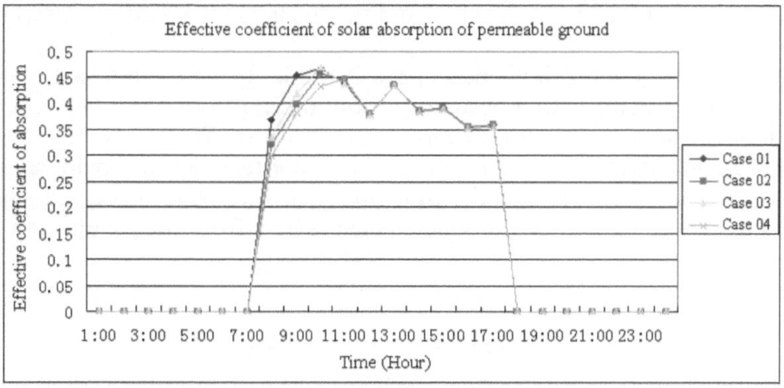

Figure 1. Effective coefficient of solar absorption for each case.

variation curve of effective coefficient of solar absorption for each case is shown as Figure 1. From the analysis of evaporation, following main points can be summarized:

1. The maximum value of evaporation surface temperature ranges from 45°C to 46°C, and occurs at 13:00~14:00. The minimum value of evaporation surface temperature ranges from 28°C to 29°C, and occurs at 6:00.
2. The maximum value of evaporation rate for each case is about 0.4 kg/(m^{2}h). Effective coefficient of solar absorption varies from 0.35 to 0.48 during the time from 8:00 to 17:00. Because of evaporation cooling effect, solar absorption can be reduced significantly and effective coefficient of solar absorption can be controlled no higher than 0.45. For commercial CFD software, like FLUENT, the parameter of absorptivity in radiation boundary conditions setting panel can be substituted by effective coefficient of solar absorption, which has taken surface evaporation in consideration.

3 APPROACHES ON IMPROVEMENT OF OUTDOOR THERMAL ENVIRONMENT

Cool surfaces and shade trees can not only create a good outdoor environment, but also can reduce the energy consumption, as stated by Akbari (2001). This paper tries to analyze the importance of reasonable building layout and design in improving outdoor environment from three viewpoints of outdoor natural ventilation promotion, modification of thermo-physical properties of underlying surface by vegetation and permeable pavement and ecological compensation with large area of wetland.

3.1 *Outdoor natural ventilation design*

To promote outdoor ventilation is one of the most effective and feasible approaches to improve the quality of thermal environment. In designing, following methods can be adopted to improve the outdoor ventilation.

1. The direction of district road should be consistent with the prevailing wind direction to make the road a passageway for the air flow. Meanwhile, arrangement of slab-type building along the wind direction can further increase air velocity. This way can not only improve the quality of thermal environment for street, but also can enhance the ventilation and cooling effect of surrounding buildings.
2. Reasonable arrangement of space distance between tall buildings contributes to the wind channel. When building distance is less than 0.8 H (H is the height of the upwind building), above 50% of the area between buildings is eddy current area. The ventilation can obviously be improved if the building distance increase to 1.0~1.5 H. To arrange the

highest and longest buildings downwind and the small and low building upwind can also promote ventilation.

3. Promote air flow by adopting open space in building. One approach is to choose open space for lobby and other accessory room in first floor. Another approach is to choose an open space for structure transformation floor. When the height of the transfer floor is higher than 3.0 meters, the ventilation effect can be obviously improved. When the area of open space of the first floor reach to 30%~50%, and space height reach to 3~6 m, ventilation for the space 2 meters high from the ground can be greatly improved.

3.2 *Improvement of thermo-physical properties of underlying surfaces*

Thermo-physical properties of urban underlying surfaces can affect air flow, air temperature, air humidity and mean radiant temperature for the space near the ground by complex heat exchange. As for the district with high building density, it is possible to decrease hot island effect by modifying thermo-physical properties of underlying surfaces. The effective approaches include increasing the area proportion of the permeable pavement or light color surface and providing sunshade to water-proof ground, as shown in following:

1. To arrange large area of vegetation land. Continuous vegetation can lower hot island effect notably. In green space planning, the width of green space should be more than 500 meters as possible, as proposed by Wilmers (1988). According to Jonsson's research (2004), with the vegetation coverage percentage increasing 10%, the maximum temperature can decrease 2.6%. When the vegetation coverage percentage reaches to 40%, hot island effect can be decreased by 75%. When the percentage is more than 60%, hot island effect can be mostly removed.
2. To avoid marble, granite and other water-proof pavement by large area. It is possible to combine water-proof pavement with porous permeable pavement, such as vegetation bricks. The permeable pavement should have larger porosity to hold more water to evaporate, which is beneficial to adjust the temperature and humidity of underlying surface and eliminate hot island effect.
3. To adopt light color surface to impermeable surface. General speaking, solar radiation absorption coefficient of dark ground surface is higher than light color one by 40%. The former can produce adverse effects on thermal environment. If the factor of solar radiation absorption is lower than 0.5, the surface temperature can be reduced by more than 10~15°C.
4. To plant trees with large area of crown, which can produce shadow effect to reduce solar radiation absorption of ground. To produce well shadow effect, the height of crown should reach to more than 2 meters, and average density of leaf area should reach to more than 10 m^2/m^3. In planning and designing, light color road with tree shading can greatly reduce surface temperature by more than 20°C.

3.3 *Ecological compensation*

Continuous water, wet land, large area of forest can realize self-adjustment of urban thermal environment. Continuous water can absorb solar radiation because of evaporation, which can reduce hot island effect greatly. Single water body with the area less than 0.25 km^2 has no obvious effect on the improvement of thermal environment. Large water body has a better effect on cooling and dehumidification. A piece of isolated water body whose area is about 2 kilometers can reduce the temperature by 0.6°C, as stated by Shuyan Li et al. (2008). Wet land plays an important role on regulating regional climate, reducing urban hot island effect and improving the urban environmental quality. Wet land evaporation rate is at least two times higher than that of water because of vegetation.

4 PRACTICE AND EXAMPLE

Urban planning of Keyuan, a small county with an area of 1.0 square kilometer in south China, is taken for consideration. At present planning, typical building type of Keyuan is the combination of new-built high rising building in the south with reconstruction of built-up multi-floor house in the north. The quality of thermal environment of Keyuan has been analyzed on the base of simplified evaporation model combined with CFD simulation.

4.1 *Outdoor ventilation simulation and analysis*

Figure 2 and Figure 3 describe outdoor wind field of the elevation with 1.5 meters in height, called reference elevation, in the Keyuan county. From the figures, we can see the space between the high-rise buildings and open space inside the buildings can be served as air flow channel, which promote air flow through the county. In the south high rising building area, negative impact caused by the presence of high rising building area is not obvious because of large open space area in high rise building area. Figure 3 also shows that there are two beams of air flow running through the old town area caused by urban street and channel formed by high rise building along the street, which improves the ventilation of old town area significantly.

Figure 4 shows distribution of wind speed of reference surface in the east district of Keyuan county. The eastern district is dominated by high rising residential buildings, and the overall layout scheme is semi-enclosure. The first floor adopts open space design scheme, and the ratio of open space is higher than 30 percent. The figure reveals that open floor and the semi-enclosed openings can be served as air flow passage, which contributes to the improvement of ventilation around semi-enclosure area. Figure 5 describes the outdoor temperature profile of the surface of reference elevation in the Keyuan county. Compared with

Figure 2. Velocity distribution of reference elevation.

Figure 3. Area with speed higher than 1.0 m/s.

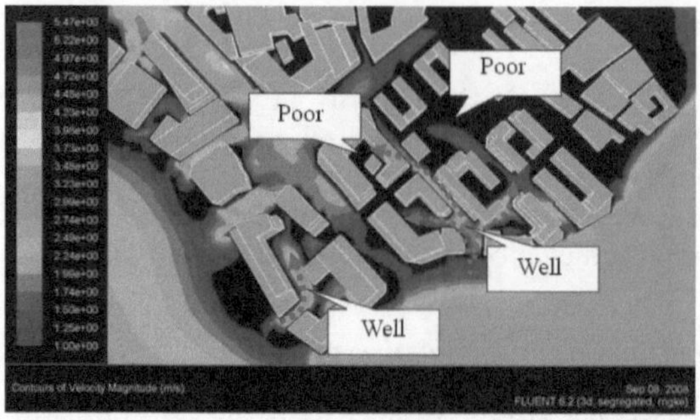

Figure 4. Detailed velocity distribution of reference elevation for east district.

Figure 5. Temperature distribution of reference elevation.

the temperature of outside the district, which is between 35°C and 37°C, the district air temperature is between 33.5°C and 36.5°C. There are four distinguished regions with lower air temperature distribution.

1. The core region covered with Keyuan park. Because of evaporation cooling effect caused by large area of trees and water body, the temperature distribution in the region is between 33°C and 34°C, which is 2 to 3°C lower than the surrounding area significantly.
2. The area enclosed by the space between the high-rise buildings. For the reason that the heat is removed outside the region by ventilation, the temperature distribution in the region is between 34 to 35°C, surpassing the temperature of incoming flow more than 1°C, which is the ventilation effect on improving thermal environment.
3. The shadow area of the high-rise buildings. The high-rise buildings lead to large shadow on the ground significantly, which can reduce the solar radiation heat gain of the ground significantly, thereby reducing local hot island effect. As can be seen from Figure 5, the air temperature of the shadow of the high-rise building is between 33.5°C to 35.5°C, which is lower than the temperature of incoming flow at least 1°C.
4. Open green space covered by tree and shrub in the old town. Public vegetation spaces with area of 1000~3500square meters can not only serve as a local entertainment space to improve relationship between neighborhoods, but also do good to improve the quality of thermal environment. The passive evaporation cooling effect can lower the groud temperature. Meanwhile, tall trees formed shadow effects. The air temperature of local public green space could be cooled more than 1°C, compared with adjacent area with no vegetation.

Figure 6 describes the temperature distribution of underlying surface and building surface, which shows temperature differences of underlying surface and building facade caused by various thermo-physical characteristics of the surface. The figure describes the difference in temperature distribution for 7 kinds of underlying surface, they are river surface, shadow

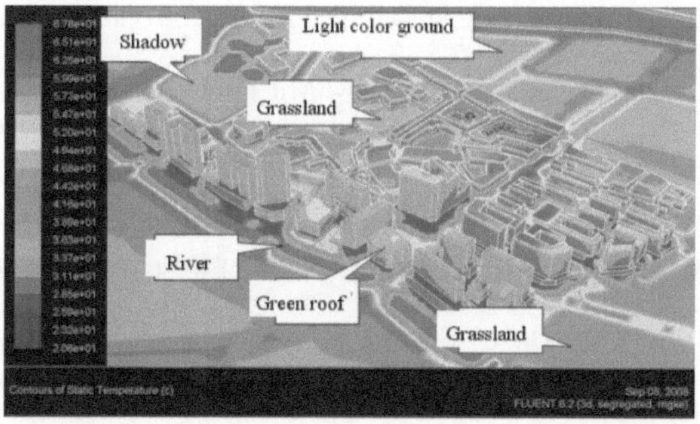

Figure 6. Air temperature distribution of underlying surface and district building facade.

Figure 7. Streamline behind high-rise building.

of Keyuan park, grassland of open space, vegetation roof of high building area, vegetation area of old urban area, light color but non-permeable ground, hardened cement roof. Surface temperature distributions are 28.5 to 30°C, 31.1 to 33.7°C, 38.9 to 41.6°C, 41.6 to 44.2°C, 44.2 to 46.8°C, 49 to 54.7°C, 59 to 65.1°C correspondingly.

Figure 7 describes air flow path around high-rise building area. From the figure, we can see, the streamline of old town area is nearly straight, which shows slight negative impact on downstream ventilation.

5 CONCLUSION

Effective coefficient of solar absorption is a simple approach to consider passive cooling effect in CFD simulation. For most commercial CFD software, effective coefficient of solar absorption makes it possible to analyze the quality of thermal environment with the consideration of evaporation, which has significant influence in hot-humid climate regions. Effective coefficient of solar absorption can not only be a constant, but also a function of time, which can reflect dynamic cooling effect caused by evaporation. Based on improved CFD simulation with evaporation consideration, it can be seen that three approaches can effectively improve the outdoor thermal environment, they are urban natural ventilation promotion, improvement of thermo-physical characteristics of underlying surface and ecological compensation.

REFERENCES

Akbari, H. & Pomerantz, M. 2001. Cool surfaces and shade trees to reduce energy use and improve air quality in urban areas. *Solar Energy* 70(3): 295–310.

Jonsson, P. 2004. Vegetation as an urban climate control in the subtropical city of Gaborone, Botswana. *International Journal of Climatology* 27: 222–237.
Luikow, A.V. 1980. *Heat and Mass Transfer*. Moscow: Mir Publisher.
Matthias, R. 2007. Review of urban climate research in subtropical regions. *International Journal of Climatology* 27: 1859–1873.
Qigao Chen. 1991. *The Fundamental of Building Thermal Physic*. Xi'an: The press of Xi'an Jiao Tong University.
Shuyan Li & Chunyi Xuan. 2008. Analysis of Microclimate Effects of Water Body in a City. *Chinese Journal of Atmospheric Sciences* 5: 20–25.
Van Bavel, C.H.M. & Hilled, D.L. 1976. Calculating potential and actional evaporation from a bare soil surface by simulation of concurrent flow of water and heat. *Agriculture Meteorology* 17: 453~476.
Wilmers, F. 1988. Green for melioration of urban climate. *Energy and Building* 11: 289–299.

Modeling and Computation in Engineering II – Xie (ed)

Fatigue test specimen of orthotropic steel deck of cable-stayed bridge with arch pylons

Chong Wu, Xinxin Wang & Huayi Wang
Department of Bridge Engineering, Tongji University, Shanghai, China

ABSTRACT: Design parameters of the fatigue test specimen of the Zhijiang Bridge in Hang Zhou City, such as boundary conditions, diaphragm height, deck plate width and the number of U-rib, were investigated by FEM. Analysis results of try test specimens were compared with that of full bridge analysis model with framed finite elements and substructure of shell elements. The influence of different design parameters on the stress distribution of fatigue specimen models was studied and the value ranges of design parameters were discussed.

1 INTRODUCTION

Recently, fatigue damages were detected in orthotropic steel decks of many bridges all over the world and affect the normal use of bridges greatly. Fatigue cracks were found at the fillet weld of U-rib-to-deck plate and at the deck plate along the welding roots in the Van Brienenoord Bridge in the Netherlands in 1997 (Inokuchi, S. 2006). On the Severn Crossing in the United Kingdom, cracks occurred at the intersections of the discontinuous ribs with the floor beams and at the rib-to-deck weld (Jong, 2004). There were also many fatigue cracks detected in orthotropic steel decks of bridges in China.

Many scholars have done lots of research on the fatigue performance of typical details of the orthotropic steel deck in theory or libratory test. Rong used tensile test to confirm the fatigue strength of details of the U-rib-to-deck plate connection and the butt weld of the U-rib supplementing segment in the orthotropic steel deck of a long-span cable-stayed bridge (Rong, 2009). Yu and Wei designed and conducted a fatigue test research on three detail categories in the orthotropic steel deck of the Sutong Bridge, which were the U-rib-to-diaphragm connection detail, the U-rib-to-deck connection detail and the detail of the free edge of cope hole in the diaphragm (Yu, 2009; Wei, 2008). Yamada and Ya Samol designed a fatigue test with small specimen, loaded by bias motor, studied the fatigue performance of U-rib-to-deck connection detail and summarized the S-N curve of such detail (Yamada, 2007). To study the effects of fabrication procedures on fatigue resistance of the rib-to-deck plate connection detail in steel orthotropic decks, Sim designed fatigue tests with big full-scale specimens. These specimens were fabricated with weld penetration ratio transitioning from 80% to 100% and some of them with precamber (Sim, 2009).

There are three types of fatigue test specimen which are small specimen, big full-scale specimen and small full-scale specimen. The big full-scale specimen is fabricated in true dimension, with more than two diaphragms and many ribs. The small full-scale specimen is also fabricated in true dimension, often with one diaphragm and 3 to 5 ribs. There will be contrast between the test result of small specimen and the test result of true bridge structure, due to differences in residual stress and stress concentration. Although presenting the fatigue performance of true bridge structure, the big full-scale specimen test has higher requirement to facilities and costs much time and money. Many steel deck fatigue tests focused on the fatigue resistance of rib-to-diaphragm connection detail, rib-to-deck plate connection detail away from diaphragm and longitudinal rib butt welded connection detail. The fatigue performance of U-rib-to-deck plate connection detail near diaphragm, however, was considered in this paper. In order to design a fatigue test specimen of the Zhijiang Bridge in Hang Zhou City, design parameters, such as boundary conditions, diaphragm height, deck plate width and the number of U-rib, were investigated by FEM. Analysis results of try test specimens were compared with that of full bridge analysis model with framed finite elements and substructure of shell elements. The influence of different design parameters on the stress distribution of fatigue specimen models was studied and the value ranges of design parameters were discussed.

2 PROJECT BACKGROUND

The Zhijiang Bridge (Fig. 1) is a cable-stayed bridge with span length of (116+246+116) m and double 90.5 m high arch pylons. The main girder (Fig. 2) is a flat steel box of 3.5 m high and 41.36 m wide, with diaphragms of 3.5 m long intervals in general segment. The thickness of deck plate is 16 mm in traffic lanes and 12 mm in sidewalk. The U-rib of deck plate is 280 mm high and 8 mm thick. The thickness of bottom plate is 14 mm in pylon zone and 12 mm in other zones. The U-rib of bottom plate is 260 mm high and 6 mm thick.

3 FINITE ELEMENT MODEL

The ANSYS was adopted for the analyses. The full bridge analysis model was shown in Figure 3 while the deck segment to be investigated was simulated by shell elements and the other members were simplified to framed elements. A pair of 10t wheel loads distributed on 600 mm × 200 mm deck was added at each lane of the bridge with six lanes. Specimens corresponding to different design parameters were simulated by FEM with shell elements (Fig. 4). A 10t wheel load distributed on 600 mm × 200 mm deck was added on specimen models.

For the full bridge analysis model and specimen models, the transverse normal stress of the deck plate and the normal stress in tangential direction along the free edge of cope hole

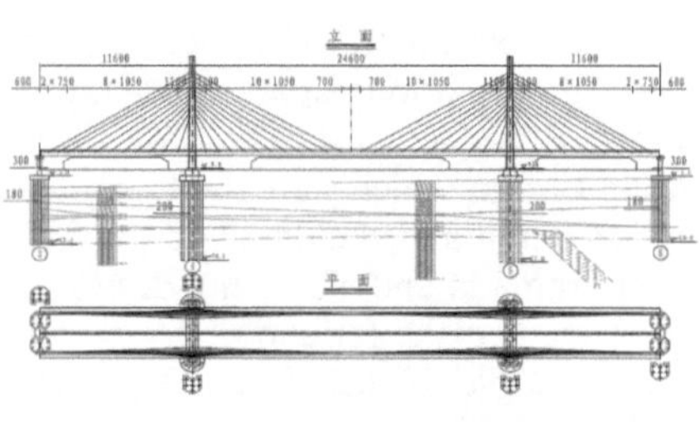

Figure 1. Zhijiang Bridge elevation (unit: m).

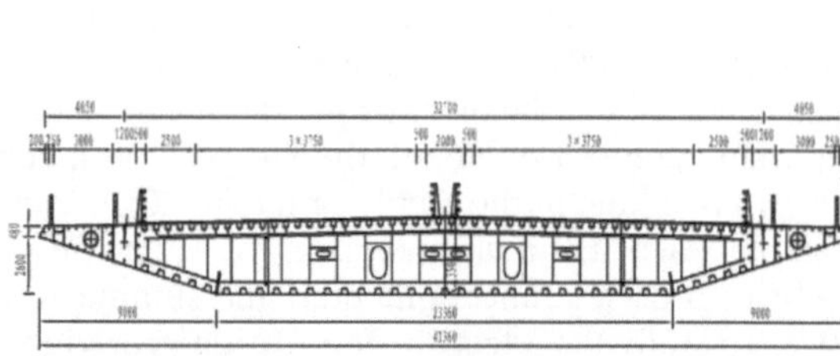

Figure 2. Steel box cross section (unit: mm).

(a) Full bridge analysis model (b) Substructure (c) Comparing part in substructure

Figure 3. Full bridge FEM.

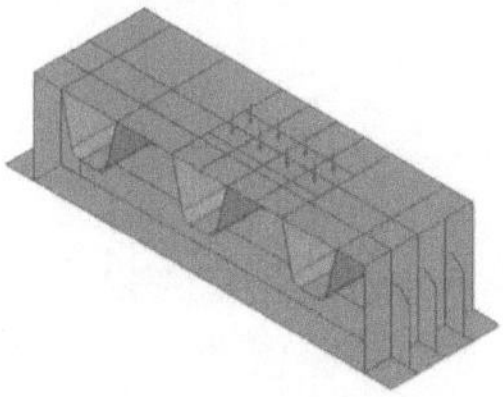

(a) FEM with 3 U-rib

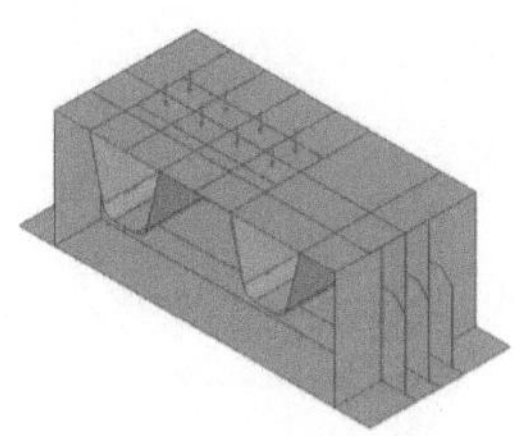

(b) FEM with 2 U-rib

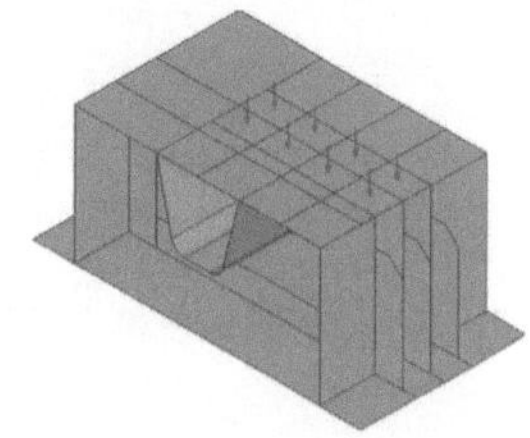

(c) FEM with 1 U-rib

Figure 4. Test specimen FEM.

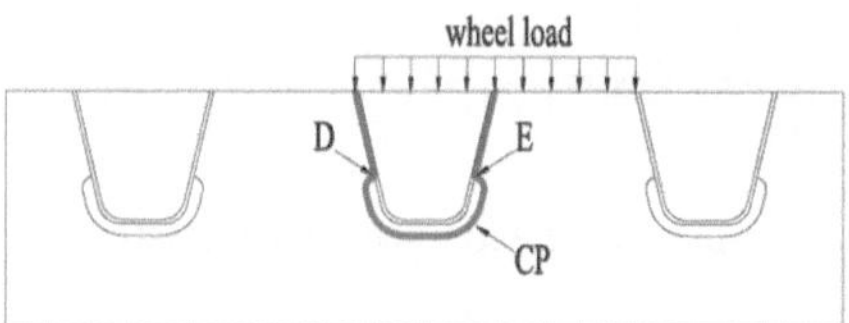

(a) Stress comparing path in the cope hole

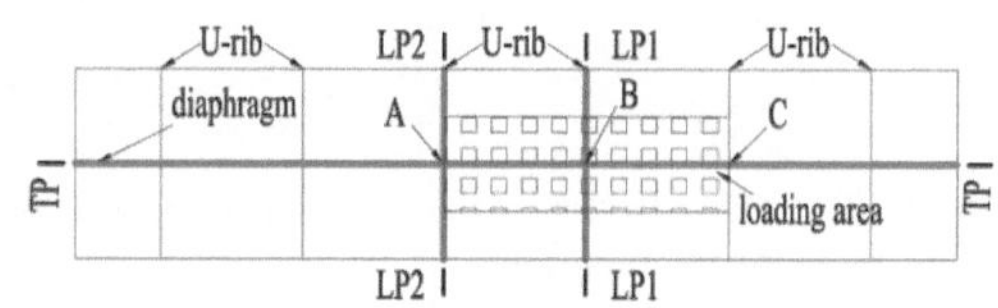

(b) Stress comparing paths in the deck

Figure 5. Sketch of stress comparing paths.

in the diaphragm were calculated and compared. The stress paths were shown in Figure 5, in which LP1 is short for longitudinal path 1, LP2 for longitudinal path 2 and CP for cope hole path.

4 EFFECT OF BOUNDARY CONDITION AND DIAPHRAGM HEIGHT

There are two types of boundary conditions which are often used in fatigue tests. One is the full supported condition (Fig. 6), and the other is the simple supported condition (Fig. 7).

For each boundary condition mentioned before, four specimens with 3 U-ribs, whose diaphragm height was 500 mm, 700 mm, 900 mm and 1000 mm respectively, were simulated by FEM.

The transverse normal stress of sections along the loading area centre line in the U-rib-to-deck weld direction of simple supported specimen models were compared with that of the full bridge analysis model (FBAM) in Figure 8. The normal stress in tangential direction along the free edge of the cope hole below the loading area of simple supported specimen models were compared with that of the full bridge analysis model (FBAM) in Figure 9.

As shown in Table 1, in the simple supported boundary condition, the transverse normal stress of the deck plate changed little with the diaphragm height increasing. The normal stress in tangential direction of the free edge of the diaphragm cope hole became closer to the FBAM with the diaphragm height increasing, especially when the diaphragm height was 1000 mm.

The transverse normal stress of sections along the loading area's edge in the U-rib-to-deck weld direction of full supported specimen models were compared with that of the full bridge analysis model (FBAM) in Figure 10. The normal stress in tangential direction along the free edge of the cope hole below the loading area of full supported specimen models were compared with that of the full bridge analysis model (FBAM) in Figure 11.

As shown in Table 2, in the full supported boundary condition, the transverse normal stress of the deck plate and the normal stress in tangential direction along the free edge of the cope hole changed weakly with the diaphragm height increasing, and were very close to the FBAM.

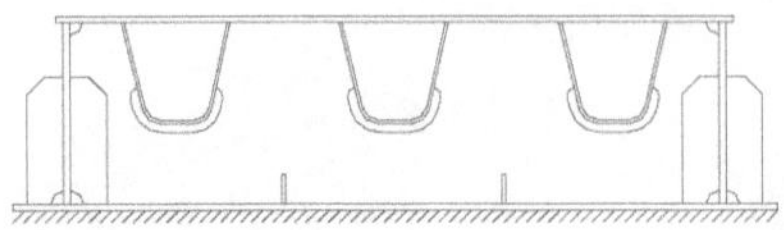

Figure 6. Sketch of full supported condition.

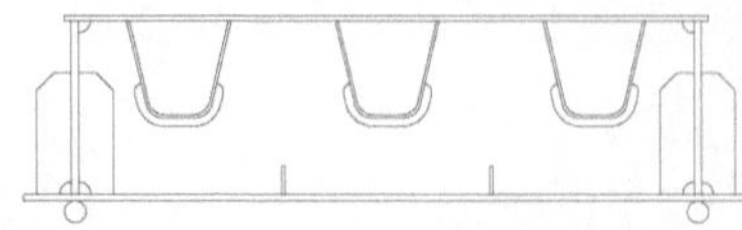

Figure 7. Sketch of simple supported condition.

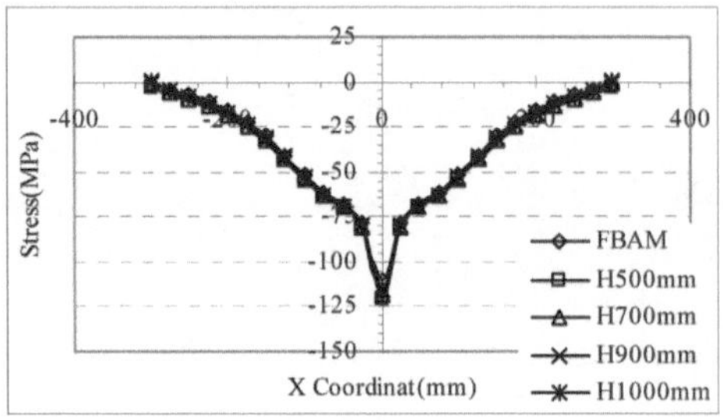

Figure 8. Transverse normal stress distribution of LP1 of simple supported specimen models with different diaphragm height.

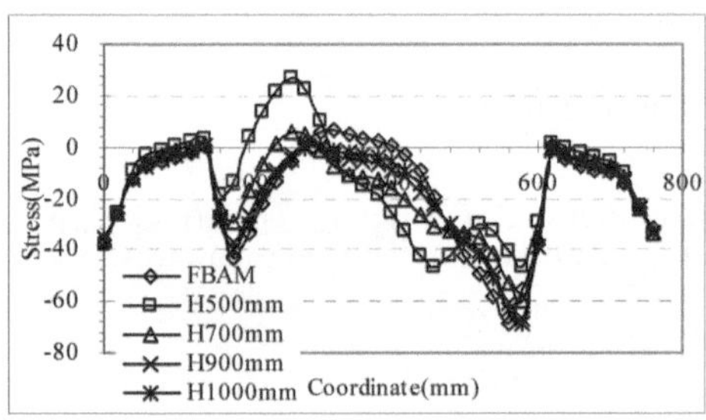

Figure 9. Normal stress distribution in tangential direction of CP of simple supported specimen models with different diaphragm height.

Table 1. The stress and relative error of simple supported specimen models with different height.

Position	Coordinate mm	Stress MPa	FBAM	H500 mm	H700 mm	H900 mm	H1000 mm
Point B*	0	Stress	110	120	118	116	115
		Relative error	0	9.1%	6.9%	5.0%	4.4%
Point E	560	Stress	68.6	40.4	53	61	63.4
		Relative error	0	–41.1%	–22.8%	–11.2%	–7.5%

* Point B is shown in figure 4, the same as below.

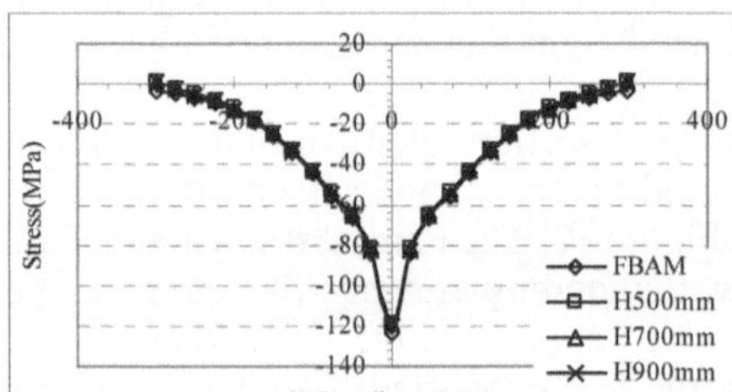

Figure 10. Transverse normal stress distribution of LP2 of full supported specimen models with different diaphragm height.

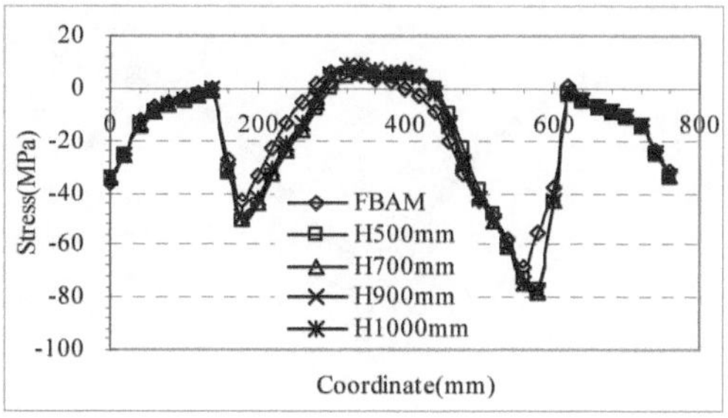

Figure 11. Normal stress distribution in tangential direction of CP of full supported specimen models with different diaphragm height.

Table 2. The stress and relative error of full supported specimen models with different diaphragm height.

Position	Coordinate mm	Stress MPa	FBAM	H500 mm	H700 mm	H900 mm	H1000 mm
Point A	0	Stress	123.7	118.4	119.2	119.9	110.2
		Relative error	0	–4.2%	–3.6%	–3.0%	–2.8%
Point E	560	Stress	68.6	73.2	74.5	74.3	74.2
		Relative error	0	6.6%	8.5%	8.3%	8.1%

5 EFFECT OF U-RIB NUMBER

Three specimens with diaphragm 500 mm high in the full supported boundary condition, whose U-rib number was 1 to 3 respectively, were simulated by FEM.

The transverse normal stress of sections along the loading area centre line in the U-rib-to-deck weld direction of specimen models were compared with that of the full bridge analysis model (FBAM) in Figure 12. The normal stress in tangential direction along the free edge of the cope hole below the loading area of specimen models were compared with that of the full bridge analysis model (FBAM) in Figure 13.

As shown in Table 3, the effect of specimen U-rib number on the transverse stress of the deck plate and the normal stress in tangential direction along the free edge of the cope hole was little. The mechanical behaviors of specimen models with 1 to 3 U-ribs were all consistent with the FBAM better and the U-rib amount could be decided according to test condition.

6 EFFECT OF DECK PLATE WIDTH

Four specimens with diaphragm 500 mm high and 3 U-ribs in full supported boundary condition, whose deck plate width was 300 mm, 400 mm, 500 mm and 600 mm respectively, were simulated by FEM.

The transverse normal stress distributions of sections along the loading area centre line in the U-rib-to-deck weld direction of specimen models were compared with that of the full bridge analysis model (FBAM) in Figure 14. The transverse normal stress distributions of sections along the loading area's edge in the U-rib-to-deck weld direction of specimen models were compared with that of the full bridge analysis model (FBAM) in Figure 15.

As shown in Table 4, the wider the deck plates of specimens were, the closer their stress distribution were to the full bridge analysis model. When the deck plate width was 400 mm, although there was a litter difference in stress at the edge of the deck plate along the U-rib-to-deck weld direction between specimen model and the FBAM, the stress coincided very well near the diaphragm.

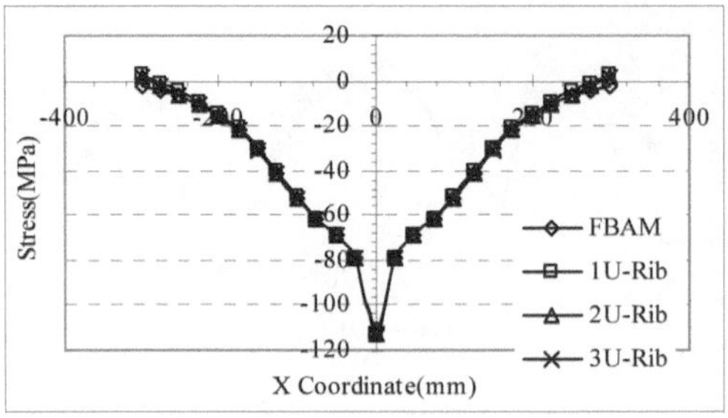

Figure 12. Transverse normal stress distribution of LP1 of full supported specimen models with different U-rib number.

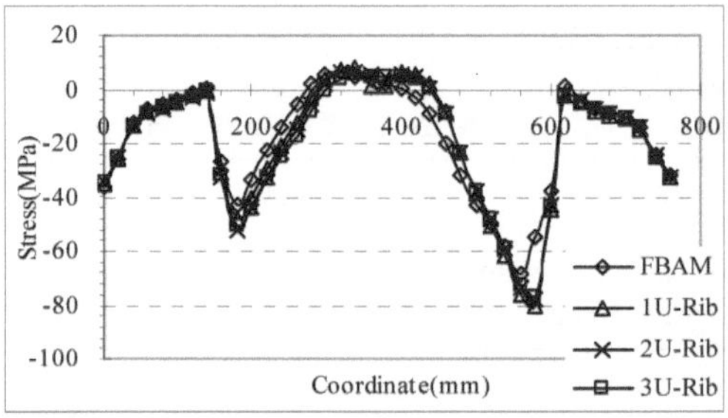

Figure 13. Normal stress distribution in tangential direction of CP of full supported specimen models with different U-rib number.

Table 3. The stress and relative error of full supported specimen models with different U-rib amount.

Position	Coordinate mm	Stress MPa	FBAM	1U-rib	2U-rib	3U-rib
Point B	0	Stress	110	112.4	112.5	113
		Relative error	0	1.8%	1.9%	2.1%
Point E	560	Stress	68.6	75.8	73.5	73.2
		Relative error	0	10.5%	7.2%	6.6%

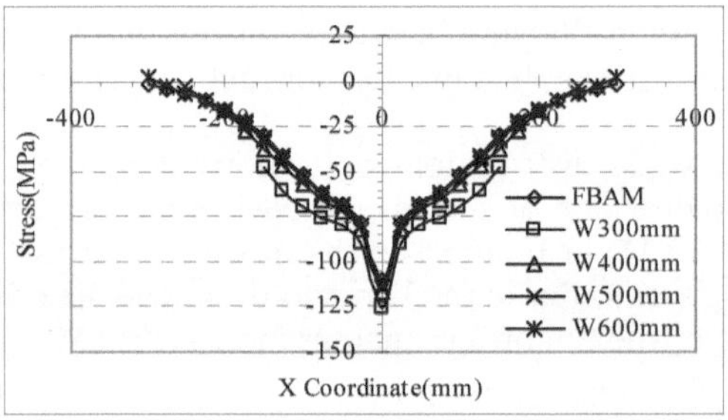

Figure 14. Transverse normal stress distribution of LP1 of full supported specimen models with different width.

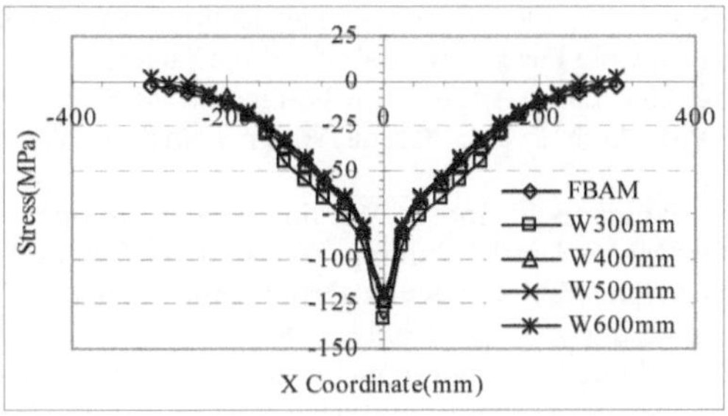

Figure 15. Transverse normal stress distribution of LP2 of full supported specimen models with different width.

Table 4. The stress and relative error of full supported specimen models with different deck plate width.

Position	Coordinate mm	Stress MPa	FBAM	W300 mm	W400 mm	W500 mm	W600 mm
Point B	0	Stress	123.7	126	117	114	113
		Relative error	0	14%	6%	3.2%	2.1%
Point A	0	Stress	124	133	123	120	118
		Relative error	0	7.7%	–0.3%	–3.2%	–4.2%

7 SUMMARY

In the simple supported boundary condition, the transverse normal stress of the deck plate of the fatigue test specimen changes little with the diaphragm height increasing, however the normal stress in tangential direction of the free edge of the diaphragm cope hole changes greatly. In the full supported boundary condition, the stress distribution of specimens changes little with the diaphragm height increasing. The effect of U-rib number on the stress of full supported specimen is small. The wider the deck plates of full supported specimens are, the closer their stresses are to the full bridge analysis model.

ACKNOWLEDGEMENTS

This research is sponsored by Science and Technology Program of Department of Transport of Zhejiang Province (No. 2011H17), National Science and Technology Support Program of China (No. 2009BAG15B02, 2009BAG15B03), Science and Technology Program of Western Transportation Construction (No. 200831849452). These supports are gratefully acknowledged.

REFERENCES

Inokuchi, S. 2006. Wheel Trucking Test For Weld Of U-Shape Rib And Deck Plate In The Orthotropic Steel Deck, Japan Bridge Association.

Jong, F.B.P.de. 2004. Overview Fatigue Phenomenon in Orthotropic Bridge Decks in the Netherlands, Orthotropic Bridge Conference, 2004: 489–512.

Rong, Z.H., Zhang, Y.L., Liu, X.G. & Tao, X.Y. 2009. Fatigue Experimental Research on Orthotropic Plate of Long-Span Cable-Stayed Bridge, Steel Construction Vol. 24, No. 120: 13–16. (in Chinese)

Sim, H.B., Uang, C.M. & Sikorsky, C. 2009. Effects of Fabrication Procedures on Fatigue Resistance of Welded Joints in Steel Orthotropic Decks, Journal of Bridge Engineering, 2009(9): 366–373.
Wang, H.Y., Wu, C. & Yang, G.T. 2011. Analysis on Fatigue Stress Range of Orthotropic Steel Deck in Cable-Stayed Bridge with Arch Pylons, China Water Transport Vol. 12, No. 11: 182–184. (in Chinese)
Wei, L.Q. 2008. Research on Fatigue Behaviors of Welding Conformations in Orthotropic Steel Bridge Deck. Engineering Master Degree Thesis, Nanjing: Wei, L.Q. (in Chinese)
Wolchuk, R. 1990. Lessons From Weld Cracks in Orthotropic Decks on Three European Bridges, Journal of Structural Engineering Vol. 116, No. 1: 75–84.
Yamada, K. & Samol, Y. 2007. Plate bending fatigue tests for root crack of trough rib of orthotropic steel deck, Proceedings of Society of Civil Engineering Vol. 63, No. 1: 35–42.
Ya, S., Yamada, K. & Ishikawa, T. 2011. Fatigue Evaluation of Rib-to-Deck Welded Joints of Orthotropic Steel Bridge Deck, Journal of Bridge Engineering 2011(8): 492–499.
Yu, B., Qiu, H.X., Wang, H. & Guo, T. 2009. Experimental Research on Fatigue Behavior and Damage Development of Welded Conformation of Orthotropic Steel Bridge Deck, Journal of Highway and Transportation Research and Development Vol. 26 No. 9: 64–69. (in Chinese)

Modeling and Computation in Engineering II – Xie (ed)
© 2013 Taylor & Francis Group, London, ISBN 978-1-138-00058-2

Uplift ultimate bearing capacity analysis of piles in sloped ground

Liu-Yong Cheng, Shan-Xiong Chen, Fei Yu & Jian Li
State Key Laboratory of Geomechanics and Geotechnical Engineering,
Institute of Rock and Soil Mechanics, Chinese Academy of Sciences, Wuhan, P.R. China

ABSTRACT: There is a strong engineering practical and theoretical reference value in solving the uplift ultimate bearing capacity of oblique slope pile foundation. With the oblique slope conditions, load transfer calculation in pile-soil system and formula of ultimate bearing capacity for pile foundation are derived in this paper. We also contrast the theoretical value derived by the formula, analog values based on FLAC3D software and the field test results. On this base, we calculate the ultimate bearing capacity of oblique slope foundation by using FLAC3D software and analyze the capacity change law and influencing factors on different conditions. The study showed that: the calculation formula of ultimate bearing capacity, FLAC3D numerical value and test calculation results can be most consistent. The slopes pile foundation bearing capacity is little influenced by the slope, provisional slope distance, but relatively large impacted by pile length. Within a certain range, we can improve its bearing capacity by increasing the length of the pile.

1 INTRODUCTION

The uplift foundation is an important type of building foundations, which is widely used in the transmission lines basis, the tall construction basis, the sluice foundation basis and so on. Compared with the pressure piles, in addition to the role of the superstructure weight, there are also the wind load and other role uplift loading on the uplift piles. So it is necessary to consider the bearing capacity when we design the uplift piles. With more foothills in China, there are more mountain transmission line project in power transmission. In order to meet the requirements of the transmission tower construction, part of the pile foundation had to be built in the oblique steep Lot even cliffs. The slope influences the transmission tower pile foundation obviously. Compared with ordinary foundations, oblique slope foundation is the most obvious different in facing empty without lateral restraint, which reduces the strength of rock and soil.

For the working mechanism of uplift pile, many scholars have done a lot researches. Kulhawy considers that equal section uplift pile's failure pattern mostly occur along the pile—soil sidewall interface, inverted cone shear break, or mixed the shear surface damage (Kulhawy etc. 1979). Through a large number of field tests and model tests, Liu Wen-Bai analyzes uplift pile pullout mechanism, failure mode, and come to the pullout capacity and uplift displacement amount computing formula (LIU 2002). Liu Zu-de describes systematically the failure mode and its mechanism of uplift piles, uplift bearing capacity calculation methods, the uplift test method to determine the pullout resistance, uplift untapereds carrying capacity, etc (LIU 1996). Since 1960s, with numerical simulation method introduced in the field of geotechnical engineering, numerical simulation of pile foundation engineering has got an immense improvement. De Nicola and Randolph use FLAC procedures of the finite difference method to simulate pile—side-friction effects under pulling and pressure loads in Gibson sand (Nicola 1993).

Yang Shen-sheng qualitatively analyzes uplift untapered the deformation and stress characteristics in soft soil with ANSYS (YANG 2003). To determine the uplift piles pulling ultimate capacity, Baars uses Plaxis establishing a numerical model founding that uplift piles pulling limit capacity based on numerical model have higher accuracy than the analytical model under empirical formulas (Baars etc. 1996).

Currently, the more mature the calculation of the theoretical system has not formed for slopes uplift piles. In this paper, we will derive pile—soil system calculated differential equations with load transfer and ultimate bearing capacity formula of pile foundation in the condition of slope, and compared with the test and FLAC3D software simulation results to determine the inversion parameter. On this base we use FLAC3D to conduct slope piles numerical simulation calculation and analysis ultimate bearing capacity variation law and influencing factors on different conditions.

2 ULTIMATE BEARING CAPACITY CALCULATION METHODS OF SLOPE PILE

2.1 *Load transfer differential equations in the pile—soil system*

When uniform-section-pile bears uplift loading, it passed load to soil through the pile, but the direction of the stress is contrast to sink friction. Firstly, there is pile tensile stress on pile top, with the increase of the upward displacement of the pile top, pile tensile stress gradually downward portion extension, when the pile top displacement reach a certain value (bit 6 to 10 mm), the frictional resistance of the interface has got to its limit, but the lower side friction is not fully used, expansion of friction occurred as the load increases, down expansion friction occurred as the load increases, the interface with peak friction expand to below. Obviously, uplift pile friction can be described as distribution function along to pile deep named z.

Influenced by the slope, friction near the pile side of the slope is not the same as the friction way from the side of the slope pile skin. If the length is L, the cross-sectional area is A, radius of slope single pile is r, the pile top vertical load is F, at the depth Z takes dz as differential pile segment.

The pile uplift ultimate bearing capacity consists of three parts: the pile friction, pile weight and the vacuum suction of the pile bottom. In practical engineering, water seepage in the soil, vacuum suction disappear over time under the long-term load, pile weight is to smaller to ignore. Base on pile force balance, a cross-section force formula at any pile internal is as following:

$$P(z)=P-\pi r\int_0^z q_{s1}dz-\pi r\int_0^z q_{s2}dz \tag{1}$$

where p = uplift force; r = radius of pile; q_{s1} = slope side skin friction of pile; and q_{s1} = the other side skin friction of pile.

Uplift displacement of pile consists of two parts: the amount of stretching of the pile and the lift displacement of pile end:

$$S_0=S_b+S_p \tag{2}$$

where s_0 = the lift displacement of pile; s_b = the amount of stretching of the pile; and s_p = he lift displacement of pile end.

Uplift amount for each section:

$$S(z)=S_0-\frac{1}{E_pA}\int_0^z P(z)dz \tag{3}$$

where E_p = elastic modulus of pile; and A = cross-sectional area of the pile.

Substituting the derivation of (3) into (2):

$$q_s(z) = q_{s1}(z) + q_{s2}(z) = \frac{E_p A}{\pi r} \cdot \frac{d^2 S(z)}{dz^2} \tag{4}$$

The above equation is differential equations of pile-soil system load transfer. We can obtain curve of skin friction through stress or displacement test components.

2.2 *Formula of ultimate bearing capacity*

Due to the relationship of the load and displacement, we know that the pile friction is deep changes in the function with the pile. Making pile tip as the origin of coordinates, we assume that piles on both sides of the friction resistance:

$$q_{s1}(z) = k_1 G_0 z e^{\frac{dS(z)}{dz}} \tag{5}$$

$$q_{s2}(z) = k_2 G_0 z e^{\frac{dS(z)}{dz}} \tag{6}$$

where k_1 = correlation coefficient of slope side; k_2 = correlation coefficient of the other side; G_0 = shear modulus of soil; q_{s1} = slope side skin friction of pile; and q_{s1} = the other side skin friction of pile.

From the above we note that:

$$q_s(z) = q_{s1}(z) + q_{s2}(z) = \frac{(k_1 + k_2) G_0 z}{1 - \frac{(k_1 + k_2)\pi r G_0}{2AE_p} z^2} \tag{7}$$

Pullout capacity:

$$Q = \pi r \int_0^l q_{s1}\, dz + \pi r \int_0^l q_{s2}\, dz \tag{8}$$

$$Q = -AE_p \ln\left[1 - \frac{(k_1 + k_2)\pi r G_0}{2AE_p} l^2\right] \tag{9}$$

where E_p = elastic modulus of pile; A = cross-sectional area of the pile; k_1 = correlation coefficient of slope side; k_2 = correlation coefficient of the other side; G_0 = shear modulus of soil; r = radius of pile; and l = pile length.

The above equation is formula of ultimate bearing capacity of slope pile foundation. As long as we know parameters and correlation coefficients of the pile and soil, we will be able to estimate uplift bearing capacity value of slope pile foundation.

3 NUMERICAL SIMULATION

The calculations use the FLAC3D software which based on finite difference theory. It analyzes the influence factors of pile foundation uplift capacity. According to the design, choosing different slope ratio, different depth of the pile, different facing slope distance and different length of the pile to calculate.

3.1 *Model*

The model is built according to the following rules: taking piles of actual slope into simulation calculation, and the slope ratio, facing slope distance and different length of the pile are based on the setting conditions of different working conditions. The concrete geometric model is shown as figure 1.

Boundary conditions: in horizontal direction, we fix the left and right side of the boundary in order to limit the horizontal displacement in the x direction; Limit the displacement in the y direction of the entire model; Fix displacement of three directions underside of the model. There are no limits on the rest boundary.

The computing model (shown as Figure 2) could be built according to the geometric model. Computing models would be different due to different conditions of different working conditions. In view of more than a few of calculation conditions, here gives only one model grid diagram of corresponding working condition.

3.2 *Brief introduction of calculation process*

Aiming to a concrete working condition, the calculation process of using FLAC3D software is shown as figure 3.

3.3 *Theoretical validation and parameter inversion*

To verify the reasonableness of the formula, referring to the test given by Wang Ai-min, which is about a field pullout test in a flat slope, the appropriate parameters could be

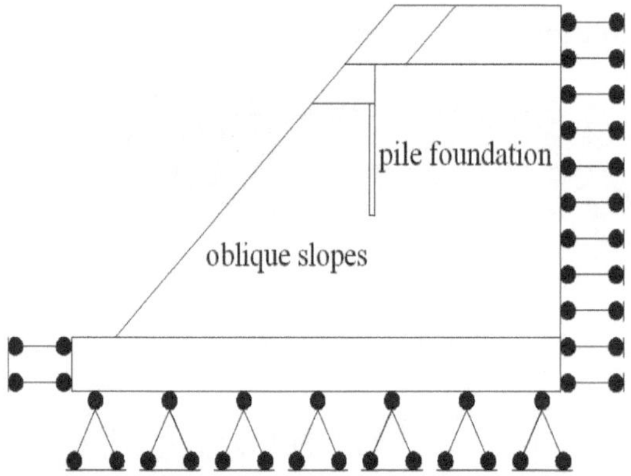

Figure 1. Schematic diagram of model.

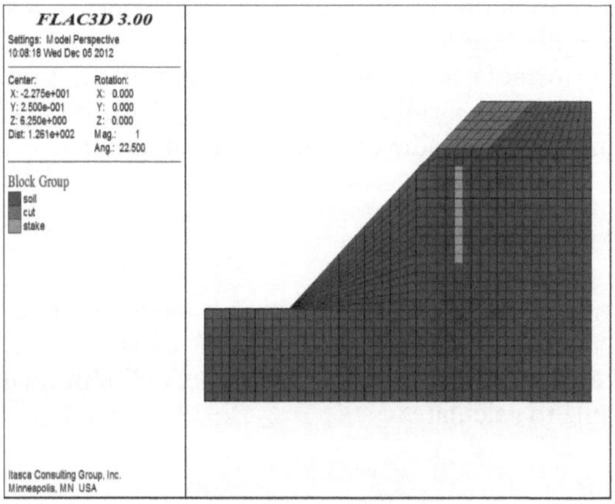

Figure 2. Model grid diagram of computing model.

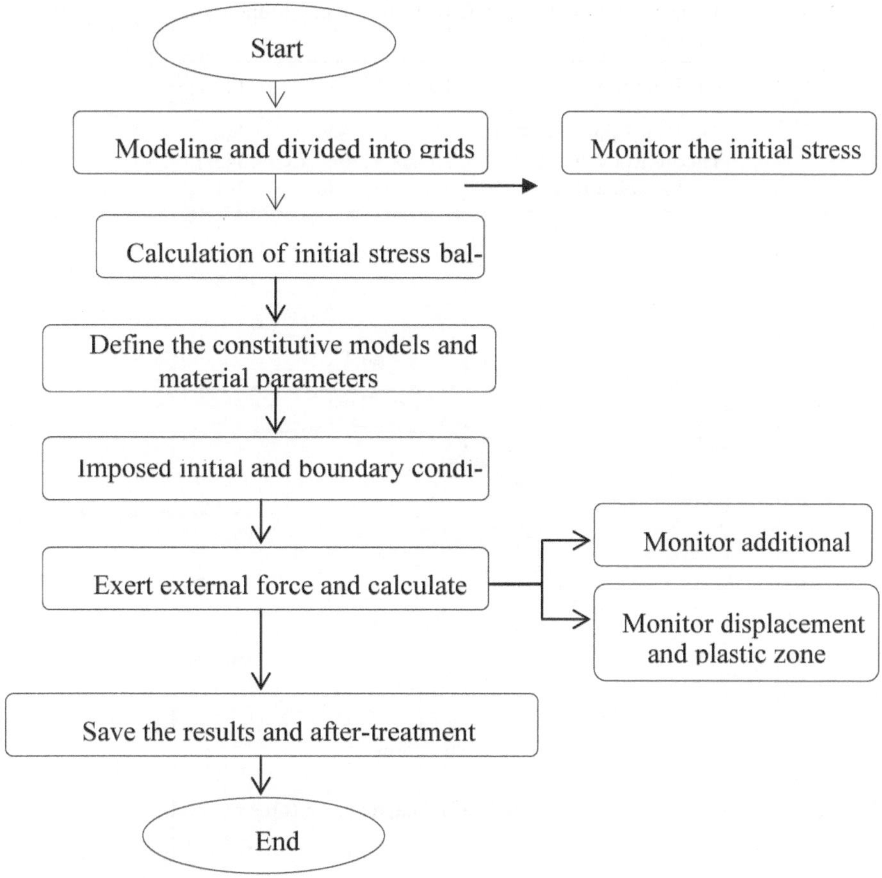

Figure 3. Schematic diagram of calculation process.

defined (WANG 2005). Taking the pile diameter as 0.3 m, the soil elastic modulus is equal to 11.5 × 106 N/m^2 the Poisson's ratio is 0.35 and the elastic modulus of concrete pile is 2.55 × 104 MPa. Taking the value of ultimate bearing capacity in test and those derived from numerical simulation and theoretical into compared. To coincide with the field test, the numerical simulation and theoretical calculation are taken flat slope of pile foundation to calculate that k1 = k2. The values on three methods are shown in table 1.

In the numerical calculation, the parameters inversion has a direct impact on the numerical simulation of the effect. The parameter inversion requires not only experience, but also need compared with the measured data, and only a combination of both could make the numerical simulation reflect the true results accurately. The contrast on three methods is shown as figure 4.

As can be seen from the table, ultimate bearing capacity of pile foundation on three methods of is different. When the pile length is 10 m, the value in numerical simulation is the maximum, with a difference of 6.2% of the minimum value in theoretical. When the pile length is 12 m, the value in theoretical is the maximum, with a difference of 1% of the minimum value in test. When the pile length is 14 m, the value in theoretical is the maximum, with a difference of 6.9% of the minimum value in test. Results show that the ultimate bearing capacity on three methods vary, but with little difference, thus proving the reasonableness of the theoretical and numerical calculations.

Table 1. Ultimate bearing capacity of pile foundation on three methods.

Pile length/m	10	12	14
Experimental value/KN	320	340	340
Theoretical value/KN	299.78	349.73	365.23
Analog value/KN	328	346	351

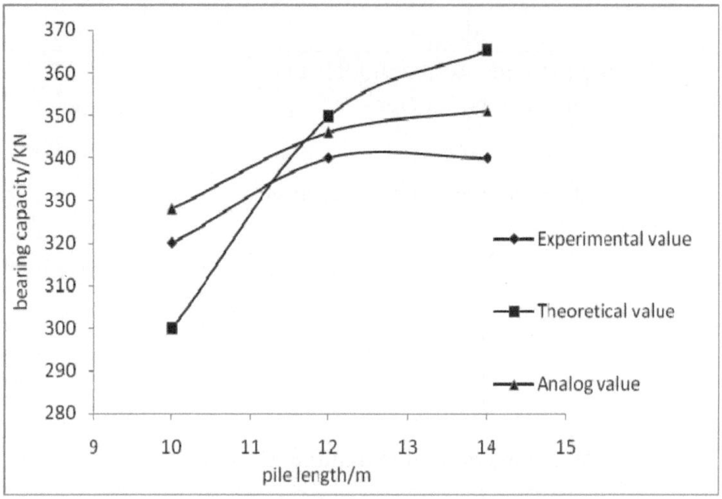

Figure 4. Comparison about ultimate bearing capacity on three methods.

3.4 *Variation of tensile bearing capacity of incline pile foundation*

According to the above parameters, we make the further calculation for uplift bearing capacity of the oblique slope pile foundation with different slopes, slope lengths, pile lengths. And we also use it analyze the capacity change law and influencing factors on different engineering conditions.

1. The influence of different gradient on ultimate bearing capacity.
   Figure 5 shows ultimate bearing capacity in different gradient.
   The picture shows tensile bearing capacity of incline pile foundation of different gradient is different, but the difference is not large. When gradient is 1:1, it is 240 KN; when gradient reduce to 1:1.5, it is 270 KN; when the size of gradient has further shrunk, it does not increase in substance. This suggests that incline has some influence on tensile bearing capacity of pile foundation, but when the gradient reduces, the influence reduce. When the slope is less than 1:1.5, the influence is little; when the size of gradient has further shrunk, the influence of incline on tensile bearing capacity can be insignificant.
   Because of the existence of incline and the loss of soil of pile foundation outer flank, it does not offer enough soil counterforce, thus tensile bearing capacity of incline is less than flat slope. As the slope of incline and the loss of soil of pile foundation outer flank reduce, the influence of in cline on bearing capacity reduces.
2. The influence of different distance to slope on ultimate bearing capacity.
   Figure 6 shows ultimate bearing capacity in different distance to slope.
3. The influence of different pile length on ultimate bearing capacity.
   Figure 7 shows ultimate bearing capacity in different pile length. This picture shows as the pile length increase, ultimate bearing capacity increase. When the pile length is 5B (5 times pile diameter), ultimate bearing capacity is 140 KN; when the pile length is 20B

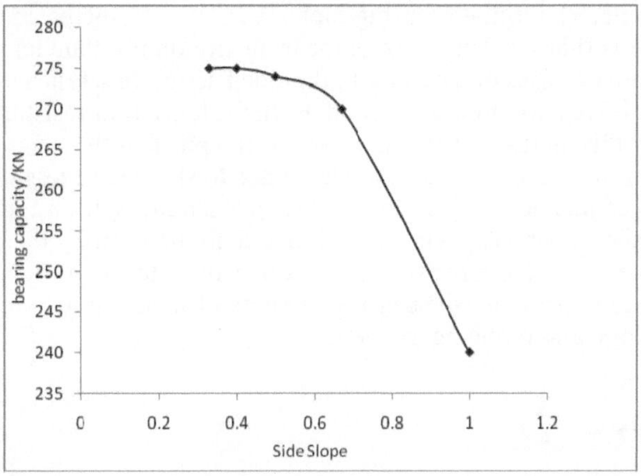

Figure 5. Ultimate bearing capacity on different gradient.

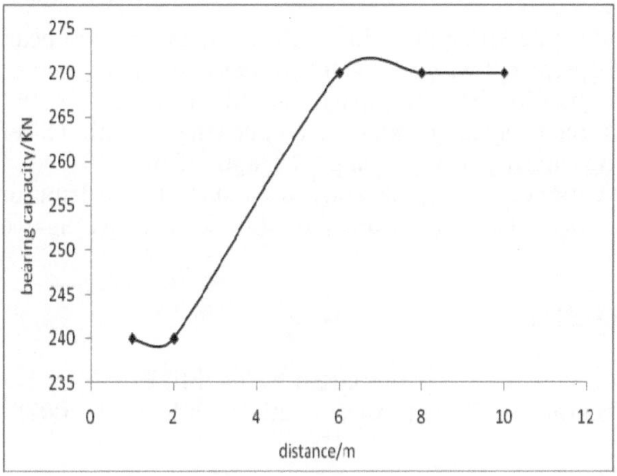

Figure 6. Ultimate bearing capacity on different distance to slope.

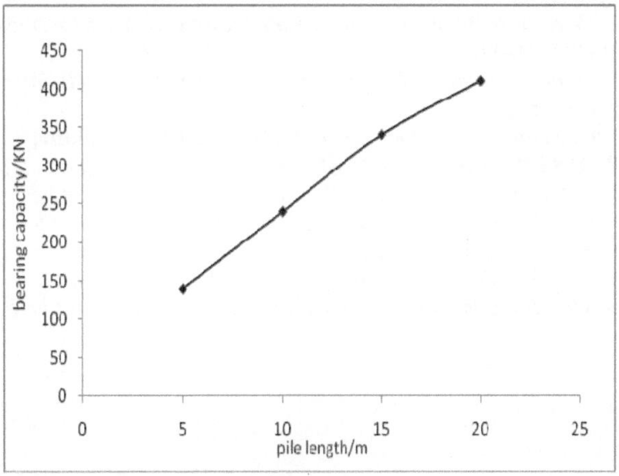

Figure 7. Ultimate bearing capacity on different pile length.

(20 times pile diameter), ultimate bearing capacity is 410 KN, increasing 270 KN, almost 2 times. This suggests that pile length has large influence on ultimate tensile bearing capacity. Because of the increases of pile length, the side friction resistance of pile increase, but the size of uplift of pile foundation is equal to the side friction resistance of pile and the weight of pile, so it is increase as the increases of the pile length.

Upon inspection, we can see gradient and distance to slope have less influence on tensile bearing capacity of incline pile foundation than pile length. Within the specific limits, we can increase tensile bearing capacity of incline pile foundation by increasing pile length. When the distance to slope is more than 2 times pile diameter or gradient is less than 1:1.5, it is not effective to improve tensile bearing capacity of incline pile foundation by increasing distance to slope and reducing gradient.

4 CONCLUSION

By the analysis on bearing capacity behavior of pile foundation on oblique slope, this paper derives the formula for the ultimate bearing capacity of uplift pile and gets numerical calculation by using FLAC3D. And we analyze the capacity change law and influencing factors on different engineering conditions, and we draw the following conclusions:

1. For the oblique slope pile with pullout loads, the formula for uplift bearing capacity in this paper can be well agreement with FLAC3D numerical calculation and the test result.
2. The slopes pile foundation bearing capacity is little influenced by the slope, provisional slope distance, but relatively large impacted by pile length. Within a certain range, we can improve its bearing capacity by increasing the length of the pile.
3. When Deducing Ultimate bearing capacity, we assume the distribution of the skin friction. Reasonableness of the assumptions should be further investigated.

ACKNOWLEDGEMENT

Research presented in this paper was supported by Twelfth Five-Year National Science and Technology Pillar Program of China (project numbers: 2011BAB10B02).

REFERENCES

De Nicola, A. & Randolph, M.F. 1993. *Tensive and Compressive Shaft Capacity of Piles in Sand*. Journal of Geotechnical Engineering. 119(12): 1952–1973.

Kulhawy, FH., Kozera, D.W. & Wilhiam, J.L. 1979. *Uplift testing of Model Drilled Shaft in Sand*, J. of Geotech. Eng. 105 (GT1): 31–47.

Liu Wen-Bai. *Development in the theory and engineering practice of pile foundation project*. 2002. Beijing: China Building Industry Press.

Liu Zu-de. *Pile foundation engineering technology*. 1996. Beijing: China building industry press.

LüYARU, LIU HANLONG & DING XUANMING. 2012. *Field tests on bearing characteristics of X-section pilecomposite foundation*. Journal of Performance of Constructed Facilities. 2012.

Van Baars, S. & Van Niekerk, W.J. 1999. *Numerical modelling of tension piles*. Symposium Beyond 2000 in Computational Geotechnics, Balkema.

Wang Aimin. 2005. *Single uplift pile stress and deformation*. Jilin University.

Yang Shensheng. 2003. *Uplift pile in soft soil foundation stress and deformation behaviors*. Hangzhou: Zhejiang.

Modeling and Computation in Engineering II – Xie (ed)
© 2013 Taylor & Francis Group, London, ISBN 978-1-138-00058-2

The application of self-designed machinery to remove ACM and DCM in the chimney

J.Y. Liu & P.F. Li
China International Water & Electric Corporation, Beijing, China

ABSTRACT: This passage described and discussed a method used for cleaning and removal of asbestos-containing materials (ACM) and dioxin contaminated material (DCM) which is confined in the chimney flues. It is based on an actual civil engineering project which is Kwai Chung Incineration Plant (KCIP) Demolition and Decontamination Works in Hong Kong. This method by using a type of self-designed machinery is unique and effective to avoid the ACM and DCM divulging to the environment and is very safe for the operator and worker. The technique can be spreaded in relevant field and applied in similar projects or works.

1 INTRODUCTION

In general, the designated demolition and decontamination project at the chimney of Kwai Chung Incineration Plant (KCIP) will involve materials of significant health and safety concerns, namely asbestos-containing materials (ACM) and dioxin contaminated material (DCM).

Although DCM is confined in the KCIP chimney flues, it is present in ash form all over the inner surface of the flues and special removal technique is needed to effectively clear it all. It will be thoroughly removed using a method in compliance with the requirements of the Environmental Permit (EP), Clause No. 121/2002 Cl 2.12 to 2.14 as well as approved Environmental Impact Assessment Report, Reg. No. AEIAR-049/2002.

2 GENERAL DEMOLITION APPROACH

2.1 *List and location of ACM and DCM to be removed*

In view of asbestos-containing materials (ACM) and dioxin contaminated material (DCM) currently present at KCIP chimney, the following is a list of items to be removed as described in subsequent sections of this demolition and decontamination plan.

ACM remaining at chimney: asbestos flue guide plates (Platform 7 to Platform 1, total: 16 nos.), asbestos sampling port gaskets (Platform 2 to Platform 1, total: 8 nos.), asbestos door sealants (G/F, total: 4 nos.) and asbestos ventilation louvers (between Platform 7 and Platform 6, total: 6 nos.). DCM inside chimney flues: ash material (approximately 7 m^3) within four metal flues from top to bottom of total height approximately 150 m above ground.

2.2 *General demolition approach*

The principle of the demolition procedure for the chimney flues is that they will be cut into small pieces of manageable size (e.g., 1.2 m × 0.8 m) by flame-cutting or any other appropriate method with the assistance of hand-held tools on the spot by operatives who will work from intermediate working platforms between chimney platform levels.

If flame-cutting procedure is carried out, fire-resistant materials (non-ACM) should be used for physical protection of both the immediate work environment and the workers in action. In particular, a piece of woven fire-retardant fabric (non-ACM) suspended by wire will also be placed within the metal flue space to collect any material or object inadvertently falling from the flame-cutting point. As the KCIP chimney is considered a confined space, a suitable risk assessment regarding flame-cutting to be conducted inside the chimney should be made by a Safety Officer in advance. Hazards including presence of flammable or toxic gases, risk of fire or explosion, oxygen deficiency and in-rush of free-flowing solid or liquid etc shall be assessed. All relevant local work safety regulations and codes of practice for hot work (e.g., welding process) are also to be observed.

Localized demolition of chimney flues may be conducted when no DCM or ACM is present in the vicinity or associated with their structure in a defined work zone. In exceptional circumstances, access to ACM can only be gained in line with demolition work process and caution should be exercised. Advice should therefore be sought from the Asbestos/Dioxin Specialist in advance.

The main site will be protected by security personnel and a high hoarding such that the public would be totally excluded from the Project. The area beneath the chimney would be cordoned off and only authorized staff involved in the demolition of the chimneys would be allowed admission into the vicinity of the chimney structure.

Concerning demolition of concrete chimney wall after clearance of ACM/DCM, the principle is that the upper portions of the chimney will be cut into pieces by hand-held tools on the spot by operatives who will work from working platforms inside the chimney. Hydraulic breakers will be an option for use for the remaining lower portions of the chimney.

The demolished concrete debris will then be broken down and removed by hydraulic excavators and loaded on to trucks for transportation to the designated disposal site. During the demolition work, water sprays will be used to suppress excessive dust generated by the processes.

2.3 *DCM removal*

To accomplish thorough DCM removal at chimney flues, the main considerations and work preparation requirements are listed as follows:

- Construction of internal temporary structures for work access;
- Ash is confined in flues space and escape of DCM is not expected under normal condition. Therefore, any cracks or holes (porous condition) present on the metal flue body should be properly sealed up using foam adhesive;
- Construction of internal scaffolding with certification on structural stability or safety;
- CCTV monitoring of the cleaning process along with video recording;
- Use of a rotary cleaner to clear DCM inside chimney flues by means of HEPA vacuuming and wet-wiping;
- Performance criteria are that ash material is not found remaining on previously contaminated surfaces through CCTV monitoring and subsequent spot-check by visual inspection;
- Treatment of removed DCM ash with cement in a containment setup; and
- TCLP test procedures required after cement stabilization process to confirm <1ppb dioxin concentration in the cement/ash solid mixture for safe disposal.

3 DCM REMOVAL STRATEGY

Since DCM removal is to be carried out prior to demolition of metal flues and ACM removal, confirmation of DCM clearance is required at various stages. The overall strategy of DCM removal is to start from Platform 7 down to Ground Floor of chimney and finally back up to chimney top (i.e. above Platform 7). Eight (8) major DCM removal zones are divided for progress of work and are graphically illustrated in Figure 1.

The eight DCM removal work zones in containment setup are defined as follows:

- Zone P7-P6 (between Platform 7 & Platform 6)—Containment C2;
- Zone P6-P5 (between Platform 6 & Platform 5)—Containment C3;

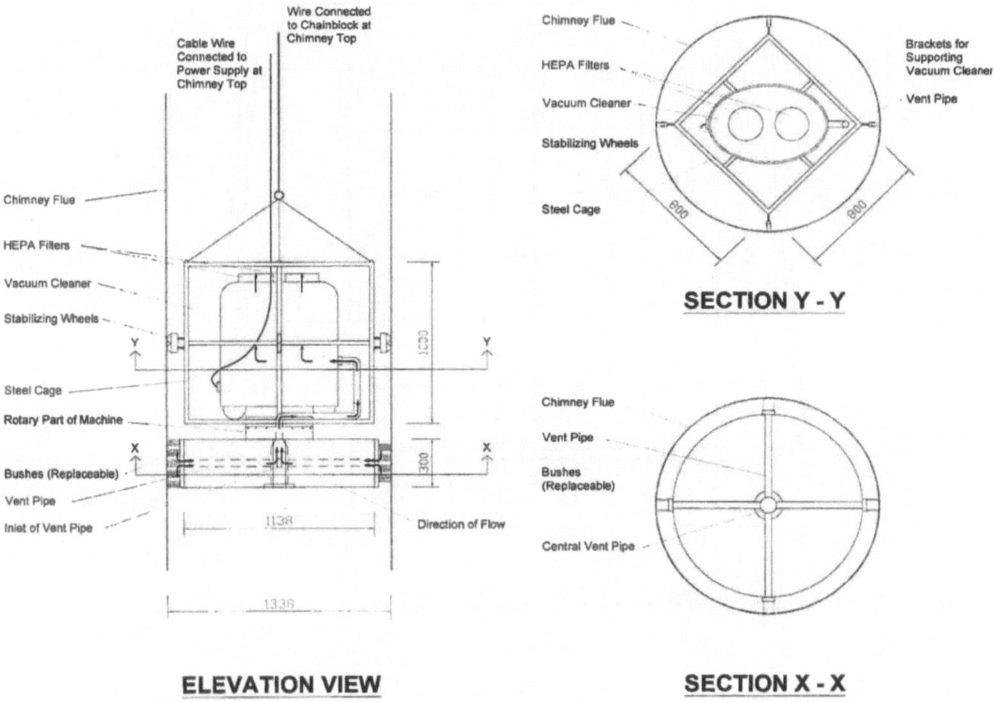

Figure 1. Details of cleaning Robot for removal of DCM in chimney fule.

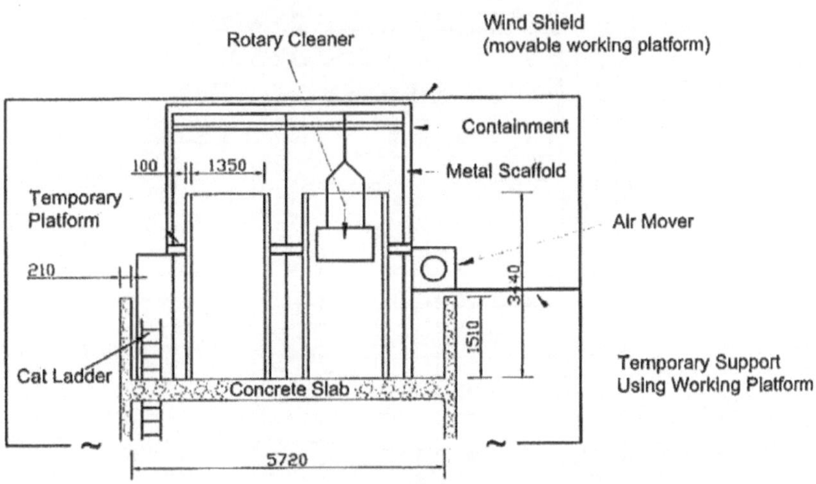

Figure 2. Use of rotary cleaner for removal of DCM.

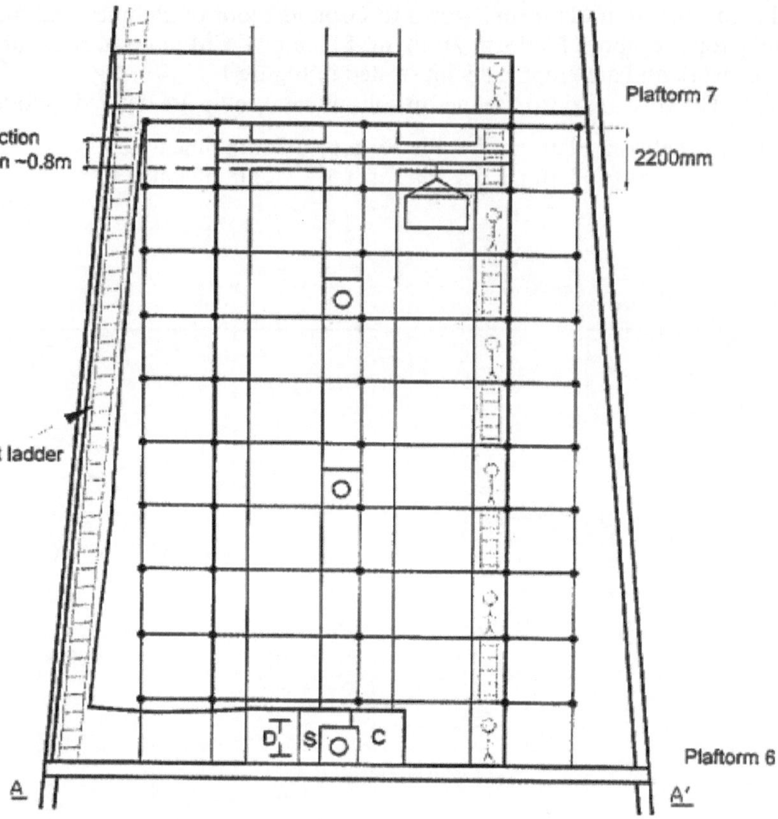

Figure 3. The procedure of removal of DCM inside the chimney.

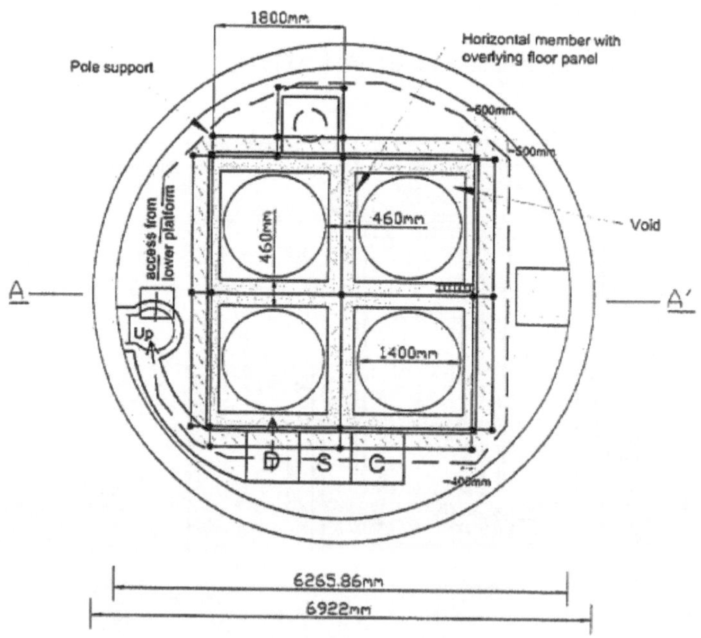

Figure 4. The section of the chimney.

Figure 5. The photo show of the rotary cleaner.

- Zone P5-P4 (between Platform 5 & Platform 4)—Containment C4;
- Zone P4-P3 (between Platform 4 & Platform 3)—Containment C5;
- Zone P3-P2 (between Platform 3 & Platform 2)—Containment C6;
- Zone P2-P1 (between Platform 2 & Platform 1)—Containment C7;
- Zone P1-G/F (between Platform 1 & Ground Floor)—Containment C8; and
- Zone Top-P7 (Chimney top to Platform 7)—Containment C9.

4 CONCLUSIONS

The approach described in this passage has been proved to be safely and effectively in cleaning and removal of asbestos-containing materials (ACM) and dioxin contaminated material (DCM) especially in the confined space such as in the chimney flues. It can ensure the safety and health of the operators and workers involved in the project. This method can be widely promoted in relevant field and applied in similar projects or works.

REFERENCES

Atkins China Ltd. 2000. Contract No. CV/2007/06 Asbestos Abatement Plan.

Atkins China Ltd. 2001. *Kwai Chung Incineration Plant Environmental Impact Assessment (Final Report).*

China International Water & Electric Corp. 2009. Demolition and Decontamination Plan for Chimney Flue DCM/ACM Removal at Kwai Chung Incineration Plant (KCIP).

Cheremisinoff, N.P. 2002. *Solid Waste Management and Waste Minimization Technologies.* Oxford:Butterworth.

De Nevers, N. 2000. *Air Pollution Control Engineering. 2nd ed.*. New York:McGraw-Hill.

EPD. 2002. Environmental Permit EP-121/2002.

EPD of the HKSAR. Air Pollution Control Ordinance (CHAPTER 311).

EPD of the HKSAR. ENVIRONMENTAL IMPACT ASSESSMENT ORDINANCE (CHAPTER 499).

Hong Kong Construction Association. 2009. *Best Practice Guide for Environmental Protection on Construction Site.* Hong Kong.

Occupational Safety and Health Branch Labour Department. 2002. Code of Practice on Control of Air Impurities in the Workplace.

Shine, H.J. & Zhao, D.C. 1990. Electron transfer to excited doublet states. Photoirradiation of 10-methylphenothiazine cation radical perchlorate in solutions of phenylacetylene and p-tolylacetylene in acetonitrile. *J. Org. Chem.*

Yuji Aso, Yukiko Miyamoto, Karen Mine Harada, Keiko Momma, Shigeyuki Kawai, Wataru Hashimoto, Bunzo Mikami & Kousaku Murata. Engineered membrane superchannel improves bioremediation potential of dioxin-degrading bacteria. *Nature Biotechnology*, 24.

Modeling and Computation in Engineering II – Xie (ed)
© 2013 Taylor & Francis Group, London, ISBN 978-1-138-00058-2

Static analysis of cable under self-weight and horizontal loads

Dong Shi, Yujun Qi & Weiqing Liu
Nanjing University of Technology, Nanjing, Jiangsu, China

ABSTRACT: This paper presents a new approach to investigate the static response of cables subjected to self-weight and horizontal distributed loads. Two innovative strategies that coordinate transformation matrix and equivalent form of deformation compatibility equation of cable are proposed. Further, the three-dimensional problem that static reasons of cable under self-weight and horizontal loads is simplified to two 2-D problems by coordinate transformation matrix. Then, based on the theory of catenaty action, the equivalent form of deformation compatibility equation is used and the static response of cable is obtained. Finally, a comparison of the theoretical method in this paper and FEM analysis is presented, and shows that the theoretical method which proposed in this paper is valid.

1 INTRODUCTION

Cable has wide application in engineering as structural components such as cable-stayed bridges, cable-nets structure, and mooring system of floating structures (Buonopane, S. & Billington, D. 1993, Mei, K.-H. & Lu Z.-T. 2002). Due to its advantages of simple stress state and light weight, the structural response of cable has attracted numerous researchers. Based on finite difference discretization, Najib Bouaanani & Mohamed Ighouba (2011) investigated the static response of the inclined suspended cables subjected to vertical concentrated and distributed loads. Static solutions of cable under self-weight were derived out in two cases about considering elasticity or not respectively (Wei Jiandong et al., 1999). Qi Yujun et al. (2011) gave the precision for the catenary and parabola solutions, and proposed the application ranges of both approximate methods. The researches above mainly focus on mechanics characteristic of cable under vertical loads. However, horizontal loads, wind and flow load for instance, are also having an effect on the design of cable in engineering structures. The cable is in 3D forced state and the geometric nonlinearity analysis will be more complex under self-weight and horizontal loads. Only few scholars pay attention to this problem. Two simplified models were developed for static analysis of stayed cable under wind load, but the models had some limitations (Liu Zhiwen et al., 2005). Chin-Sheng Kao et al. (2006) analyzed the static instability of long-span cable-stayed bridges due to wind loading by 3-D analysis. For this model, a 4-node isoparametric cable element is used which is approximate, and it can not be used to analysis the real state of the cable.

In this paper, three-dimensional model is simplified into two planes by introducing coordinate transformation matrix. The existing deformation compatibility equation can not be directly used for the cable in two coordinates. In view of this, equivalent form of it needed to be derived out.

2 MECHANICAL MODEL AND SOLUTION

2.1 *Problem description*

There are two mechanical states named state I and state II, and the geometry of the problem is shown in Fig. 1. The load is only self-weight q_{z0} of cable and the Cartesian coordinate OXYZ was employed in the state I. (0, 0, 0) and (l_x, 0, c) are coordinates of cable's ends called A and B (see Fig. 2). An additional arbitrary distributed load q_f is loaded and the Cartesian coordinate OX′Y′Z′ was employed in state II, at same time q is the resultant force of q_{z0} and q_f (see Fig. 3). The problem is to get static solution of cable in state II.

2.2 *Equivalent form of deformation compatibility equation*

From state I to state II, geometrical elongation of cable is (Shen Shizhao 1997)

$$\Delta S_g = \int ds - ds_0 = \int \sqrt{1+\left(\frac{dz}{dx'}\right)^2}\, dx' - \int \sqrt{1+\left(\frac{dz_0}{dx}\right)^2}\, dx \tag{1}$$

where z_0 and z are curve equations of cable in Cartesian coordinate OXYZ and OX′Y′Z′. Physical elongation can be expressed as Eq. (2) by introducing non-stress state of the cable

$$\Delta S_f = \int\left(\frac{T}{EA}\right)ds - \int\left(\frac{T_0}{EA}\right)ds_0 \tag{2}$$

where A and E are section area and Young modulus of cable, T and T_0 are the internal force of cable in state I and state II, respectively. First item of Eq. (2) is the physical elongation from non-stress state to state I, and the second is from non-stress state to state II. Deformation compatibility equation is abbreviated as

$$\Delta S_g = \Delta S_f \tag{3}$$

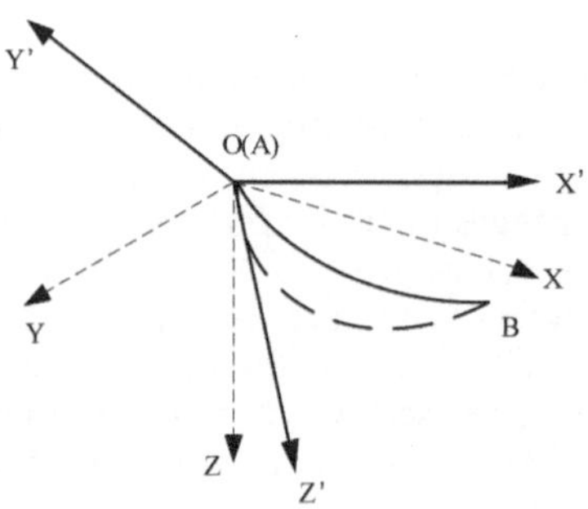

Figure 1. Curve of cable in two states.

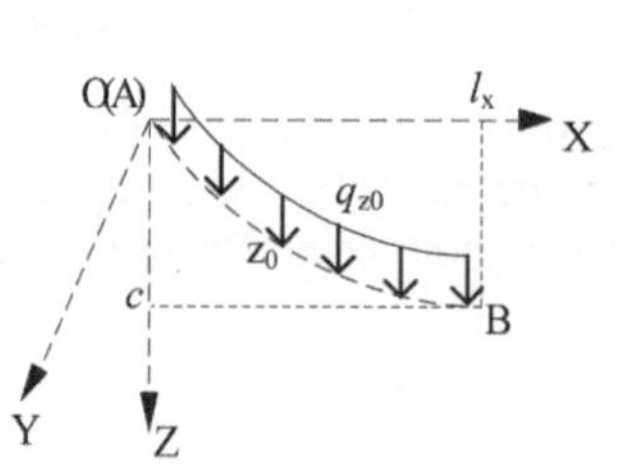

Figure 2. Cable in state I.

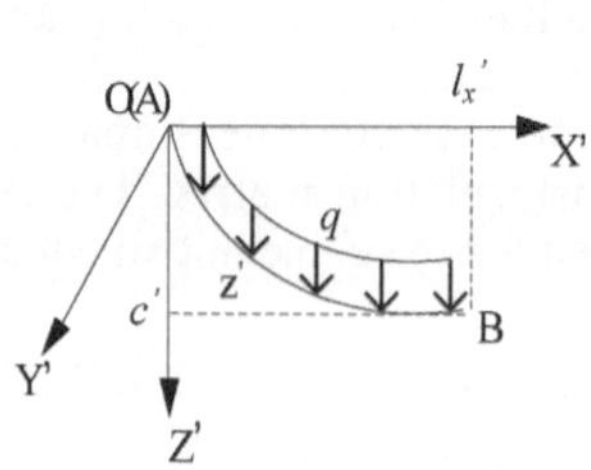

Figure 3. Cable in state II.

Substituting Eqs. (1) and (2) into Eq. (3), and transposing terms, Eq. (3) becomes

$$\int\sqrt{1+\left(\frac{dz}{dx'}\right)^2}dx' - \int\left(\frac{T}{EA}\right)ds = \int\sqrt{1+\left(\frac{dz_0}{dx}\right)^2}dx - \int\left(\frac{T_0}{EA}\right)ds_0 = S_w \tag{4}$$

where S_w is the non-stress length of cable. Accurate solution of S_w is provides by Feng Jiechun & Dai Yingzhang (2012) as

$$S_w = \int_0^l \frac{ch\left(x/\xi + c_1\right)}{\eta ch\left(x/\xi + c_1\right)+1}dx = \frac{\xi}{\eta}\left[\frac{l}{\xi} + g(c_1) - g\left(\frac{l}{\xi}+c_1\right)\right] \tag{5}$$

where $g(t) = \frac{1}{\sqrt{1-\eta^2}}\ln\frac{1+\eta e^t - \sqrt{1-\eta^2}}{1+\eta e^t + \sqrt{1-\eta^2}}, \xi = H/q, \eta = H/EA$ and $c_1 = arsh\left[(h/2\xi)/sh(l/2\xi)\right] - l/2\xi$

.

2.3 *Coordinate transformation*

Here, q_x, q_y and q_z are defined as coordinate components of q_f, as is shown in Fig. 4. So

$$q = \sqrt{q_x^2 + q_y^2 + q_z'^2} \tag{6}$$

where $q_z' = q_z + q_{z0}$.

There are three basic coordinate transformations from Cartesian coordinate OXYZ to OX′Y′Z′ by introducing two Cartesian coordinates OXY_1Z_1 and OX_1Y_1Z' (see Fig. 4), and they can be outlined as follows:

1. From OXYZ to OXY_1Z_1: X-axis is the public axis, Z_1-axis is parallel to projection of q in YOZ plane and the angle between Z-axis and Z_1-axis is denoted by θ_1;
2. From OXY_1Z_1 to OX_1Y_1Z': Y_1-axis is the public axis, Z′-axis parallel to q and the angle between Z-axis and Z_1-axis is denoted by θ_2;
3. From OX_1Y_1Z' to OX′Y′Z′: Z′-axis is the public axis, line AB is in plane OX′Z′ and the angle between X-axis and X′-axis is denoted by θ_3.

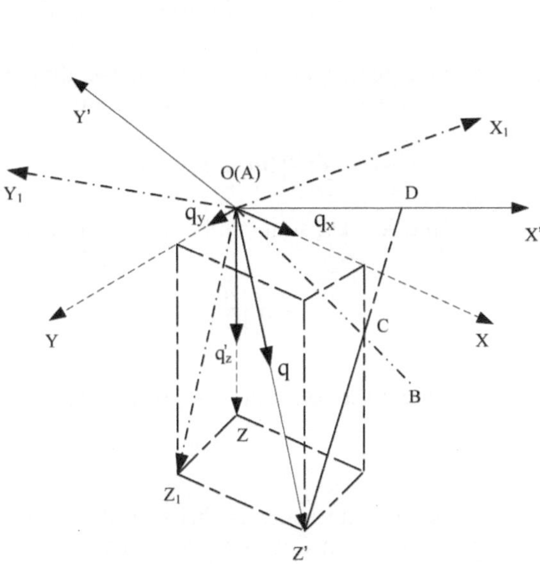

Figure 4. Coordinate transformation of cable.

The relationship of coordinates between Cartesian coordinate OX′Y′Z′ and OXYZ is

$$[x' \quad y' \quad z']^{\mathrm{T}} = [D][x \quad y \quad z]^{\mathrm{T}} \tag{7}$$

where [D] is the coordinate transform matrix. And it can be decomposed by Ke Jingpu (2008) as

$$[D] = \mathrm{R_Z}(\theta_3)\mathrm{R_Y}(\theta_2)\mathrm{R_X}(\theta_1) \tag{8}$$

in which

$$R_x(\theta_1) = \begin{bmatrix} 1 & 0 & 0 \\ 0 & \cos\theta_1 & -\sin\theta_1 \\ 0 & \sin\theta_1 & \cos\theta_1 \end{bmatrix}, R_y(\theta_2) = \begin{bmatrix} \cos\theta_2 & 0 & -\sin\theta_2 \\ 0 & 1 & 0 \\ \sin\theta_2 & 0 & \cos\theta_2 \end{bmatrix}, R_z(\theta_3) = \begin{bmatrix} \cos\theta_3 & -\sin\theta_3 & 0 \\ \sin\theta_3 & \cos\theta_3 & 0 \\ 0 & 0 & 1 \end{bmatrix}$$

Above all, θ_1, θ_2 and θ_3 are needed to be discussed to determine [D].

Based geometrical relationship, θ_1 and θ_2 can be calculated directly by Eqs. (9) and (10).

$$\tan\theta_1 = \frac{q_y}{q_z'}, \sin\theta_2 = \frac{q_x}{\sqrt{q_x^2 + q_y^2 + q_z'^2}} \tag{9}$$

The equation of plane X_1OY_1 is

$$q_x x + q_y y + q_z' z = 0 \tag{10}$$

and equation of plane S_1, composed of Z′-axis and line AB, is represented as

$$cq_y x - (cq_x - l_x q_z')y - l_x q_y z = 0 \tag{11}$$

Using Eqs. (11) and (12), the direction vector of X′-axis which is the intersection line of plane S_1 and X_1OY_1 can be obtained as

$$\vec{n_1} = (l_x q_z'^2 - cq_x q_z' + q_y^2 l_x, -cq_y q_z' - l_x q_x q_y, cq_y^2 - l_x q_x q_z' + cq_x^2) \tag{12}$$

From the steps (1) and (2) of coordinate transformations above, direction vector of X_1-axis is given by $R_y(\theta_2)\, R_x(\theta_1)\, (x, y, z)^{\mathrm{T}}$ when $x = 1$ and $y = z = 0$, namely

$$\vec{n_2} = (\cos\theta_2, -\sin\theta_1\sin\theta_2, \cos\theta_1\sin\theta_2) \tag{13}$$

θ_3 is the angle between X_1-axis and X′-axis, so it can be calculated by Eq. (15)

$$\cos\theta_3 = \frac{\vec{n_1}\cdot\vec{n_2}}{|\vec{n_1}||\vec{n_2}|} \tag{14}$$

By the method above, we can achieve the solution of θ_1, θ_2 and θ_3.

2.4 *Mathematical formulation*

The cable curves are 2-D curves in state I and II, because it can only bear axial tension force. So the curve equation z′ is given in Cartesian coordinate OX′Y′Z′ by Shen Shizhao (1997) as:

Table 1. Theory solutions (THES) and FEM solutions (FEMS) of cable-end internal force.

Internal force (kN)	$q_z = 0.7975$ kN/m, varying q_x			$q_z = 0.7975$ kN/m, varying q_y	
	0	0.06 kN/m	0.6 kN/m	0.06 kN/m	0.6 kN/m
T_A THES	5169.836	5170.526	5237.196	5171.180	5194.005
T_A FEMS	5169.313	5170.036	5236.307	5170.637	5193.195
T_B THES	5081.724	5082.414	5149.084	5095.724	5232.448
T_B FEMS	5081.389	5082.110	5148.384	5095.291	5231.582
T_A Difference	1.012e-4	0.948e-4	1.697e-4	1.050e-4	1.559e-4
T_B Difference	0.659e-4	0.599e-4	1.359e-4	0.849e-4	1.655e-4

$$z' = \frac{H'}{q}\left[\cosh\alpha' - \cosh\left(\frac{2\beta' x}{l_x'} - \alpha'\right)\right] \tag{15}$$

where $\alpha' = \sinh^{-1}\left[\frac{\beta'(c'/l_x')}{\sinh\beta'}\right] + \beta'$ and $\beta' = \frac{q l_x'}{2H'}$, in which c' and l_x' are projections of l_{AB} on axes of plane X′OZ′. And they are acquired by Eq. (7) as

$$\begin{bmatrix} l_x' \\ c' \end{bmatrix} = \begin{bmatrix} D_{11} & D_{13} \\ D_{31} & D_{33} \end{bmatrix} \begin{bmatrix} l_x \\ c \end{bmatrix} \tag{16}$$

As H_0 is known in state I, we can obtain S_{w1} by Eq. (5). Using Eqs. (5) and (4), H' is complied in state II in coordinate OX′Y′Z′. Thereafter, vertical component of tension can be expressed in term of

$$\frac{dz'}{dx} = \frac{V'}{H'} = -\sinh\left(\frac{q}{H'}x - \alpha'\right) \tag{17}$$

Now that the curve and the internal force of the cable have been known in plane X′OZ′, we can get the solution of internal force in Cartesian coordinate OXYZ using

$$\begin{bmatrix} F_x & F_y & F_z \end{bmatrix}^T = [D]^{-1}[H' \;\; 0 \;\; V']^T \tag{18}$$

2.5 *Example*

In this section, an example is given to verify the accuracy and feasibility of the method above. Dates of the cable's parameters come from the research developed by Mei Kuihua & Sun Shengjiang (2007). The values of horizontal distributed loads are 0, 0.06 and 0.6 kN/m, and the directions are negative X-axis and Y-axis. The internal force of the cable ends are summarized in Table 1 and compared to FEM solutions. Table 1 illustrates the differences are lower than 2×10^{-4}. When the horizontal loads are correspond to self-weight, the internal force of the cable changes apparently.

3 CONCLUSIONS

Coordinate transformation matrix and equivalent form of deformation compatibility equation are used to investigate the static response of cables subjected to horizontal distributed loads and self-weight. Equivalent form of deformation compatibility equation indicates that the non-stress length is invariable in different states. Coordinate transformation matrix is introduced to simplify 3D model into tow planes. From the example given in the paper, when considering the horizontal loads, the internal force of the cable changes a lot. So the horizontal loads are also playing an important role in static analysis of the cable.

REFERENCES

Buonopane, S., Billington, D. 1993. Theory and history of suspension bridge design from 1823 to 1940. *Struct. Div., ASCE*, 119(3): 954–977.

Chin-Sheng Kao, Chang-Huan Kou, Xu Xie. 2006. Static Instability Analysis of Long-Span Cable-Stayed Bridges with Carbon Fiber Composite Cable under Wind Load. *Tamkang Journal of Science and Engineering*, 9(2): 89–95.

Feng Jiechun, Dai Yingzhang. 2012. Calculation of non-stress length of cable in cable-stayed bridge. *Highway*, 5: 143–147.

Ke Jingpu. 2008. Three-dimensional coordinate conversion principle and application. *Jiangxi Surveying*, 74(3): 17–20.

Mei, K.-H, Lu Z.-T. 2002. Application Prospect of CFRP to Super Length Suspension Bridge and Cable-Stayed Bridge. *Bridge Constructer*, 2: 75–78.

Mei Kuihua, Sun Shengjiang. 2007. The static design calculation of the stayed cables. *Journal of China & Foreign Highway*, 27(4): 21–26.

Najib Bouaanani, Mohamed Ighouba. 2011. A novel scheme for large deflection analysis of suspended cables made of linear or nonlinear elastic materials. *Advances in Engineering Software*, 42: 1009–1019.

Qi Yujun, Feng Peng, Ye Lieping. 2011. Comparison of theoretical Solutions and ultimate span length between FRP cables and steel cables. *Civil construction and environmental engineering*, 33(4): 52–58.

Shen Shizhao. 1997. *Cable structure design*. Beijing: China Architecture Industry Press, 75–78.

Wei Jiandong, Zhao Renda. Che Huimin. 1999. Static design of cable in cable-stayed bridge. *Bridge construction*, 2: 21–26.

Modeling and Computation in Engineering II – Xie (ed)
© 2013 Taylor & Francis Group, London, ISBN 978-1-138-00058-2

Thermal analysis of bonded-in anchor in steel-to-concrete connection exposed to fire

Qun Xie
School of Civil Engineering and Architecture, University of Jinan, Jinan, China
Shandong Provincial Key Lab of Appraisal and Retrofitting in Building Structures, Jinan, China

Hao-Xue Ju
Shandong Institute of Commerce and Technology, Jinan, China

Yu Zhao
China Construction Eighth Engineering Division. Corp. Ltd., Jinan, China

ABSTRACT: A serial of numerical simulation by the software ABAQUS has been carried out to investigate the thermal response of boned-in anchor at elevated temperatures. The 3D model of steel-to-concrete connection with fire resisting coating is employed in this research and the development of temperature is supposed to follow the ISO834 standard time-temperature curve. Several variables including thickness of fire resisting coating, edge distance and embedment depth have been considered. The result shows that ambient temperatures have great impact on the thermal field of the whole connection and the temperature of adhesive anchor would reach above 160°C after a fire rating of 90 minutes even in the case of fire protection method has been used on the connection surface. Based on the comparison of parametric analysis, some advices are given for the fire-resistance design of bonded-in anchor.

1 INTRODUCTION

Adhesive anchor is a widely-used technology with the advantage of reliable performance and high strength in structural retrofit and reinforcement, however, the feature that bond stress of adhesive will decrease in elevated temperatures tends to reduce the capacity of anchorage if the bonded-in anchor is subjected to fire [Harmathy & Sultan 1987]. Limited information is currently known about the fire resistance for adhesive anchor and fire tests are so expensive and time-cost, thus the usage of numerical approach is becoming a convenient and efficient way to investigate the performance of anchors in fire, especially for the parametric studies [Wang & Xue 2010]. A 3D FE model in this paper is employed to study the thermal behavior of anchor in elevated temperature by heat transfer analysis [Wang 2012]. The object of this research is the tensile anchor in a steel-to-concrete connection with fire resisting coating exposed to fire, and the purpose is to figure out the temperature distribution and development of the whole connection.

2 HEAT TRANSFER MODEL

In practical engineering a steel-to-concrete connection with adhesive anchor is a typical application, and a steel attachment is a necessary component in the connection to transfer the load from new steel element to the existing concrete base [Wan & Zheng 2012]. What's more, because the structural steel will lose its strength in the case of being heated to high temperature, the surface of structural steel is generally required to be covered by fire resisting material after they have been constructed [Robinson & Said 2011]. The ISO834 standard

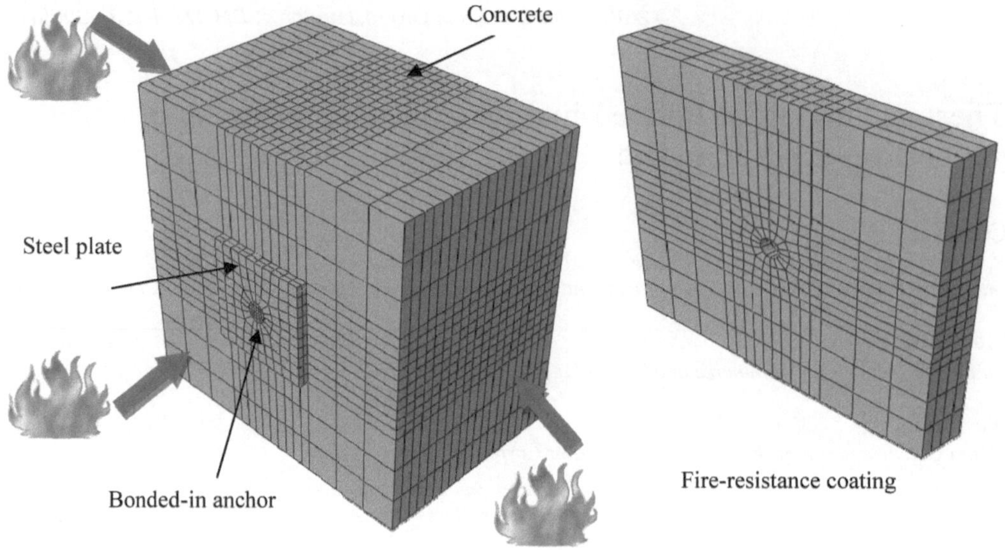

Figure 1. Model details.

time-temperature curve is used in this research. Based on the above description, the simulation model for heat transfer analysis includes reinforced concrete base, anchor, steel attachment and fire-resisting coating. Details about the model are shown in Figure 1. The model is measures 300 mm × 300 mm in x-y plane, and 200 mm height in z-axis direction. The diameter of anchor is 20 mm and embedment depth is 160 mm, and the steel plate locates the central area of the concrete surface with 10 mm thickness. The fire-resisting coating with thickness 40 mm just covers the whole base surface where steel plate and anchor locate. Finally the HRB335 grade reinforcing steel is used in the model which is conventional steel in civil engineering.

The whole model is meshed with cubic cells in a side length of 0.01 m, In order to obtain accurate temperature field, the anchor cross-section has more refined mesh, and the purpose is to assure that there is full heat transfer between all surfaces for individual parts in the model [Cook & Klingner 1992]. There are three surfaces of model that are assumed to expose to fire as shown in Figure 1, The structural elements should support reliably the designed load without loading capacity reduction in a fire rating of at least 90 minutes. A room temperature 20°C is specified as an initial condition and it is assumed that the initial temperature is uniformly distributed throughout the whole model.

3 MATERIAL PROPERTIES AT ELEVATED TEMPERATURE

The properties of material including concrete, steel and fire resisting coating would be greatly influenced by elevated temperature and be quite different from the properties in room temperature. The thermal properties of all material in model are listed in Table 1. For the reason of simplified analysis, the thermal coefficients, conductivity and specific heat, are all assumed to be constants.

4 RESULTS OF HEAT ANALYSIS

Finite element analysis results are presented in the following sections. The temperature field and temperature-time relationship for each component in the model will be discussed respectively.

Table 1. Material property at elevated temperature.

Material	Concrete	Steel	Fire-resistance coating
Emissivity	0.9	0.8	0.95
Poisson's ratio	0.17	0.3	–
Thermal conductivity	1.92	30	0.12
Specific heat (J/kg •°C)	1000	500	600
Density (kg/m^3)	2400	7850	450
Thermal expansion	0.000008	0.000012	–

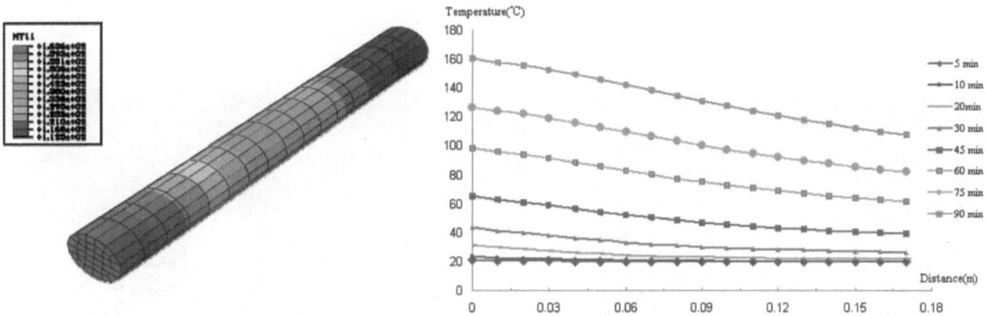

(a) Temperature contour

(b) Temperature gradient along anchor length

Figure 2. Thermal results of anchor.

4.1 *Anchor*

Figure 2 describes the anchor temperature contour at elevated temperatures. The temperatures have a relatively identical distribution along anchor length at the early phase of heating. With the increase of time, the temperature near anchor top has a faster growth and reaches the maximum value of 163°C, after the fire duration of 90 minutes. The minimum temperature 110°C occurs at other end of anchor, so the final temperature gradient isn't uniform.

4.2 *Concrete*

Figure 3 shows the temperature contour for concrete obtained from the FEM heat transfer analysis. The figure shows that the concrete temperature gradients have huge variation along the surface, with somewhat high temperatures up to 1000°C in regions directly exposed to fire, and low down to 150°C at central part.

4.3 *Steel plate*

Steel plate not only has the function of connecting steel to concrete but also plays a role on heat transfer during the process of heating. Figure 4 plots the temperature gradients of steel plate in both surfaces. In Figures 4(a) and (b) top surface means the surface contacting with coating and bottom surface is the surface contacting with concrete.

The surface temperatures in top surface are obviously higher than the corresponding temperatures in bottom surface and for every curve in Figure 4(c) there is a sharply temperature drop in central position where the anchor is welded with steel-plate because of the heat transfer between the two components.

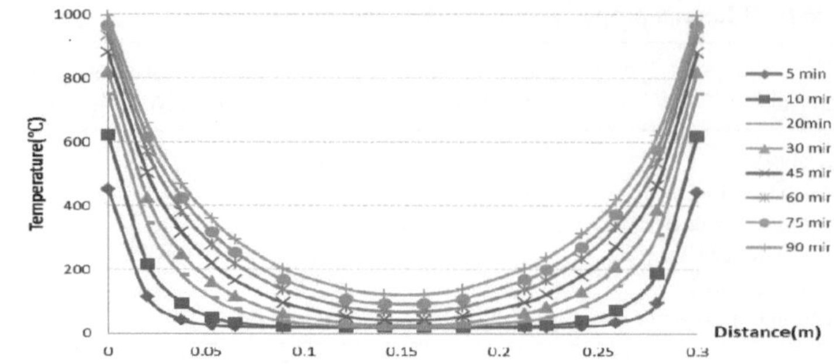

Figure 3. Thermal results of concrete.

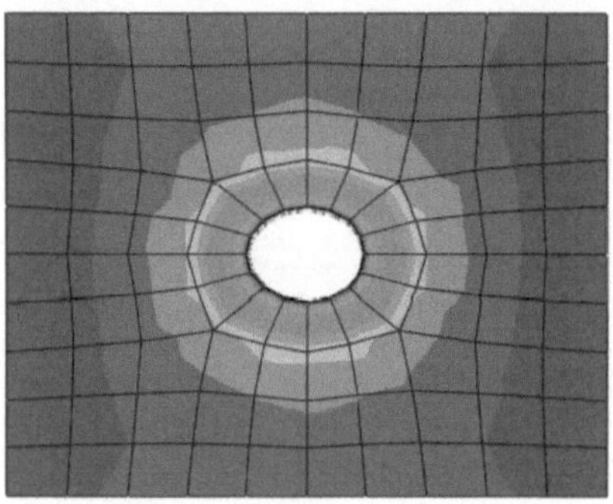

(a) Temperature contour of top surface

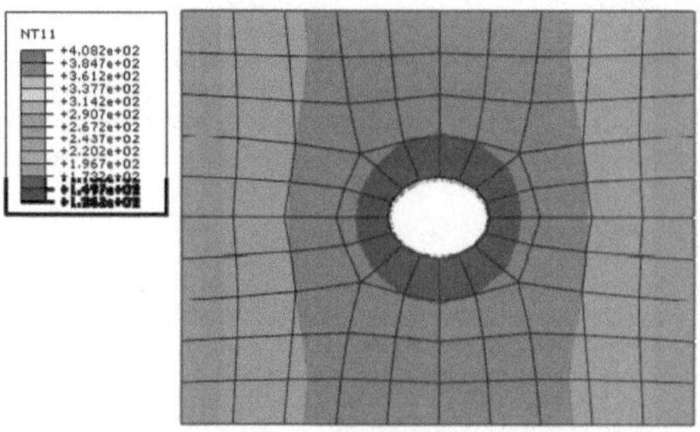

(b) Temperature contour of bottom surface

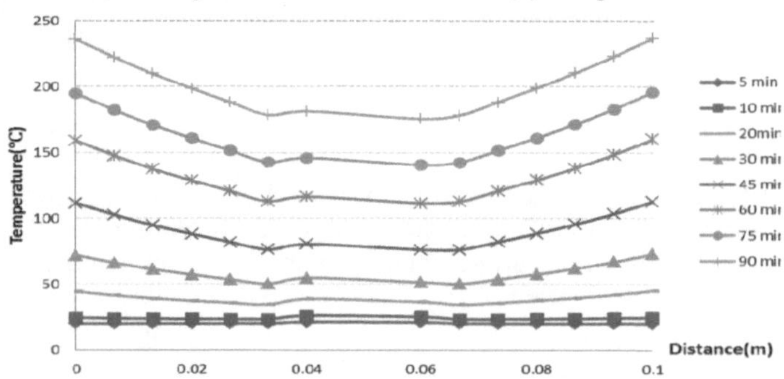

(c) Temperature gradient of top surface

Figure 4. Thermal results of steel plate.

4.4 *Fire resisting coating*

Figure 5 shows the temperature contour in both surfaces of fire resisting coating at elevated temperature. The temperatures for the whole surface directly exposed to fire reach 1000°C and there is an obvious temperature distribution for inner surface.

4.5 *Interface points*

In order to verify the compatibility of node temperature for different components, the time-temperature relationships have been compared for the interface nodes in model as shown in Figure 6. The curves of common nodes for different parts match each other well which shows that the interface nodes have satisfactory compatibility to transfer heat as expected.

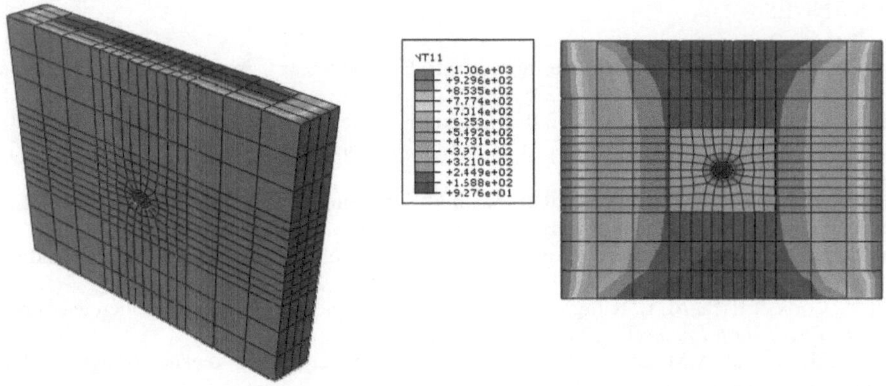

(a) Temperature contour of surface directly exposed to fire (b) Temperature contour of opposite surface

Figure 5. Temperature results of the fire-resistance coating.

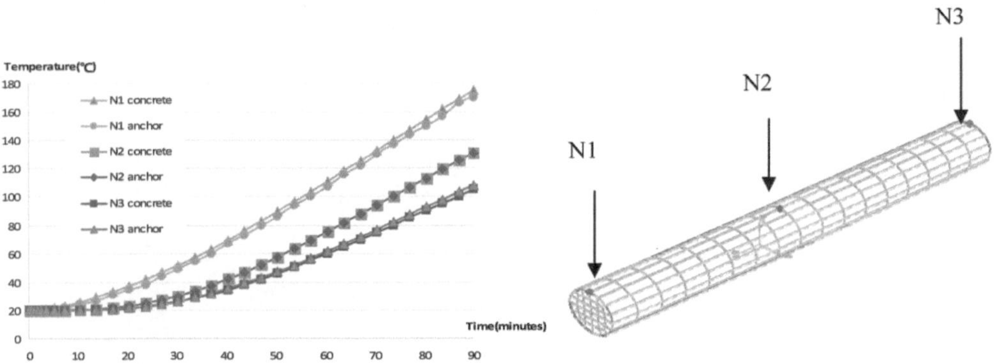

Figure 6. Temperature comparison of interface node.

5 CONCLUSIONS

A 3D FE model is employed to investigate the thermal response of steel-to-concrete composite connecting with bonded-in anchor at elevated temperatures. The analysis of temperature distribution for different components is carried out and some conclusions could be obtained as following:

1. The temperature distribution along anchor length isn't uniform during the process of heating. The maximum value could reach 160°C after fire duration of 90 minutes even under the protection of fire resistance material at connection surface.
2. There is obvious temperature gradient along the thickness direction for steel plate and fire resisting coating.
3. The temperatures for interfacial node between different element parts have good compatibility which verify that the mode mesh is reliable to transfer heat.

ACKNOWLEDGEMENTS

This research was financially supported by the Open Fund for Shandong Provincial Key Lab of Appraisal and Retrofitting in Building Structures (JG01).

REFERENCES

Aaron, J. & Wang. 2012. Numerical Investigation into Headed Shear Connectors under Fire. *Journal of Structural Engineering* 1 38(1): 118–122.

Fu-Xiong, Wan & Wen-Zhong, Zheng. 2012. Thickness Value of Fireproof Coating for Slabs Strength with Carbon Fiber Sheet Bonded with An Inorganic Adhesive. *Journal of Harbin Institute of Technology* 44(2): 11–16.

Guang-Yong, Wang & Su-Duo Xue. 2010. Method Analyzing Reinforced Concrete Frames in Fire Based on ABAQUS. *Journal of Beijing University of Technology* 36(3): 316–320.

Harmathy, T.Z. & Sultan, M.A. 1987. Comparison of Severity of Exposure in ASTM E119 and ISO 834 Fire Resistance Tests. *Journal of Testing and Evaluation* 15(6): 371–375.

Ronald, A. Cook & Richard, E. Klingner. 1992. Ductile multiple-anchor steel-to-concrete connections. *Journal of Structural Engineering* 118(6): 1645–1665.

Robinson, S.E. & Said, A.M. 2011. Effect of Different Parameters on the Tensile Strength of Adhesive Anchors. *Structures Congress ASCE*: 1850–1861.

Modeling and Computation in Engineering II – Xie (ed)
© 2013 Taylor & Francis Group, London, ISBN 978-1-138-00058-2

Effect of GGBS on chloride diffusion coefficients of cement mortar

S. Zhang, Y. Fan & N. Li
Institute of Road and Bridge Engineering of Dalian Maritime University, Dalian, Liaoning, China

ABSTRACT: The effects of Ground Granulated Blast Furnace Slag (GGBFS) and fly ash on the chloride diffusion behavior of cement mortar at early and later ages are examined in this paper. A kind of rapid chloride-ion diffusion tests was applied to evaluate the diffusion coefficients of cement mortar incorporating with GGBFS or fly ash. It is indicated that the chloride ion diffusion coefficients of cement mortar containing GGBFS decreased significantly at 14 and 28 days. Fly ash will further reduce the chloride diffusion coefficient of GGBS modified cement mortar. The optimum content of GGBS is 50 wt.% of cement.

1 INTRODUCTION

With the construction and application of a large amount of coastal concrete structure, the problem of concrete cracking and steel corrosion caused by chloride ions has attracted more attention. How to improve the resistance to chloride ion erosion performance of reinforced concrete, and strengthen the durability of concrete structures has become to be a hot topic. GGBFS has a higher potential activity; there have some studies about the effect of GGBFS on the chloride diffusion performance of concrete since the 1980s (Leng, et al., 2000). It is shown that GGBFS can effectively improve the resistance to chloride diffusion performance of concrete at 28 days, and it will defer the corrosion destruction of steel bar and significantly improve the durability of the reinforced concrete structure (Kyong, et al., 2005, Sujjavanich and Cheng, 2005, Li, 2012). However, with the impact of loads and harmful substances at early age of engineering construction, concrete materials will inevitably generate the initial destruction and reduce the durability of concrete structure. So it is significant to research the affect of GGBFS on the durability of concrete, especially the performance of concrete at the early age.

To study the effect of ground granulated blast furnace slag (GGBFS) on the resistance to chloride penetration of cement mortar at various curing periods. In this paper, an experimental investigation on chloride ion diffusion coefficient of cement mortar with GGBFS and GGBFS-fly ash modified cement mortar were carried out.

2 EXPERIMENTAL PROGRAM

2.1 *Materials*

Portland cement (P.O 42.5) was applied in this paper. The GGBS is S95 with a specific surface area of 425 m^2/kg, and the density is 2.90 g/cm^3. The chemical composition of cement and GGBS are shown in Table 1. Grade I fly ash was studied in this paper. The X-ray diffraction (XRD) spectrum of GGBS, and the scanning electron microscopy (SEM) images of GGBS and fly ash are shown in Figure 1.

Table 1. The chemical composition of cement and ground granulated blast furnace slag (mass content/%).

Chemical agent	CaO	SiO_2	Al_2O_3	Fe_2O_3	MgO	SO_3	Others
Cement	59.30	21.91	6.27	3.78	1.64	2.41	4.69
GGBS	26.59	34.18	13.81	14.55	8.95	1.05	0.87

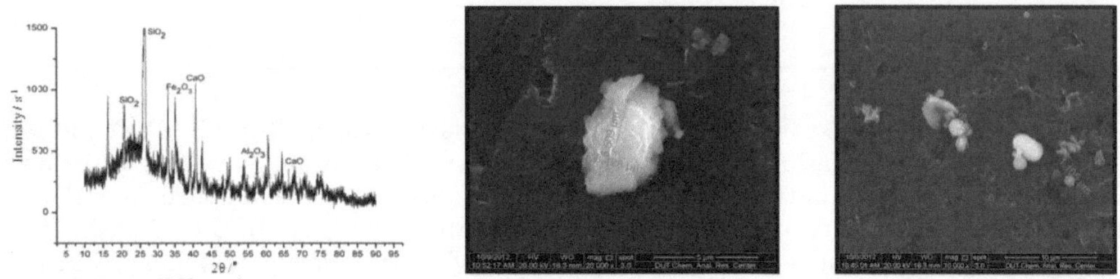

a. The X-ray diffraction spectrum of ground granulated blast furnace slag b. The scanning electron microscopy image of ground granulated blast furnace slag c. The scanning electron microscopy image of fly ash

Figure 1. The XRD spectrum, and the SEM of ground granulated blast furnace slag and fly ash.

2.2 *Specimen preparation*

The blended cement mortar was prepared by the mixture of ordinary Portland cement and GGBS. The content of GGBS modified cement paste was illustrated in Table 2. The cement to sand ratio of the cement mortar was 1:2, and water to binder ratio is 0.5. The composition of the GGBS and fly ash was illustrated in Table 3.

To ensure the homogeneity of the paste, the ingredients were firstly homogenized in an electric mixer. Then, the mortar pastes were molded into Φ100 mm × 100 mm cubes for chloride ion diffusion test. The molds filled with mortar were vibrated for one minute to remove the air bubbles. The cement samples were kept in the molds at 100% relative humidity for 24 hours, and then cured in water for 3, 7, 14, 28 days before the diffusion test.

2.3 *Test method*

The rapid chloride migration (RCM) was selected to measure the chloride ion diffusion coefficient, which was an electrically accelerated migration test.

The specimens for chloride permeability tests were obtained through sawing the cylinder specimens with the dimension of Φ100 mm × 100 mm into two slices of about 50 mm thickness. To expel the air from the internal voids that all the voids will filled with water, the slices were washed thoroughly and then put into an ultrasonic cleaning machine shown in Figure 2a. Then, the slices are taken out and fixed in a rubber tube with two stainless steel hoops shown in Figure 2b. Then, put the rubber tube into an electrolyzer and injected 335 ml of 0.2 mol/L KOH solution into the rubber tube and 0.2 mol/L KOH solution containing 5% NaCl into the electrolyzer. Finally, properly connected to the positive and negative (shown in Figure 2c) to electromigration process. Chloride ion diffusion depth is measured and the equation as follows is used to calculate the chloride diffusion coefficient.

$$D_{RCM} = 2.872 \times 10^{-6} \frac{Th\left(x_d - a\sqrt{x_d}\right)}{t} \tag{1}$$

Table 2. Dry mix composition of ground granulated blast furnace slag modified cement.

Mix designation	Cement (%)	GGBS (%)
SC0	100	0
SC1	70	30
SC2	60	40
SC3	50	50

Table 3. Dry mix composition of ground granulated blast furnace slag and fly ash modified cement (mass, %).

Mix designation	Cement/%	GGBS/%	Fly ash/%
S5F0	50	0	0
S5F1	40	50	10
S5F2	30	50	20
S5F3	20	50	30

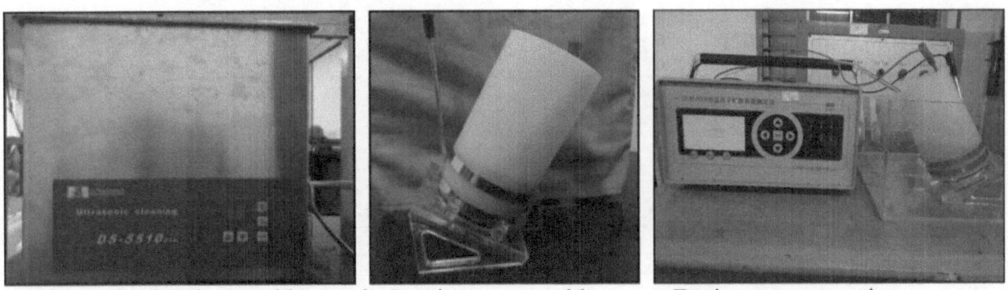

a. Ultrasonic cleaning machine b. Specimens assembly c. Equipment connection

Figure 2. Instrument used in the chloride permeability test.

$$a = 3.338 \times 10^{-3}\sqrt{Th} \quad (2)$$

where, D_{RCM} is the chloride ion diffusion coefficient (m^2/s); T is the average temperature of the initial and final of the anolyte (K); h is the height of the specimen (m); x_d is the chloride ion diffusion depth (m); t is the test time (s).

3 TEST RESULTS AND DISCUSSION

3.1 *Test results of chloride permeability of GGBS modified mortars*

The test results for rapid chloride ion permeability of GGBS modified cement mortars at various curing period are presented in Figure 3 and Table 4.

From Figure 3 and Table 4, it can be drawn that:

1. Compared with the ordinary cement mortar, the chloride diffusion coefficients of GGBS modified cement mortars at 3 days increase with the increasing content of GGBS. Therefore, GGBS can reduce the resistance to chloride ion penetration of cement mortar.
2. The chloride diffusion coefficients of cement mortars with the addition of 30 wt.%, 40 wt.% GGBS at 7 days take on an increase. However, the chloride ion diffusion coefficients

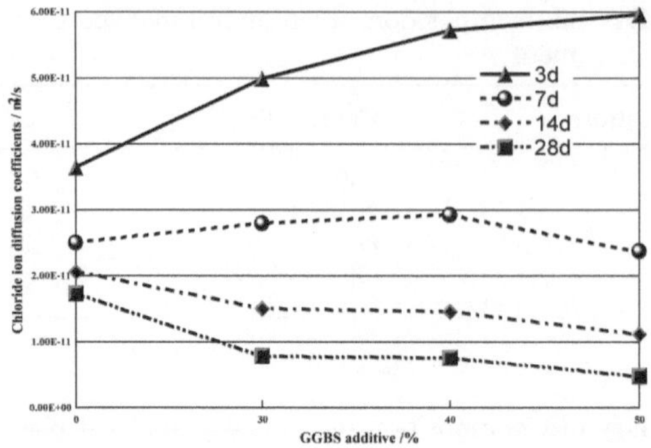

Figure 3. Relationship between the chloride ion diffusion coefficients with ground granulated blast furnace slag additives.

Table 4. Increase extent of the chloride ion diffusion coefficient.

Mix designation	3d (%)	7d (%)	14d (%)	28d (%)
SC0	0	0	0	0
SC1	36.70	11.73	–27.01	–54.47
SC2	56.74	16.93	–29.17	–56.28
SC3	63.33	–5.04	–45.70	–72.47

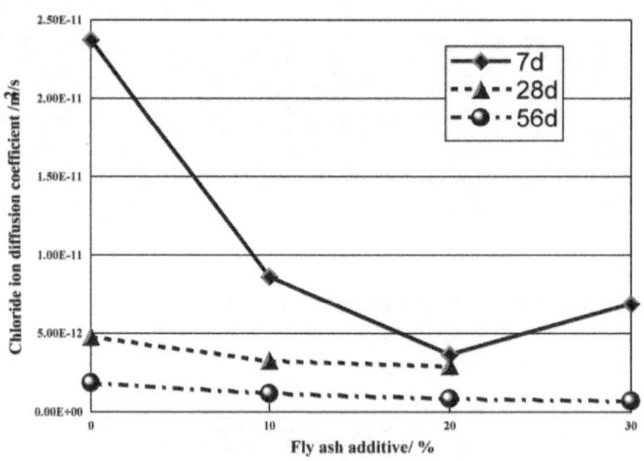

Figure 4. Relationship between the chloride ion diffusion coefficients with fly ash additives.

of cement mortars with the addition of 50 wt.% GGBS will decrease. Compared with the ordinary cement mortar, the chloride diffusion coefficient has a slight decrease (about 5.04%).

3. For the cement mortars with the addition of GGBS, the chloride diffusion coefficients decrease significantly with the amount of GGBS at 14, 28 days. The reduction of cement mortar with the addition of 50 wt.% GGBS is about 72.47% at 28 days. It can be concluded that GGBS can significantly improve the resistance of the cement mortar to the chloride ion penetration of cement mortar at long-term period, which keep in agreement with the research result (Li, et al., 2011).

3.2 *Test results of chloride permeability of GGBS and fly ash modified mortars*

The test results for rapid chloride permeability of GGBS modified cement mortars containing fly ash at various curing periods are presented in Figure 4 and Table 5.

Table 5. Degradation of the chloride ion diffusion coefficient/%.

Mix designation	7d/%	28d/%	56d/%
S50	0	0	0
S50F10	63.64	31.36	37.27
S50F20	84.59	38.44	54.66
S50F30	70.69	–	62.84

From Figure 4 and Table 5, it can be seen:

1. With the addition of GGBS and fly ash, the chloride ion diffusion coefficients of cement mortars at 7 days decrease, the reduction amount is about 84.59% for the mortar with the addition of 50 wt.% GGBS and 20 wt.% fly ash. Obviously, GGBS and fly ash enhance the performance of the resistance to chloride ion diffusion of cement mortar significantly at early age.
2. The chloride diffusion coefficients of cement mortars with the addition of GGBS and fly ash decreased with the amount of fly ash at 28 and 56 days. The cement mortar with the addition of 50 wt.% GGBS and 30% fly ash at 56 days has a decrease of 62.84%. We can reach the conclusion that GGBS and fly ash can further improve the resistance to chloride ion penetration of cement mortar at long-term age. Similar conclusions have been achieved in the former researches (Pan, et al., 2009).

4 CONCLUSIONS

This paper presents the effects of ground granulated blast furnace slag (GGBFS) and fly ash on the anti chloride penetration of mortar at various curing ages. Based on the experiments executed in this paper, the following conclusions can be drawn:

1. GGBS can improve the ant-chloride ion diffusion property of mortar at early age except 3 days; the optimal content of GGBS is 50 wt.% of cement.
2. GGBS can significantly enhance the capability of the resistance to chloride penetration of cement mortar at long-term age.
3. The resistivity to chloride penetration of cement mortar with the addition of 50 wt.% GGBS and fly ash can further strengthen at various curing ages, the optimal content of fly ash is about 20 wt.% to 30 wt.% of cement.

ACKNOWLEDGEMENTS

This research was financially supported by the National Natural Science Foundation of China (Grant No. 51178069), Program for New Century Excellent Talents in China, and the Program for Excellent Yong Talents in Liaoning Province.

REFERENCES

Cheng, A.H., Ran, W. & Chen, J.K. 2005. Influence of GGBS on durability and corrosion behavior of reinforced concrete. *Materials Chemistry and Physics* 93(2): 404–411.

Kyong, Y.Y. & Eun, K.K. 2005. An experimental study on corrosion resistance of concrete with ground granulate blast-furnace slag. *Cement and Concrete Research* 35(7): 1391–1399.

Leng, F.G., Feng, N.Q. & Lu, X.Y. 2000. An experimental study on the properties of resistance to diffusion of chloride ions of fly ash and blast furnace slag concrete. *Cement and Concrete Research* 30(6): 989–992.

Li, K.L. 2012. Ocean engineering concrete using high-volume GGBS. *Applied Mechanics and Materials* 23(8): 1279–1286.
Li, K.L., Huang, G.H. & Lin, J. 2011. Durability of High-Volume GGBS Concrete, *Advanced Materials Research* 26(3): 338–343.
Pan, Y.C., Liu, J.B. & Xun, Y. 2009. Experimental research on durability of ocean-engineering high-performance concrete affected by fly ash, *Journal of Harbin Institute of Technology* 41(6): 179–181.
Sujjavanich. S. & Voradej. S. 2005. Chloride permeability and corrosion risk of high-volume fly ash concrete with mid-range water reducer. *ACI Materials Journal* 102(3): 177–182.
Wang, H.Y. 2008. The effects of elevated temperature on cement paste containing GGBFS. *Cement and Concrete Composites* 30(10): 992–999.

Modeling and Computation in Engineering II – Xie (ed)
 ISBN 978-1-138-00058-2

In-situ test for RC frame structure lightly destroyed by the 5.12 great earthquake

ChunYang Lin
Institute of Engineering Mechanics of China Earthquake Administration, Harbin, HeiLongJiang, China

ABSTRACT: The paper tests different responses of RC frame under different excitation-frequency cyclic loading. The dynamic drive comes from motor eccentricity momentum on the tope of the RC framed building. In course of In-situ test the structure itself is a full size model. All structural natural characteristics are test perfectly. The tested RC frame structure is established models to analyses the dynamical responses by finite element method. One model is established with traditional methods considering infill masonry walls' effect by period reduction method. Another model uses plane element to simulate the effect on structural stiffness contribution caused by masonry infilling walls. The paper contrasts the analysis results and the test results. By the test results and the analysis results the thesis confirms and verifies the RC structural dynamical characteristic. Structural frequency and damping ratio can be deduced. The data from in-situ test are useful for further researches.

1 INTRODUCTION

In-situ test plays important roles in civil area especially in structural test just like bearing-test in soil mechanics, the In-situ test firstly is used in soil mechanics and rock engineering.

In-situ test mentioned in the paper can be used to test RC frame dynamical characteristic and research rules of cracked frame structure. Taking advantages of In-situ test, researchers can fully and truly test dynamical responses. All the data from In-situ test can really reveal the structural characteristic because model's the boundary is full size without any sale size and simplification. The full size model (the real structure) is not simplified calculation model with some inaccuracy which is not as good as supposed perfect boundary condition.

The paper firstly uses the In-situ test to verify structural dynamic characteristics of cracked RC frame. Considering the structural coupling between foundation and upper structural frame, the dynamical characteristics is influenced by filler wall of RC frame is obvious. The RC frame structural is destroyed by WenChuan the 5.12 great earthquake.

2 INTRODUCTION FOR RC FRAMED BUILDING FOR IN-SITU TEST

The three-floor RC framed structure is a building with long lean conveyer draper room which is supported by reinforced concrete columns. The single span building transverse span is 6.9 meters long. Beams and slabs are deposited by the cast—in—place concrete. Slab is deposited with in-situ reinforced concrete and slab thickness is 80 mm. The top floor is deposited as in-situ concrete slab whose thickness is 70 mm. The masonry infilling wall in the reinforcement concrete structure is brick material without any plaster cover. There is no any separation wall in first floor in the building. The plane of RC framed structure has an arc shape in plane. Masonry conveyer-belt pass is 2.6 m width and 1.2 m height. The base of the

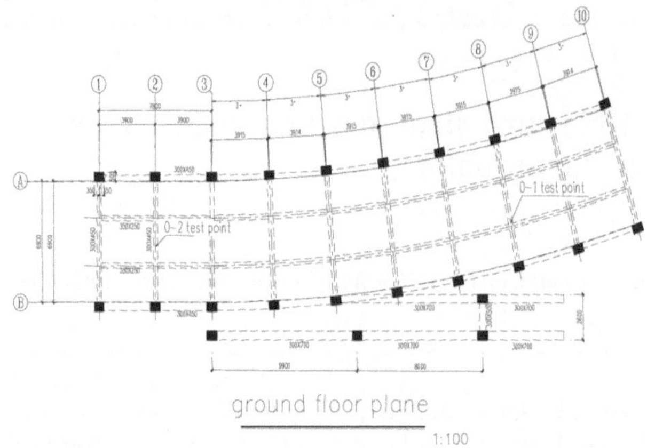

Figure 1. Floor plan.

Figure 2. Front elevation.

Figure 3. Back elevation.

building is used as storehouse for coal, where is fully filled with coal when In-situ test is being test. The building is supported by stone-columns which stand on stable rock. Some column in the first floor were lightly destroyed by 5.12 great earthquake so that there is a little of crack at bottom of the columns.

The RC framed structure was built up in sixties in last century. The site class of the structure building is II class. According to the conversion value of concrete compressive strength the strength grade of concrete of framed beams and columns is no more than C25. Hot rolled Steel bar was used in beams and columns. The masonry filling wall and enclosure wall were mortared with strength grade of M1.0 mortar-mix materials.

Plan of structure is described as Figure 1, Figure 2, and Figure 3.

3 MOTOR AND SENSOR INSTRUMENTS USED IN IN-SITU TEST

Horizontal Forced vibrator used in In-situ test is a large forced vibrator named CSD-60 which has two largely eccentricity mass at both end of the vibrator. Exciting frequency of the vibrator is from 2 Hz to 10 Hz.\ Power of motor in vibrator is 55 kilowatt and the mass of eccentricity block is 800 Kg and the weight adds up to 361 KN. The picture followed indicates the motor's location on the tope of the building.

941B and 891 sensor were used to test vibration signals which were excited by electrodynamic-forced vibrator. All the accelerator and velocity signals collected by the sensor on the floor were transmitted into computer disk. Sensor location on the floor is described as Figure 5, where accelerator and velocity Sensor sites along the longitudinal and transverse

Figure 4. Vibration-motor working on the roof.

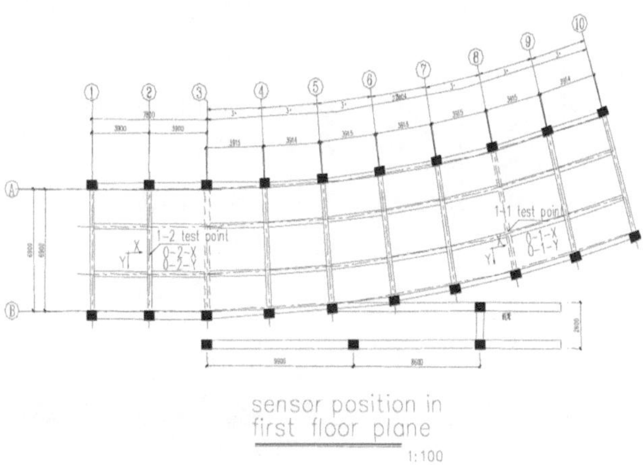

Figure 5. Location of sensor on the floor.

axis direction of the building. Eight FBA-31 types of accelerator sensor are used in structural prototype-test.

There are eight sites of sensors location in the whole building. There are two sensors sites in ground floor, two in first floor, and one in third floor. There are thirty two passes respectively which records the accelerator and velocity value (Piao 2003). The first floor was selected as example to mark the sensor situation on the floor. Test points of structure are described as Figure 5.

4 TEST CONTENT

4.1 *Natural frequency of vibration test*

Buildings in state of being undisturbed or slightly being disturbed it should do free vibration. Different channels record natural vibration velocity value and acceleration value of buildings when the building is vibrating without any external excited load. After respectively filtering, Fourier transforming, different channel frequency amplitude spectrum curve can be plot which is shown in Figure 7. After selecting the typical velocity signal of three sensors for processing the paper gets the history results shown in Figure 6. After analysis, the first natural frequency of structural vibration is about 4.5 Hz, and first fundamental period of structure is 0.22 seconds.

4.2 *Frequency response curve plot*

Horizontal Forced vibrator can changes itself vibration frequency by changing motor revolving velocity, which can be realized by using frequency conversion device as power supply. When revolving speed gets more highly the eccentricity rotor inputs more power so that forced vibration frequency increases obviously. After history curves are analyzed the different vibration frequency are found out easily, which includes all the test point in structure along X, Y direction displacement. By analysis of signals the frequency amplitude response curves were plot perfectly. The fundamental vibration frequency of the structure is about 4.5 Hz.

By exciting frequency being changed the response of the buildings, such as measured quantity (acceleration, velocity, displacement), become different. According to test point records different exciting frequency of the measured quantity of the steady state response

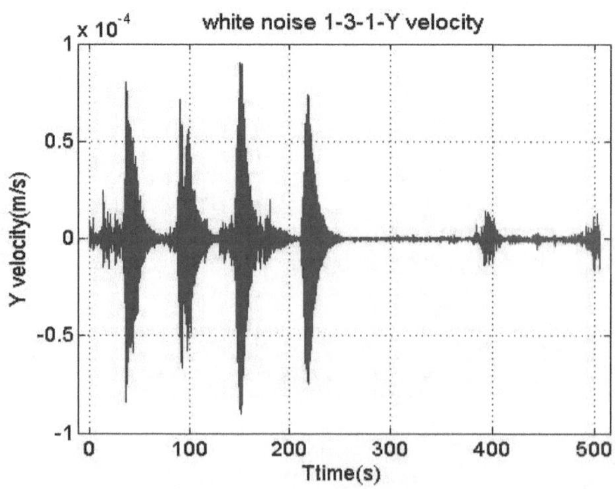

Figure 6. History graph of transverse displacement on the floor of the frame vibrating in freedom.

maximum can be found. The smooth curve was plotted by these discrete points, which are the measured quantity of the history response curve of forced vibration on structure building's floor. Different measured quantity of resonance frequency and resonance zone can be found out by the frequency response curve. The typical frequency response curves which were achieved from the test are plotted in the Figure 8 and Figure 9.

4.3 *Damping ratio calculation*

Damping refers to vibration amplitude of gradual decline characteristics for the effect of the system itself or outside, which appears in any system in the process of vibration. In the process of free vibration motion decline-ratio mainly depends on the damping ratio. Damping ratio is a character of vibration system which depends on the mass and the stiffness of the vibration system. This experiment is used to test the damping ratio ξ by using half power band width method (Hu 2006 & clough 1995).

$$\xi = \frac{\omega_b - \omega_a}{2\omega_n}$$

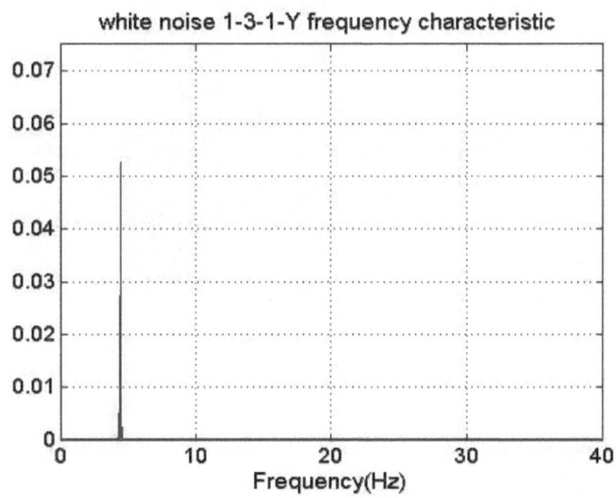

Figure 7. Frequency response curve of transverse velocity of the frame.

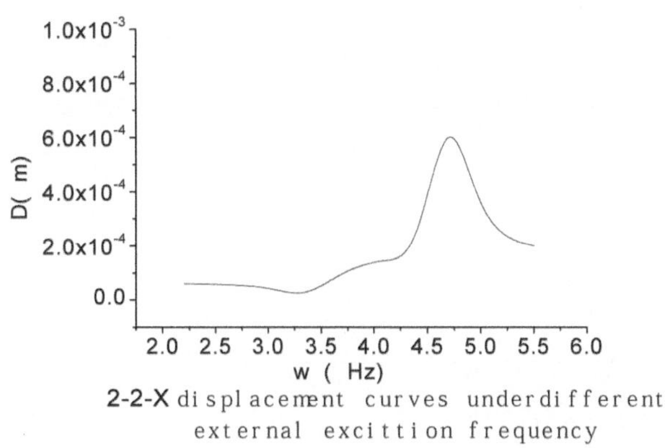

Figure 8. Frequency response curve of the frame in second sensor along transverse vibration.

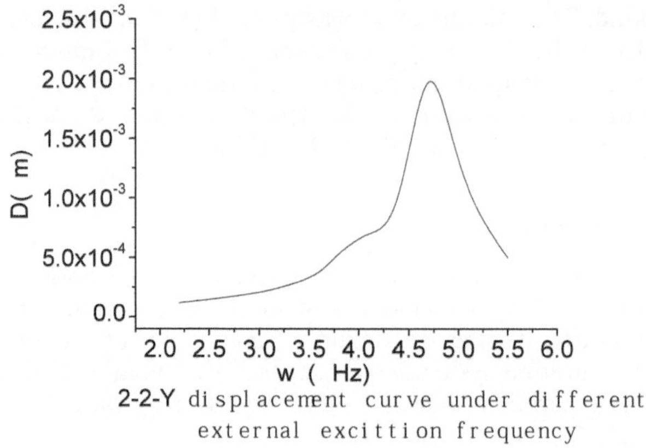

Figure 9. Frequency response curve of the frame in second sensor along longitudinal vibration.

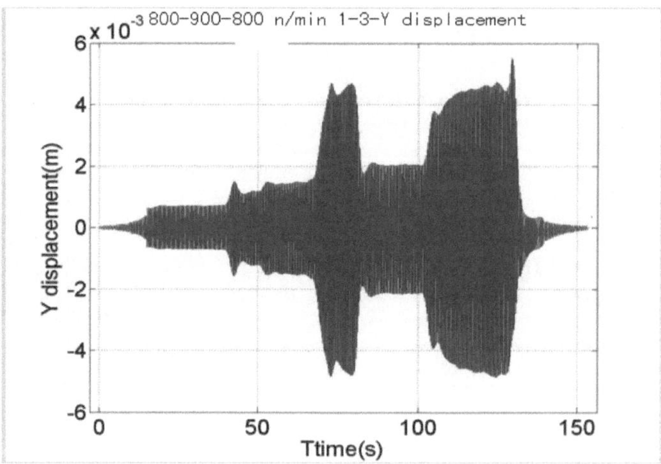

Figure 10. History displacement curve of the frame in location on third sensor along longitudinal vibration.

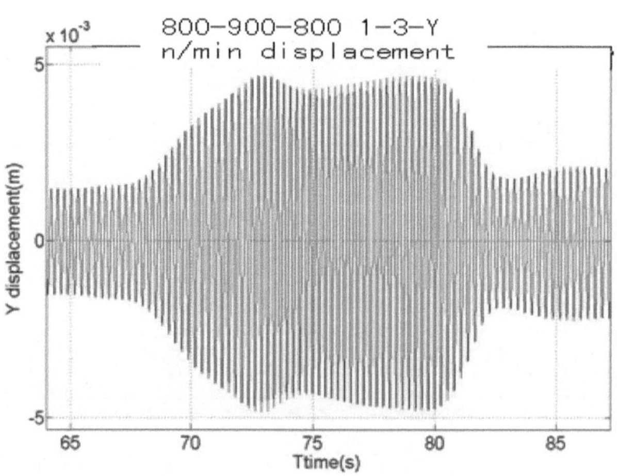

Figure 11. History displacement large part curve of the frame in location on third sensor along longitudinal vibration.

ω_n is the resonance frequency in frequency amplitude curve graphics, ω_a, ω_b is equal to 0.707 times of the vibration amplitude of the corresponding frequency 5 Hz. From frequency response curves of the RC frame structure in second sensor along longitudinal vibration, formula was used to solve ξ. The result is 0.03. The results verify the damping ratio of reinforced concrete structures is between 0.03 and 0.08 is the reasonable value interval range (industry Publish Press 2010).

4.4 *History analysis curves*

In course of test, there are 32 channels works for all displacement, accelerator and velocity signals. 8 test points respectively record the accelerator and velocity history-signals when the structure building is subjected to periodically lateral load. By changing the rotational velocity of motor different data were recorded in computer memory. After being plot all the history curves, the maximum and minimum were found and marked in the Figure. When the excitation frequency is about 4.5 Hz zone the maximum response value was discovered. Ten of full waves of test point are plot in Figure 10 and Figure 11.

5 COMPARING TEST RESULTS WITH ANALYSIS RESULTS OF SOFTWARE

The tested RC frame structure was established models to analyses the dynamical responses by finite element method which is general element software spatial analysis program.

Two models were considered, one model was established with traditional methods considering infill masonry walls effect by period reduction method. In another model plane element was used to simulate the structural stiffness contribution caused by masonry infilling walls.

Eventually we compare the calculated fundamental period with the tested fundamental period in In-situ test. The Two models are shown in Figure 12 and Figure 13. In the first case, considering slabs' stiffness contribute to the beam stiffness, the model enlarge the stiffness of beams as twice large as itself stiffness. The fundamental period Calculated by finite element analysis is 0.2677 (seconds) (Chang 2009).

Because of infilling walls the fundamental period is reduced to ranges from 0.212 (seconds) to 0.227 (seconds), which coincides with the results of test of 0.22 (seconds). In another case, the model considers the contribution of the infilling walls' stiffness to the whole structure stiffness. The model assumes the infilling walls tightly connected with structural columns and beams so that reduction of the fundamental period is unnecessary. The calculated fundamental period is 0.148 (seconds).

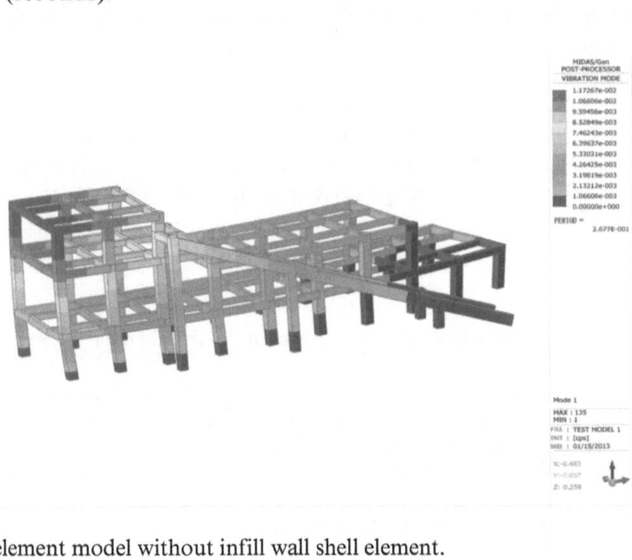

Figure 12. Beam element model without infill wall shell element.

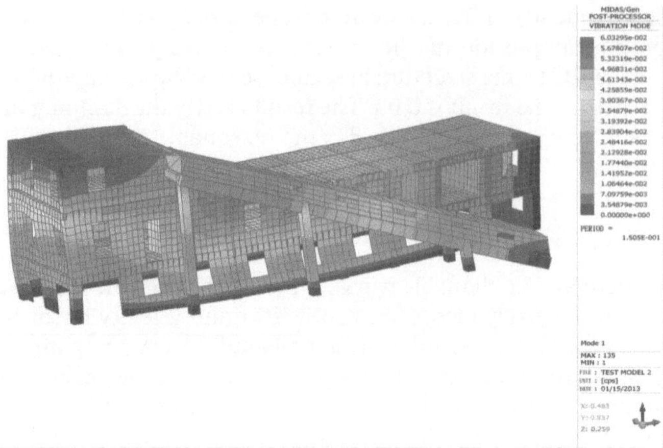

Figure 13. Beam element model with infill wall shell element.

The fundamental period calculated by finite element method is smaller than the test fundamental period. The reason why tested period is larger than period calculated by finite element software is described as followed. Firstly when the masonry filling wall was constructed (1961), there is no reliable connection between the filling walls and the members of beams and columns. It is difficult to accurately analysis the effect of the filling walls on the whole stiffness of the model. Secondly the gaps between filling wall members and frame column members make it more different to simulate the filling walls effect completely because two materials can not be coupled displacement effectively. Finally, the structure is slightly damaged by the 5.12 WenChuan great earthquake.

6 CONCLUSION

The in-situ test on building perfectly tests and identifies the Structural, dynamical characteristic, frequency and damping ratio. The test results of response can be contrast with the results of numerical analysis. The in-situ test on building perfectly reflects dynamically responses of building which is in ideal boundary conditions and really boundary conditions.

The in-situ test on building exactly reflects Dynamical accelerator and velocity spatial responses on horizontal floor diaphragm subjected to periodically lateral load. The framed structural building used in In-situ test had been lightly destroyed in the 5.12 WenChuan great earthquake. Before being test some columns have been slightly cracked. After the vibration test is over cracked columns were found no new cracked columns and beams on two ends of the members. The In-situ test proves that the cracked columns in building which were slightly destroyed in earthquake have no strong impact on the framed structure.

REFERENCES

Chang, T.Y.P. 2009. Elasticity, Plasticity & Finite Element Analysis. Department of bridge engineering Tongji university.

Chinese architectural industry Publish Press. 2010. Code for seismic design of buildings (GB 50011-2010). The Ministry of housing and urban-Rural of development of china.

Jinxian Hu. 2006. Earthquake seismic engineer. Earthqake Publish Press.

Lexing Piao. 2003. Technology and method of building structure test. Zhong Nan university Publish Press.50–73.

Ray W. Clough. 1995. Dynamics of structures. Berkeley.

Modeling and Computation in Engineering II – Xie (ed)
 ISBN 978-1-138-00058-2

Numerical simulation of smoke temperature distribution of a moving train with fire in tunnel

Jun Mao, Yanhong Xi, Guangyu Ming & Xianglei Li
Beijing Jiaotong University, Beijing, China

ABSTRACT: In subway tunnel fire, the greater possibility is the fire on the train, the emergency response is to run the train with the fire to the front of the station as far as possible. The key points to run the on-fire train in the tunnel are the intensity of fire and the train velocity. This paper adopts a fire prototype to perform a model experiment of the train fire in a 1:8 subway tunnel model. By using Computational Fluid Dynamics (CFD), the moving train with fire in the center is simulated in the tunnel by the moving mesh function of STAR-CD software. Through studying the influence of the different train speed and the heat release rate on the temperature distribution in the tunnel, the rule of smoke temperature distribution of the moving on-fire train in the tunnel is pointed out.

1 INTRODUCTION

In subway train fire, it is difficult to stop the on-fire train in the tunnel on the spot to implement the evacuation and rescue. For this case, the current national standard prescribes that the on-fire train should run to pull out the tunnel to carry out the rescue. But in this process, the fire source also moves with the train. The spread of high temperature smoke may make poisonous gases enter in compartment through the ventilation system vents or other gaps. This will endanger passengers' safety. Therefore, we need to conduct a more detailed analysis of the rule of smoke temperature variation inside the tunnel. But at present in already had, many researches generally assume that the train and source of fire are in a quiescent state, streams of smoke flow to the train, which are different to the actual process of the train carrying the fire and running in the tunnel. Therefore, this paper will use the moving mesh methods to simulate subway train moving fire in the tunnel, analyze the characteristics of the temperature distribution around the train, and provide a reference for the further study of the characteristics of a moving fire.

2 COMPUTATIONAL MODEL

A mode experiment is designed according to the prototype tunnel in Beijing Subway. A model tunnel was built in a ratio of 1:8, and its sectional size is also in a ratio of 1:8 except for its length, which can be extended by a pressure control valve so as to ensure the similarity between the full size and model. The size of tunnel model is 15 m × 0.5 m × 0.5 m, and the size of train model is 3 m × 0.26 m × 0.35 m, as shown in Figure 1.

The subway tunnel is modeled for simplicity as follows: ignore the influence of the cable inside the tunnel, and rails on the ground as well as mechanical ventilation on the smoke spread; treat the combustion as a heat source and ignore the changes of chemical reaction and smoke composition in the combustion process; consider the air flow and smoke in the tunnel as incompressible viscous fluid; ignore the influence of the radiation heat; ignore the influence of thermal diffusion, viscous diffusion and pressure on smoke flow. The size of

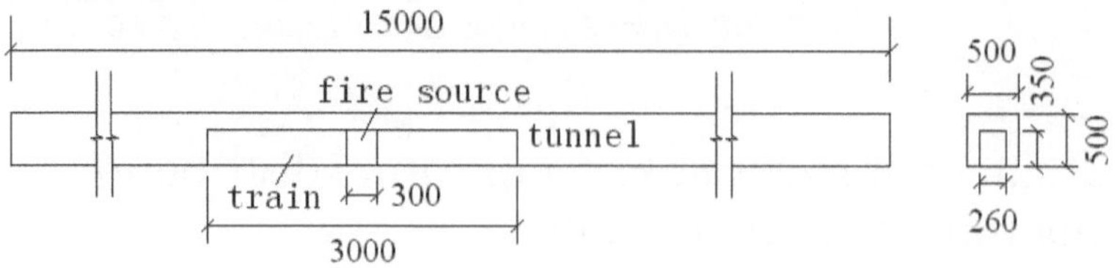

Figure 1. Longitudinal section and cross sectional view of the tunnel model (unit: mm).

computational domain is 24 m × 0.5 m × 0.5 m. For fire scene, generally, there are four steps in the progress of burning in the tunnel: free combustion, relatively stable combustion, anoxic combustion and burning out. However, it is difficult to find out the exact time and the changes of the heat release rate of the fire during each step, so we mainly focus on the period of relatively stable combustion when the temperature can reach the highest level and regard the fire as constant fire in order to analyze the characteristics of the smoke temperature in the most disadvantageous circumstances. In this paper, the fire is located in the middle of the train, the fire area is $S = 0.3 \times 0.26$ m^2. According to the result of the model experiment, the numerical simulation includes four projects which heat release rates (HRR) are 31.3 kW, 35.64 kW, 40.14 kW, and 41.08 kW.

In the paper, we use transient simulation for the subway train fire. According to the heat release rate mentioned above, we take the train speed to 0.82 m/s, 1.83 m/s, 2.65 m/s and 3.54 m/s. The initial time is when the train tail has just entered the tunnel entrance, the end time is when the train tail has left the tunnel exit, the traveling distance is 15 m, the total simulation time is 18.25 s, 8.25 s, 5.75 s and 4.25 s.

3 METHOD OF MOVING MESH NUMERICAL SIMULATION

The existed studies for the train fire inside the tunnel generally focus on a quiescent train, and the air always flows from a given upstream direction. It is quite different to the actual situation, because the flow field of relatively static conditions and the flow field of absolute motion are significantly opposite. Therefore, in order to make the result accurately close to the actual situation, we use moving fire to simulate the real situation. STAR-CD, a computational fluid software based on the finite volume method for unstructured mesh, is researched and developed by Professor Gosman in the British Empire University. STAR-CD has many characters such as a rich physical model, minimal memory footprint, and easy-to-use features. STAR-CD is a powerful CFD tool for the analysis of the heat flow phenomenon, not only for its key technologies on the completely continuous mesh, sliding mesh, mesh repair, mesh adaptability, calculated stability and convergence, but also for its strong movement mesh function (including dynamic mesh separation/connection, adding/deleting, translation/rotation, stretch/compression) in the numerical simulation. STAR-CD offers a fully implicit time-step, at the same time provides a format of the time discrete second order accuracy to solve the problem of high-precision dynamic calculation. By using the moving mesh feature of the STAR-CD, we achieve the simulation of a subway train carrying the moving fire running in the tunnel.

4 NUMERICAL SIMULATION MODEL

In this paper, the model is in a Cartesian coordinated system, and the train is running toward the positive X axis. The number of grids in the mesh in the three directions of X, Y, Z is 400 × 25 × 20 in calculation region. It will have a strong swirl around the train due to the change of the structure. Therefore, we should increase the division of the mesh in the area close to the tunnel wall and the train. Finally, the entire calculation mesh is divided into 373,060 grids.

About initial and boundary conditions, the initial time is t = 0, and boundary conditions are $p = p_a$, $u = 0$, $v = 0$, $w = 0$, t = 20°C. Due to the structural wall in the tunnel is no-slip adiabatic wall, the value has been set as the default value as given in the software; the import and export of the tunnel is the pressure boundary which the relative atmospheric pressure is defined as 0 Pa; the surface of the train is moving which speed is the train speed; slip boundary is between the moving mesh and the stationary mesh.

The two-equation model of k-ε is used in the turbulence model, the SIMPLE algorithm, momentum equations and k-ε equations are in the format of the discrete upwind difference scheme. Time is equably divided into 1000 steps, corresponding to the speed of the simulated conditions, the time of steps are 0.01825 s, 0.00825 s, 0.00575 s and 0.00425 s.

5 RESULTS AND ANALYSIS

5.1 *Longitudinal temperature*

Figure 2 is the distribution of the temperature in longitudinal sectional (unit is K) in a condition of project 1 (31.3 kW, 0.82 m/s) under 3.65 s, 10.95 s in the tunnel. As we can be seen from the temperature distribution in figures, the smoke above the fire source significantly flows to the tail of the train, when the train is moving into the tunnel. The maximum temperature is located downstream fire source at each height. In the process of running in the tunnel, the maximum temperature is in the center of the fire source, the highest temperature reaches to 356°C at 10.95 s. At 18.25 s, the tail of the train has completely run out of the tunnel. It can be found that the highest temperature is 62°C in the ceiling of the tunnel. The temperature gradually decreased with height until reduced to room temperature.

Figure 3 is the temperature curves which are 25 mm beneath the ceiling of the tunnel and represent the speed of the train at 0.82 m/s, 1.83 m/s (HRR is the same), HHR of the fire source are 31.3 kW. By comparison, all of the highest temperature is located downstream the fire source. As the speed of train increases, the distance between the center of the fire and the highest temperature point also increases. No matter how fast the train runs, the air temperature in front of the fire is always at about 20 °C as same as the initial temperature. At a speed of 0.82 m/s, the temperature at 0.3 m downstream the fire source rises the fastest, and reaches 130°C before the train runs out of the tunnel; The temperature at 0.9 m ~ 1.8 m downstream the fire source rises first and gradually decline, after 1.8 m the temperature rises again, and the temperature tends to be at about 40°C downstream the fire source. The temperature change law is the same as the train speed increases to 1.83 m/s, the temperature downstream the fire source gradually increases over time, and the farther away from the source of fire, the faster the subsequent temperature rises. When the train speed is at 1.83 m/s, the highest point of the temperature is at about 2.1 m downstream the fire source, the highest point of subsequent temperature is close to the center of the fire, and the maximum temperature is about 52°C.

When the train speed is constant but the heat release rate increases, the temperature distribution is similar. Table 2 shows the various projects on the train travelling to the middle of the tunnel and the maximum temperature on both sides of the fire source in the three groups as well as its range. The analysis shows that the temperature upstream the fire source

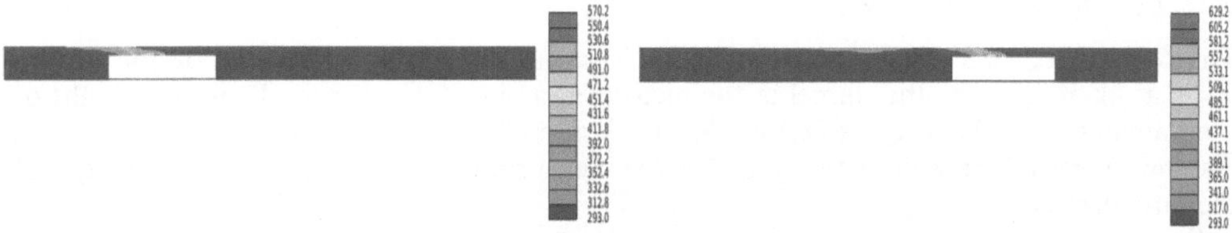

(a) t=3.65s (b) t=10. 95s

Figure 2. Temperature distributions in longitudinal section in the center of the tunnel.

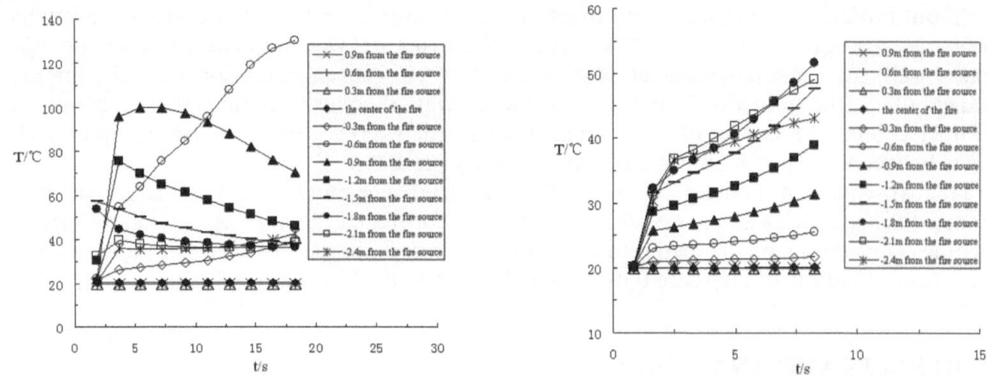

(a) U=0. 82m/s (b) U=1. 83m/s

Figure 3. Temperature curves at different train running speed on both sides of the fire source.

Table 2. Maximum temperature and its range of different projects in different regions.

		The first group		The second group		The third group	
Train speed (m/s)	Heat release rate (KW)	Maximum temperature (°C)	Distance upstream of the fire source (m)	Maximum temperature (°C)	Downstream distance of the source of fire (m)	Maximum temperature (°C)	Downstream distance of the source of fire (m)
0.82	31.3	20	0~0.9	281	0~0.6	97	0.6~2.4
	35.64			291		98	
	40.14			305		98	
	41.08			304		99	
1.83	31.3	20	0~0.9	127	0~1.2	62	1.2~2.4
	35.64			131		64	
	40.14			136		65	
	41.08			137		66	
2.65	31.3	20	0~0.9	94	0~1.8	36	1.8~2.4
	35.64			96		36	
	40.14			99		37	
	41.08			99		37	
3.54	31.3	20	0~0.9	72	0~1.8	36	1.8~2.4
	35.64			77		37	
	40.14			77		37	
	41.08			77		37	

is 20 °C, the range of high temperature in the second group rises with the speed of train increases. When the heat release rate increases to 40.14 kW, the maximum temperature substantially doesn't increase.

5.2 *Longitudinal temperature change law*

There are three measuring points T1, T2, T3 in the longitudinal direction of the tunnel, T1 is at the midline of the tunnel at the measuring point Z = 0.475 m; T2 is at a height of the human eyes at the measuring point Z = 0.1825 m of the midline in the tunnel; T3 is for a distance 6 cm off the wall of the tunnel at the measuring point Z = 0.1825 m at a height of the human eyes.

We can see the longitudinal distribution of the temperature of the three measuring points when the train is running to the middle of the tunnel (heat release rate is 31.3 kW) in figure 4a for the project 1. Figure 4a has shown that, the temperature remains the initial state upstream the fire source, and in a height of the human eyes the temperature downstream the fire source is substantially around 20°C (in the height of human eyes the temperature in the longitudinal direction of the central section of the train model is room temperature), the temperature slightly increases at 4 m the downstream of the fire source at a height of human eyes. The maximum temperature at the midline of tunnel ceiling in a distance at 1 m downstream the fire source is 105°C. Ranging from 1 m to 4 m downstream the fire source, the temperature drops sharply to 35°C, and then increases to 50°C, finally, with the distance from the center of the fire increases, the temperature gradually decreases to 20°C. The longitudinal temperature distribution is shown in figure 9b in the project that the heat release rate increases to 41.08 kW with the same speed of the train (project 4). By comparison, the rise of the heat release rate does not make the temperature of the measuring point increase. The maximum temperature is only 4°C higher than that of project 1, so the influence between the change of the train speed and the temperature can only be discussed in the case of 31.3 kW.

Figure 5 shows that comparison with the longitudinal temperature distribution of each measuring point for different train speeds. As shown in figure 5a, the train speed of 0.82 m/s, the longitudinal maximum temperature reaches 105°C at T1 at 1 m downstream the fire source. When the train speed increases to 1.83 m/s, the longitudinal maximum temperature at 2 m downstream the fire source at T1 rapidly reduces to 40°C. When the train

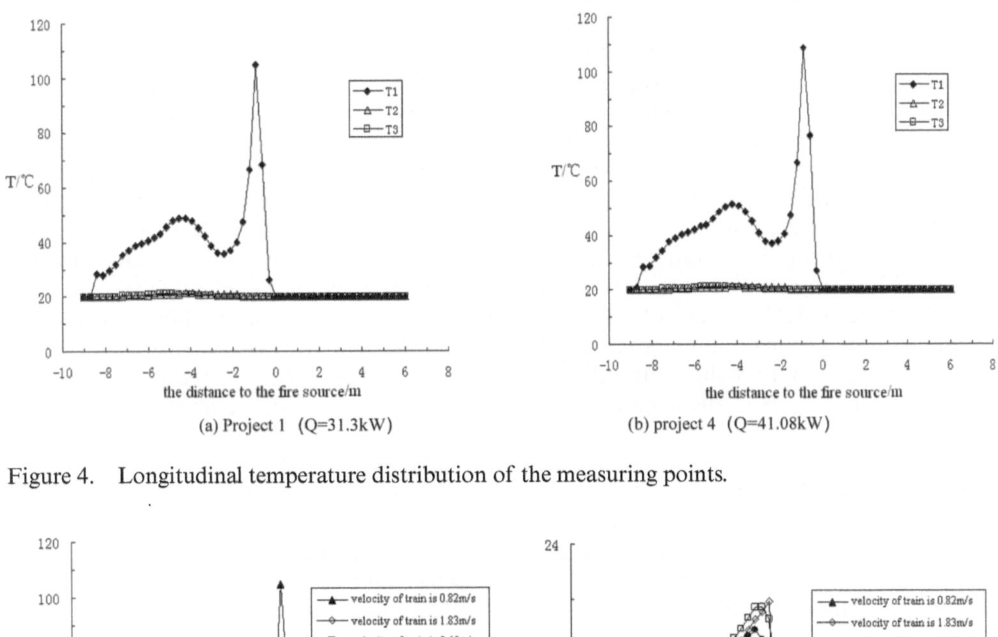

(a) Project 1 (Q=31.3kW) (b) project 4 (Q=41.08kW)

Figure 4. Longitudinal temperature distribution of the measuring points.

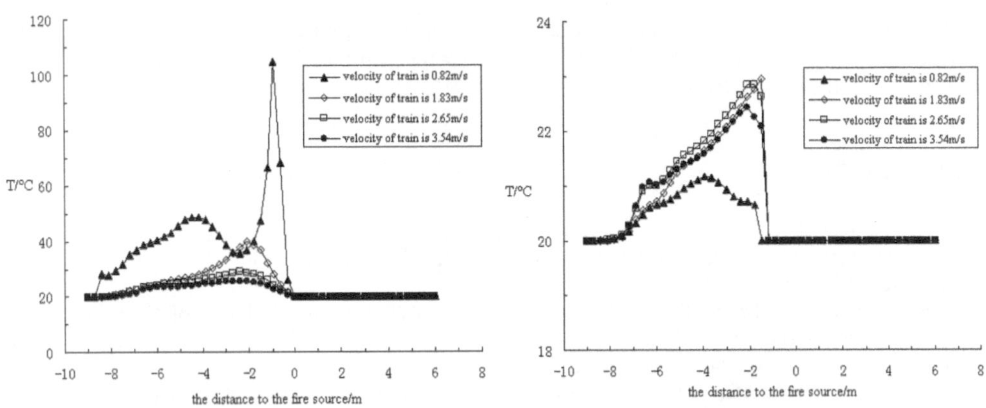

(a) The measuring points T1 (b) The measuring points T2

Figure 5. Longitudinal distribution of the temperature of the measuring point in the different train speed.

speed continually increases to 2.65 m/s and 3.54 m/s, the longitudinal temperature at 2.4 m downstream the fire source at T1 is about 26°C. As shown in Figure 5b, when the train speed is 0.82 m/s, the longitudinal temperature of each point at T2 is lower than the others, and the maximum temperature is at 4 m downstream the fire source. When the train speed is more than 1.83 m/s, the longitudinal temperature of each point at T2 rarely changes, the distance between the maximum temperature and the center of fire source reduces to 2 m. Overall, no matter what the train speed is, the longitudinal temperature of T2 is lower, and the maximum temperature is no more than 23°C. We find out that at different train speeds, the smoke mainly congregates in the ceiling of the tunnel, and there is almost no increase at a height of the human eyes.

6 CONCLUSION

1. The maximum temperature at a distance of 25 mm below the ceiling locates downstream the fire source when the fire source is in the middle of the train, and as the train speed increases, the distance between the point with maximum temperature and the center of fire source also increases. The temperature in front of the fire source is almost equal to the initial temperature.
2. According to the temperature distribution on different points away from the fire source in the vertical direction, we can divide three different regions in the X-axis direction. The first region is upstream the fire source, in which the temperature is at about 20°C (the initial temperature); the second region is from the center of the fire source to the 0.6 ~ 1.8 m downstream it, the length of the region extends as the train speed increases, in which the temperature increases at first and then decreases near the ceiling in the vertical direction. The third region locates the area of behind the second region, the change law of the temperature is the same as the second one; but the location of the maximum temperature is closer to the ceiling, and the average temperature is lower than that in the second region.
3. As the train is running at different speeds, smoke is mainly gathered near the ceiling of the tunnel. The temperature at the height of the human eyes will not evidently increase, and the increase of HHR does not make the temperature of the measuring points increased significantly. The maximum temperature of the cross-section at 1.5 m downstream the fire source is no more than 46°C, and the temperature from top to bottom reduces gradually, and the temperature at the height of the human eyes keeps at room temperature.

ACKNOWLEDGEMENT

The authors wish to acknowledge the research is supported by the National Natural Science Foundation of China (Grant No. 51078301) and Science and Technology Projects of Beijing in China (Grant No. 8101004).

REFERENCES

Bettis, R.J., Jagger, S.F. & Wu, Y. Interim validation of tunnel fire consequence models; summary of phase 2 tests [R]. The Health and Safety Laboratory Report IR/L/FR/93/11, *The Health and Safety Executive*, UK, 1993.

Bettis, R.J., Jagger, S.F., Macmillan, A.J.R. & Hambleton, R.T. Interim validation of tunnel fire consequence models; summary of phase 1 tests [R]. The Health and Safety Laboratory Report IR/L/FR/94/2, *The Health and Safety Executive*, UK, 1994.

Hwang, C.C. & Edwards, J.C. The critical ventilation velocity in tunnel fires-a computer simulation. Fire Safety Journal, 2005, 40: 213–244.

Carvd, R.O., Beard, A.N. & Jowitt, P.W. The influence of tunnel geometry and ventilation on the heat release rate of fire [J]. *Fire Technology*, 2004, 40(1): 5–26.

Fletcher, D.F., Kent, J.H., Apte, V.P. & Green, A.R. Numerical simulations of smoke movement from a pool fire in a ventilated tunnel. *Fire Safety Journal*, 1994, 23(3): 305–325.

Mao Jun. Research on smoke flow law and emergency control in subway fire. *Beijing, University of Science and Technology Beijing doctor's degree thesis* [D], 2008.

Mao Jun, Xi Yanhong & Fan Hongming. Temperature Characteristics of Flame Ceiling Jet Generated by Fire on Train in Subway Tunnel. *Journal of South China University of Technology*, 2010, 38(3): 17–24.

O Vauquelin. Parametrical study of the back flow occurrence in case of a buoyant release into a rectangular channel. *Experimental Thermal and Fluid Science*, 2005, 29: 725–731.

O Vauquelin & O Megret. Smoke extraction experiments in case of fire in a tunnel. *Fire Safety Journal*, 2002, 37: 525–533.

Woodbum, P.J. & Britter, R.E. CFD simulation of a tunnel fire-Part 1 [J]. *Fire Safety Journal*, 1996, 26 (1): 35–62.

Woodbum, P.J. & Britter, R.E. CFD simulation of a tunnel fire-Part 2 [J]. *Fire Safety Journal*, 1996, 26 (1): 63–90.

Xi Yanhong, Mao Jun & Fan Hongming. Safety velocity of on-fire train moving in subway tunnel. *Journal of South China University of Technology*, 2010, 38(3): 25–30+36.

Xu Zhisheng & Zhou Qin. Model experiment and numerical simulation of moving train in the tunnel fire [J]. *Journal of The China Railway Society*, 2004, 2, 26(1): 124–128.

Yu Li, Wang Mingnian & Guo Chun. Numerical simulation in temperature field of fire in super-long tunnel. *Journal of Civil Engineering*, 2007, 40(60): 64–68.

Zhao Yi. Numerical simulation and ventilation control in subway fire [D], Dalian, *Dalian Jiaotong University master's degree thesis*, 2009, 6.

Zhou Xiaoqing, Zhao Chenjun & Zhao Xiangxiang. Numerical simulation and analysis of some controlling in subway tunnel fire. *Journal of Guangzhou University*, 2008, 7(5): 66–70.

Modeling and Computation in Engineering II – Xie (ed)
© 2013 Taylor & Francis Group, London, ISBN 978-1-138-00058-2

A BIM-LCA framework and case study of a residential building in Tianjin

W. Yang & S.S. Wang
School of Architecture, Tianjin University, Nankai District, Tianjin, China

ABSTRACT: The growth and operation of the Chinese housing stock has caused considerable resource consumption and environmental impacts. Yet there are so far no quantitative criteria or calculation methods for the life cycle impacts and costs of residential buildings in the national and local standards. Conventional LCA relies on quantity survey and inventory data of building materials, being timing consuming and losing original building components' information. A BIM-LCA framework is presented in which the information of all assembles and materials are deduced from BIM as input data for the LCA & LCC model. The framework enables a dynamic process of optimization and evaluation of the project, and helps to integrate ecological and economic analysis. A high-rise residential building in Tianjin is analyzed. The results demonstrate that the BIM-LCA model is an effective tool, and that the calculation method based on building assemblies is more appropriate for optimization process.

1 INTRODUCTION

The LCA methodology has already been applied in calculating and evaluating the built environment since the late 1990s. In China, LCA if building materials has been increasingly developing, while there is much to do for a life cycle ecological and economical assessment of buildings based on components and assemblies. A major obstacle is that the LC Inventory database for building materials and assemblies is not well developed. With such a database and a corresponding building information model, the LCA method will be smoothly applied in every stage of a building (e.g., early design, construction, maintenance, renovation and demolition). Therefore, it is vital to find out the methodology for building life cycle assessment which can be easily applied in the rapid designing and evaluating process of the Chinese building industry, and to assist decision makers to reduce the long term environmental impacts and costs of the project.

Integrated LCA and LCC is important for sustainable building design and assessment. There has been several integrated assessment systems in practice, such as Sue-MoT 2003, LEnSE 2007 (European sustainability tool), and the German DGNB/BNB system. Saad Dawood et al. (2009) mentioned "a framework, methodologies and technologies that will facilitate the integration of Environmental Impact Assessment, Whole Life Cycle Cost and Life Cycle Assessment using 3D and BIM technologies." Yan Yuan et al. (2011) discussed about the theoretical method of the integration of BIM and LCA in China.

There is a need to present an integrated BIM-LCA framework which can be easily applied to the designing and evaluating process of the Chinese buildings. The methodology could enhance the efficiency and accuracy of the assessment. It also helps the decision makers select the best scheme.

2 DEVELOPMENT OF BUILDING LCA IN CHINA

2.1 *Relevant regulations and standards*

Energy saving codes have been developed for residential buildings in different climate zones since 1980s, and for commercial buildings since 2005. Recently, the fourth-step energy codes for heated residential buildings in north China had been issued (Design standard for energy efficiency of residential buildings in severe cold and cold zones, JGJ 26-2010), which requires 65% energy saving on the 1980 baseline. It had been adapted to local standards, such as Tianjin (DB 29-1-2010). The Green Building Design and Evaluating Standards (GB/T 50378-2006, DB/T29-204-2010) have extended the scope to building material, water and land conservation besides energy efficiency. Since 2009, carbon emission has been another important issue for the Chinese building sector. Hence the study on building LCA has developed rapidly in recent years. However, there are still no effective quantified assessment tools. A major obstacle is the lack of database for building material and assemblies. Moreover, it is time consuming to do a complete LCA calculation. It is important to introduce life cycle ecological and economical analysis in the Chinese sustainable building evaluation standards.

2.2 *Recent research on LCA in China*

The inventory database for Chinese building material and energy products is being developed in the recent years, although much has to be done in comparison to the European databases. Chen Y. et al. (2010) studied the LCA method for residential buildings. Zhang Z.H. (2010) presented the building life cycle carbon emission method and provided the inventory data for major Chinese building materials. However, these calculation methods are mainly based on quantity survey data which lost the construction and position of building assemblies. The result is difficult for architects and decision makers to interpret and to feedback in the designing process. Hence, it is important to develop a method which integrates the geometrical information, quantity information and life cycle impacts in the same framework, and allows the dynamic feedback between LCA results and the design scheme.

3 METHODOLOGY OF AN INTEGRATED BIM-LCA

3.1 *LCA tools and databases*

There have been several building LCA tools developed in industrialized countries. For example, the Athena eco-calculator has user friendly interface and contains a practical building assembly database for North America, which makes it easy to describe the buildings and to calculate the impacts. The LEGEP software developed for Germany is an effective tool of integrated LCA and LCC. It not only calculates the environmental impacts of the building, but also considers the economic factors in the same physical framework. However, the methods and databases of those tools are highly relevant to specific context, which makes them difficult to be adapted to Chinese buildings. Therefore, we try to build an integrated LCA and LCC model using Chinese data with Excel. In the model, the basic functional unit is assemblies with the information of construction details and material composition. Buildings are described as assemblies with the information of quantity and position.

With the standard drawings of building assemblies, the inventory data for major building material and energy products, and the costs for building materials and assemblies, a simplified database for Chinese buildings can be built up. Since the Chinese inventory data for building material are incomplete, international databases such as Ecoinvent and Ökobau.dat are referred to for supplementary data.

3.2 *BIM of building industry in China*

BIM is ideal for building life cycle management. It integrates the 3D geometry model with quantity and quality information, with which the parametric information of the building and of each component can be efficiently and precisely derived.

The application of BIM is developing fast in China. Many architectural design institutes have started to use BIM instead of 2D CAD drawings. The national BIM standard will be issued soon. As the most widely used BIM tool is Autodesk Revit, we use it for our BIM-LCA model.

China has publishes a series of standard drawings for building assemblies for different climate zones as the common reference for building industry, such as 05 J for five provinces in Northeast China. But the BIM database of 05 J is not available so far. Each design institute has to rebuild their own family database without sharing.

As the first step of the BIM-LCA model, major assemblies for residential buildings are chosen from 05 J standard drawings and are described as families of different categories (e.g., walls, windows, doors, floors, roofs) in Revit, which contain the information of material, construction, costs and U values. On this basis, buildings can be described and information for energy calculation and the input data for LCA can be deduced.

3.3 *LCC and Chinese construction project cost*

In China, the quantity survey of buildings is regulated by the National Cost Engineer qualification system. Referring to ISO standards and the Tianjin local regulation of construction project cost (08 construction cost in Tianjin), the database for costs of all the building materials and assemblies is built. It can be updated on the basis of local cost standard. The cost information can also be edited in BIM software, which helps architects to make choices.

In China, the LCC tool for buildings has not been well developed. In this study we adopt a simplified LCC approach in which the discount rate is chosen under the assumption that the required rate of return equals to the annual inflation rate, making the discount rate factor irrelevant to our calculation. The increase of energy price is assumed as 1% per year.

3.4 *The operation of energy consumption of the buildings*

For the operational stage of the building, Design builder for Energyplus is used to simulate the energy consumption. The result is related to the inventory data for energy system and energy products to calculate primary energy consumption.

3.5 *Conception of BIM-LCA framework*

Using the assembly category and material schedules derived from Revit, the inventory analysis of building LCA can be largely simplified. Moreover, the LCA results could be easily feedback to the designing process, which supports the modification and optimization of the scheme. Therefore, the BIM-LCA functions well in the early design stage, so that the architects can work more efficiently and find out the best scenario with minimum life cycle environmental impacts.

The assembly database is the bridge that links BIM to LCA. The information surveyed through the BIM software and core idea of LCA is the key point of the success of their connection, while the combination of the two powerful programs has been seldom concerned before.

The integrated BIM-LCA model can calculate the impacts based on either assembly schedules or material schedules derived from BIM. The results can be verified by comparing results from two different methods.

3.6 *Case study*

A high-rise residential building in Tianjin is analyzed as a case study of the BIM-LCA model. Considering the limit of time, the up-stream stage, the operation stage and the refurbishment process are considered in the study. The transportation, construction, cleaning, demolition and recycling processes are not included so far. The building lifespan is set to be 50 years. And the discount rate presumed to be equal to the rate of inflation. In the material based LCA we consider the construction waste rate factor, whereas in the assemblies approach, this part is not included.

The total floor area of the building is 17337 square meters. The building has 33 floors above ground and 3 floors under ground. It locates in Tianjin. The heating season is from Nov. 15 to Mar. 15 next year.

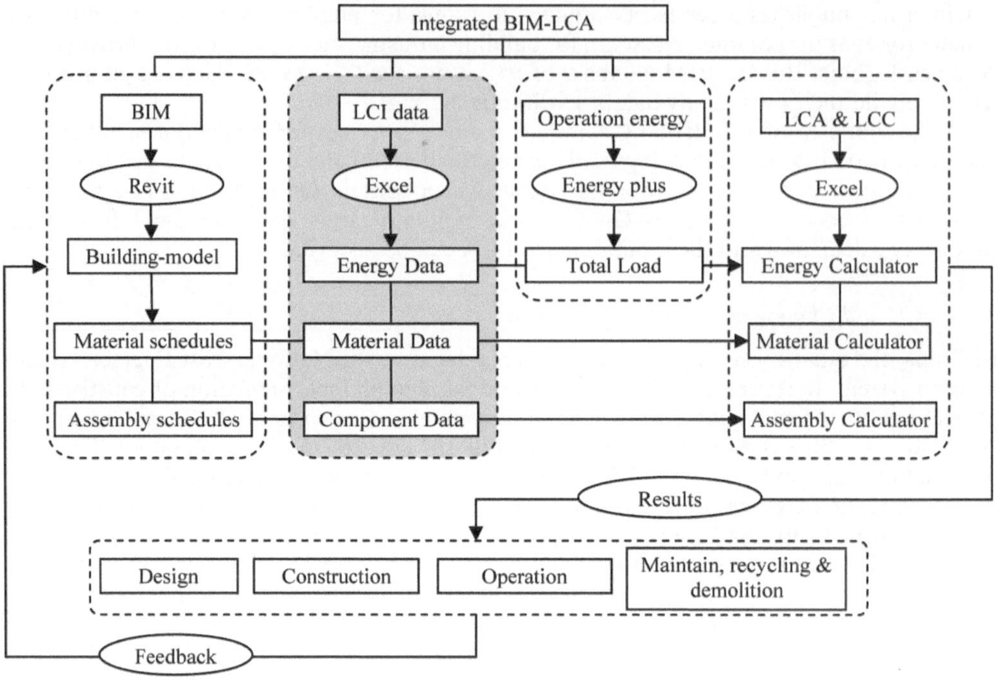

Figure 1. Integrated BIM-LCA framework.

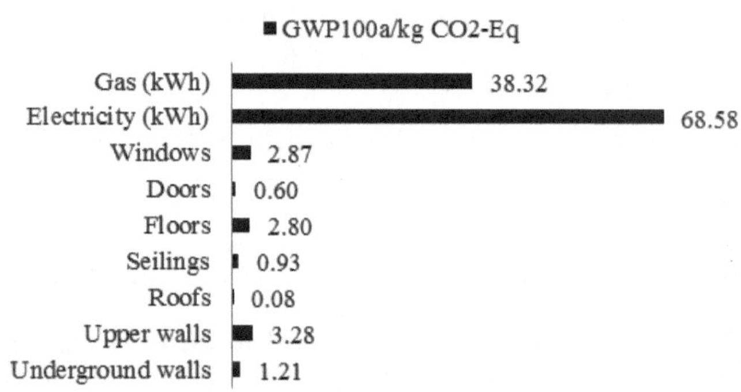

Figure 2. Comparison of CO_2 emission from major building components and from operational energy consumption (Unit: kg CO_2 /m^2/year).

The structure of building is Reinforced concrete share walls. The insulation materials are extruded polystyrene board (XPS) and FTC energy-saving materials from the thermostat phase transition. The major wall materials are 200 mm reinforced concrete and 200 mm aerated concrete block. The construction time is 2008 and the energy system for district heating is gas boiler.

The CO_2 emission from building components consists of the construction, maintenance and renovation stages. The above figure shows that the CO_2 emission of operation stage accounts for approx. 90% of total emission. One reason could be lower material intensity and more space of high-rise buildings. Another reason is that aerated concrete is used as a major wall material, which reduces the embodied energy and carbon emission from building material. However, the part of emission from operational energy is obviously higher than low energy buildings in Europe, which demonstrates that the insulation of the envelop could be further improved. Therefore, better insulation material and windows with reasonable embodied energy is the major problem to be solved.

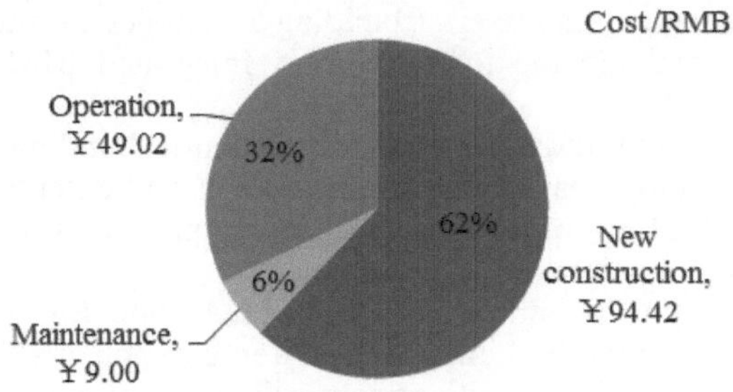

Figure 3. Comparison of cost of three life stages per square meter per year.

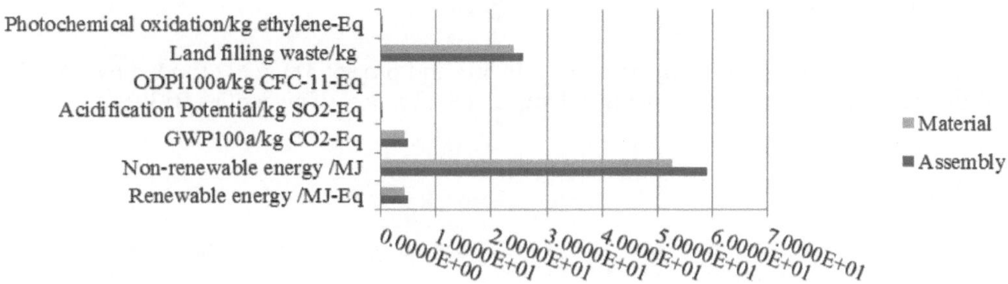

Figure 4. Comparison of material based & assembly based LCA-LCC values (kg CO_2 /m^2 year).

As figure 3 shows, the new construction cost accounts for nearly two thirds of the total, which means that in residential buildings the construction cost would be the predominant part. The building with low energy demand standard would take lower operation phrase costs.

As figure 4 shows, in the construction stage, the values of the results of the material based approach are a little bit lower than that of the assembly based approach. The main reason is that during the assemblies producing process, there are additional material and energy consumption. The current building LCA in China is mainly based on material approach, which is not precise enough. Moreover, the information of building construction is lost during the calculation process, which makes it difficult to support the optimisation process of architectural design. Therefore, it is important to set up a BIM-LCA framework based on assemblies.

4 DISCUSSION AND CONCLUSION

This paper introduced an integrated BIM-LCA methodology which combines several conventional approaches in building design and assessment, including geometrical model, quantity survey, operational energy simulation, LCA and LCC. It can support the design, assessment and management of the building from early design stage to the construction, operation, refurbishment and demolition and recycling.

According to the study of the BIM-LCA methodology and the case study, the following issues should be addressed:

1. There is so far no standards and benchmarks for life cycle environmental impacts for buildings in China. Moreover, the current energy calculation tools are mainly used after detailed design, not in the early design stage. The result is therefore difficult to be used for optimisation of the schemes.
2. The current achievements on building LCA in China are mostly based on quantity survey data and CAD drawings, which is either incomplete or time consuming. There is a need for an a BIM-LCA framework to enable an interactive optimisation process.

3. The parametric information databases of building assemblies, inventory data for building materials and their costs are the foundation of Integrated BIM-LCA application in China.
4. The first results from a case study of a high-rise residential building in Tianjin shows that the BIM-LCA methodology based on Autodesk Revit and excel program for LCA and LCC can be a useful tool in supporting the designing process. The calculation approach based on building assemblies are more accurate.
5. There is a large potential to improve the BIM-LCA tool. E.g., building the friendly user interface, completing the building material and components databases, and setting benchmarks for different types of buildings according to national and local standards.

ACKNOWLEDGEMENT

This contribution is supported by project 51108303 and project 51178292 funded by NSFC, and by the Programme of Introducing Talents of Discipline to Universities B13011.

REFERENCES

CHEN, Y. & ZHU, Y. 2010. Residential building life cycle energy consumption and emission model. Journal of Tsinghua University 50(3): 325–329.

GONG, Z. 2004. A Quantitative Method to the Assessment of the Life Cycle Embodied Environmental Profile of Building Materials. Tsinghua University M.S. thesis paper.

Holger, K., Kohler, N., Kreissig, J. & Lützkendorf, T. 2010. A life Cycle Approach to Buildings. München, Institut für international Architektur-Dokumentation GmbH & Co. KG (DETAIL Green Books).

Ignacio Zabalza Bribia´n, Alfonso Aranda Uso´n & Sabina Scarpellini. 2009. Life cycle assessment in buildings: State-of-the-art and simplified LCA methodology as a complement for building certification. Building and Environment 44: 2510–2520.

Kohler, N. & Hassler, U. 2002. The building stock as a research object. Building Research and Information 30(4): 226–236.

Kohler, N. & YANG, W. 2007. Long-term Management of the Building Stock. Building Research and Information 35(4): 1–13.

Saad Dawood, Richard Lord & Nashwan Dawood. 2009. Development of a visual whole life-cycle energy assessment framework for built environment. Proceedings of the 2009 Winter Simulation Conference M.D. Rossetti, R.R. Hill, B. Johansson, A. Dunkin and R.G. Ingalls, eds.

SHANG, C. & ZHANG, Z. 2010. Building life cycle carbon emission calculation. Journal of project management 24(1): 7–12.

Tianjin Construction and Transportation Committee. 2010. Tianjin energy efficiency design standards for residential buildings (DB29-1-2010).

YANG, W. & Kohler, N. 2008. Simulation of the Evolution of the Chinese Building and Infrastructure Stock. Building Research and Information 36(1): 1–19.

YANG, Q. 2009. The quantitative evaluation of building products full life cycle environmental effect. Tianjin University PhD thesis paper.

YANG, W. 2010. The modeling of building stocks—the building development strategies of rural and urban China in the next 50 years. Nanking: Southeast University Press.

Modeling and Computation in Engineering II – Xie (ed)
© 2013 Taylor & Francis Group, London, ISBN 978-1-138-00058-2

Optimal windbreak design for wind-erosion in high-speed railway

Y. Xi, J. Mao & L. Gao
The Civil Institute of Beijing Jiaotong University, Beijing, China

G. Yang
Institute of Mechanics, Chinese Academy of Sciences, China

ABSTRACT: It is also very important to improve resistant capabilities of wind resistance and structural stability of the windbreak while reducing train aerodynamic load. This article proposes a new type of windbreak with double-layer structure forming a cavity chamber with hole for energy dissipation of crosswind. Numerical simulation was used to analysis the wind protection effect of two kinds of windbreak and their own wind-resistance performance. The results showed that the two types of windbreaks can significantly reduce the train aerodynamic load under the action of crosswind, and the force act on the windbreak with cavity chamber structure is small than on another one with single layer structure. The new windbreak has better effect in aerodynamic load shedding for the train, stronger resistant capabilities of wind resistance and safety of structure. This new windbreak with cavity chamber will provides useful references for design of the wind protection system for high speed train.

1 INTRODUCTION

Crosswind results mainly in the disturbance of the flow field around the train and causes a rise in additional aerodynamic forces and moments. The resulting combination of forces and moments impairs the rollover threshold of the train, thus causing a major threat to operational safety (Baker, 1990; Baker, 1993; Coleman, 1994; Quinn, 2007). In addition to the magnitude of the relative velocity, the magnitudes of the aerodynamic forces and moments depend predominantly on the wind speed of incidence. Windbreaks have been used to shelter high-speed train operation safety from wind damage. They reduce wind speed and alter the characteristics of airflow around them, inducing changes in the surrounding atmospheric (Cleugh, 1998). The interaction between the windbreak and the airflow is complicated by the turbulent characteristics of the wind and by the complex behavior of natural obstacles. Direct effects are due to adsorption of momentum from the wind flow, thus decreasing wind erosivity. They result in reductions of wind speed and turbulence intensity within a certain distance in the lee.

Besides windbreak height and porosity, the actual form of the wind speed curve depends on other important characteristics of the airflow—windbreak system. These are the approach flow characteristics, such as wind speed, wind direction, turbulence intensity, and atmospheric stability, and external windbreak properties, such as windbreak shape, width, and length (Heisler and Dewalle, 1988). Airflow—windbreak interactions have been described and studied in great detail in some numerical simulation models (Wilson, 1985; Wang and Takle, 1995). These models give insights into airflow around porous barriers. In the Wind Erosion Prediction System (WEPS), windbreak effects are incorporated by a reduction factor of friction velocity that takes into account the distance from the barrier, porosity, and width of the barrier porosity, and width of the barrier.

(a) Windbreak in Nanjiang line (b) Windbreak in Lanxin line

Figure 1. Windbreak in China.

At present, there are windbreaks mainly in Nanjiang line with arc characteristics and in Lanxin line with flat feature in China, as shown in Figure 1. Considering the comprehensive properties of wind resistance and structure optimization design of the wind shield effect, method and so on, the main objective of this study is to design a windbreak with double corrugated plate and air chamber. And the effect of wind resistance of single windbreak and double will be discussed in the following.

2 CALCULATION MODEL AND CONDITION

2.1 *Calculation model*

Numerical simulation of a complete train with a length of about 200 m requires computational resources beyond those available on existing computers and hence only the flow fields of simplified train configurations can be reliably computed at present. Based on observations made by Cooper (Cooper, 1979), who suggested that the flow structure downstream of a certain distance from the nose (less than one coach length) is more or less constant. Therefore, a decrease in length does not alter the essential physical features of the flow as long as the total length remains above the limit suggested by Cooper. So, as a simplification, the length of the train was limited to that of a train assembly consisting of two identical driving coaches and one trailers (Fig. 2). The total length of the train is 76.125 m with a length to width to height ratio of 1:0.039:0.051.

Figure 3 shows the dimensions of the flow domain. This domain size was consistent with that used by Schlichiting (Schlichiting, 2000) and was deemed adequate to capture the changes in the flow field during the interaction while giving sufficient distance between the models and the boundaries.

Figure 3 shows a grid system. The finest grid size is near the train surfaces and at the leeward side. The grid size is fine too around the pantograph. The total number of grid, double-layer windbreak with chamber mode and the single-layer mode are 6000 million and 5100 million, respectively.

2.2 *Calculation condition*

Flow around three cases such as no windbreak, single layer windbreak and double layer windbreak with chamber with the high-speed train with 350 km/h under the influence of a crosswind have been studied using numerical technique.

In order to prove the wind resistance performance of the double-layer windbreak with chamber, a series of numerical simulations of the flow past the previously described geometry were carried out. The total simulation conditions are listed in Table 1.

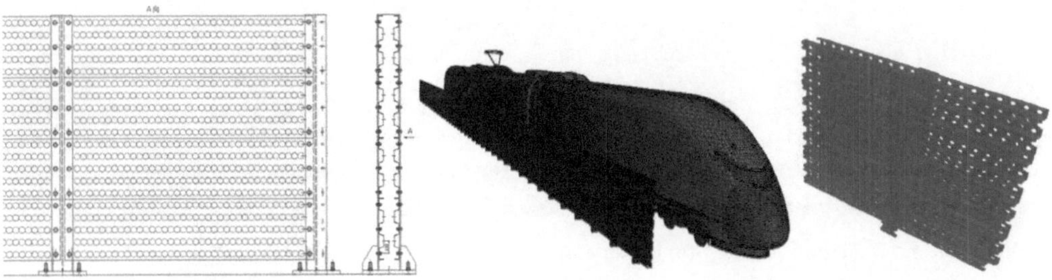

(1) Schematic diagram of windbreak (2) The relative position of the train (3) single windbreak

Figure 2. Model.

(1) Computational domain (2) Grid around the train (3) Grid around the windbreak

Figure 3. Computational domain and grid system.

Table 1. Simulation conditions.

Scenario	Train speed/ (km/h)	Crosswind speed/ (m/s)	Crosswind angle	Pore diameter/ (mm)
No windbreak	350	10, 15, 20	90°	0
Single windbreak	350	10, 15, 20	90°	50
Double windbreak	350	15, 20	90°	50

3 RESULTS AND DISCUSSION

3.1 *Pressure distribution around the train*

The flow structure for the wind speed with 20 m/s is shown for the case by the two-dimensional vorticity and pressure distribution around the train in Fig. 4. The direction of the crosswind is form lift to right.

From the perspective of vortex motion, the pressure increased in the most surface of the train windward side and the vortex separation generated in the leeward side due to the wind acted directly on the train without energy attenuation in the no windbreak case, as shown in figure 4.

3.2 *Mechanism of the windbreak with chamber*

From the analysis above, part of wind energy is consumed and the direction of the wind deflected because of the windbreak blocking. Thus the pressure of the train is changed and the aerodynamic forces are reduced. The mechanism of the double layer windbreak with chamber is same as the single one. But the load reduction effect of the double layer windbreak with chamber is more obvious.

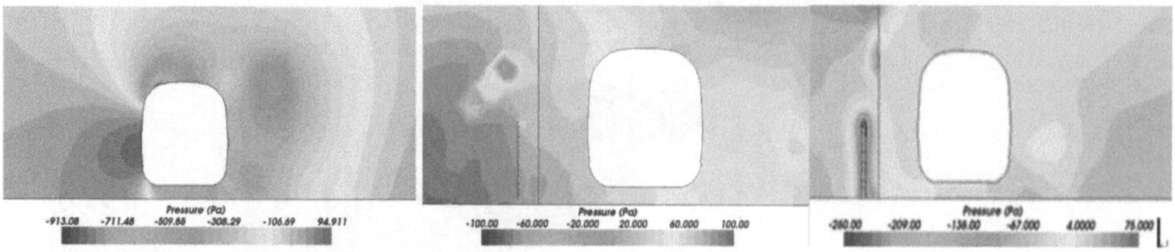

(a) Pressure, no windbreak (b) Pressure, single windbreak (c) Pressure, windbreak with chambers

Figure 4. Vorticity and pressure distribution around the train.

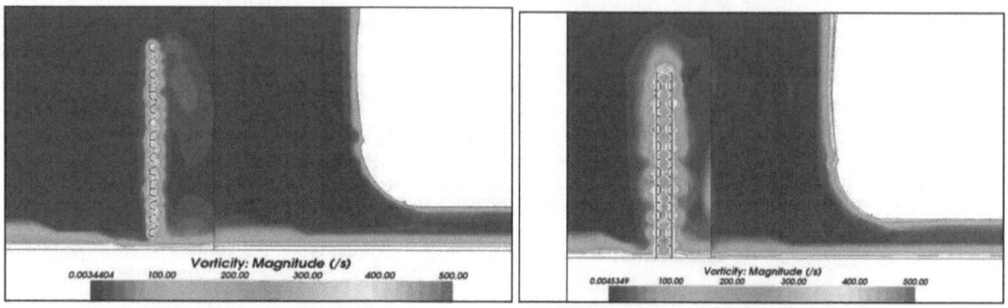

(a1) Vorticity, single windbreak (b1) Vorticity, windbreak with chambers

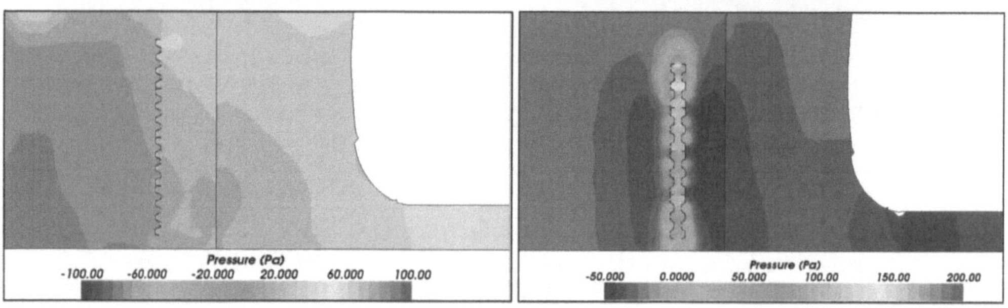

(a2) Pressure, single windbreak (b2) Pressure, windbreak with chambers

Figure 5. Vorticity and pressure distribution around the windbreak.

The reason is as follows: in the progress of the crosswind flows through the double-layer windbreak, the wind energy loss happened when the air flow through the first layer of the corrugated board. Moreover, the direction of the airflow is changed which flowing out though the second layer of the corrugated board. Furthermore, larger separation eddy and the second energy loss happened. So, the energy dissipation mechanism of the double chamber is much larger than the single windbreak, as shown in Fig. 5.

3.3 *Aerodynamic force/moment*

As can be seen from Fig. 4 and Fig. 5, a non-symmetric pressure field is created, resulting in the generation of lateral force and overturning moment. The associated forces and moments act together in the same sense, trying to turn the train over the leeward track.

As shown in Table. 2, the lateral force of the head train and the middle train decreased due to windbreak. But the tail train is complicated which is relation to the wind speed and the type of windbreak. And the effect of the aerodynamic force reduction of double layer windbreak with chamber is better than the single one. Moreover, the aerodynamic moments are all reduced in any crosswind speed with two windbreaks.

Table 2. Aerodynamic forces/moments (Force/N; Moment/N·m).

Force	Location	Crosswind speed—10 m/s		Crosswind speed—15 m/s			Crosswind speed—20 m/s		
		No windbreak	Single windbreak	No windbreak	Single windbreak	Windbreak with chambers	No windbreak	Single windbreak	Windbreak with chambers
Lateral force	Head	29370.6	2167.09	47271.83	2720.54	844.33	66708.2	2072.39	1676.40
	Middle	11658.7	1453.85	19505.65	760.47	38.19	30296.4	−158.78	423.20
	Tail	−1900.3	6604.5	331.49	4794.9	3776.75	4534.77	4596.64	810.76
Overturning moment	Head	−9425.2	1903.07	−15109.9	1701.21	833.48	−21053	863.69	351.58
	Middle	2566.59	−586.03	4015.18	−440.24	−362.70	3751.64	−414.21	478.22
	Tail	237.137	−334.05	−639.20	359.85	−412.13	−2728.87	−777.15	120.58

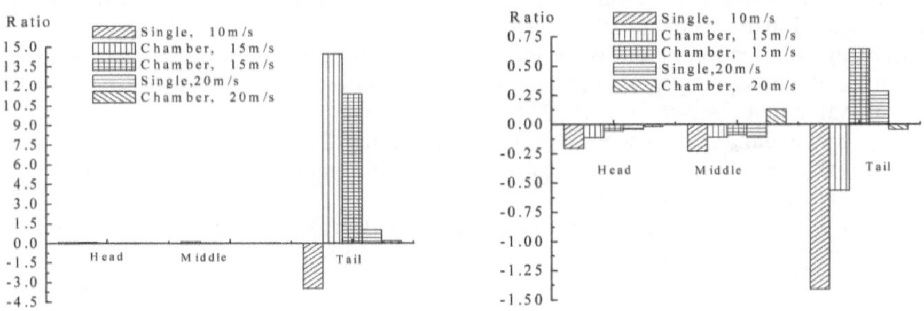

(a) Lateral force

(b) Overturning moment

Figure 6. Comparison of load reducing with two kinds of windbreaks.

Figure 6 is comparison results of load reducing effect with two kinds of windbreaks. The value of the ordinate is the ratio of the windbreak used to no windbreak. From Table 2 and Figure 7, the load reducing effect with windbreaks is very obvious, which changed with the crosswind speed. Compared with single windbreak, double windbreak with chamber can better improve the aerodynamic load.

4 CONCLUSION

A certain high-speed train with 350 km/h under the influence of crosswind has been studied using numerical technique by solving the steady three-dimensional RANS equations. The body of the train includes most important details including bogies, inter-car gaps, pantograph and rotating wheels on the rail. The dependence of the flow structure and the aerodynamic characteristics on windbreaks have been studied. Based on the analysis of the results, the following can be deduced:

1. A non-symmetric pressure field is created, resulting in the generation of lateral force and overturning moment on the one hand, and an increase of the other components of forces and moments on the other. The associated forces and moments act together in the same sense, trying to turn the train over the leeward track.
2. Two types of windbreaks can significantly reduce the train aerodynamic load under the action of crosswind; and the windbreak with chamber has better effect in aerodynamic load shedding for the train, stronger resistant capabilities of wind resistance.
3. Mechanism of the windbreak with chamber is as follows: in the progress of the crosswind flows through the double-layer windbreak, the wind energy loss happened when the air

flow through the first layer of the corrugated board. And the direction is changed; and larger separation eddy and the second energy loss happened.

REFERENCES

Baker, C.J. Assessment of Wind Tunnel Testing Techniques for Ground Vehicles in Cross Winds. Journal of Wind Engineering and Industrial Aerodynamics, 1990, 33(1): 429–438.

Baker, C.J. Behaviour of Road Vehicles in Unsteady Cross Winds. Journal of Wind Engineering and Industrial Aerodynamics, 1993, 49(1): 439–448.

Cleugh, H.A. 1998. Effects of windbreaks on airflow, microclimates and crop yields. Agroforestry Systems 41, 55–84.

Coleman, S.A. Experimental Study of the Aerodynamic Behavior of High Sided Lorries in Cross Winds. Journal of Wind Engineering and Industrial Aerodynamics, 1994, 54(3): 401–429.

Cooper, R.K. The effect of cross-wind on trains. In: Proceedings of Aerodynamics of Transportation, ASME-CSME Conference, 18–20 June, Niagara, 1979, pp. 127–51.

Heisler, G.M. & Dewalle, D.R. 1988. Effects of windbreak structure on wind flow. Agriculture, Ecosystems and Environment, 22–23, 41–69.

Quinn, A.D. An Investigation of the Wind-induced Rolling Moment on a Commercial Vehicle in the Atmospheric Boundary Layer. Proceedings of the Institution of Mechanical Engineers, Part D: Journal of Automobile Engineering, 2007, 21(11): 1367–1379.

Schlichiting, H. Boundary Layer Theory, Springer. 8th printing, 2000.

Wang, H. & Takle, E.S. 1995. A numerical simulation of boundary-layer flows near shelterbelts. Boundary-Layer Meteorology 75, 141–173.

Wilson, J.D. 1985. Numerical studies of flow through a windbreak. Journal of Wind Engineering and Industrial Aerodynamics 21, 119–154.

Modeling and Computation in Engineering II – Xie (ed)
© 2013 Taylor & Francis Group, London, ISBN 978-1-138-00058-2

Research progress on effective utilization of natural wind in highway tunnels

T. Yan
Key Laboratory of Transportation Tunnel Engineering, Ministry of Education, School of Civil Engineering, Southwest Jiaotong University, Chengdu, Sichuan, China
Department of Civil Engineering, Southwest Jiaotong University, Emei, Sichuan, China

M.N. Wang, J. Yin & C. Guo
Key Laboratory of Transportation Tunnel Engineering, Ministry of Education, School of Civil Engineering, Southwest Jiaotong University, Chengdu, Sichuan, China

ABSTRACT: With the construction of highway tunnels at high altitude, an increasing number of tunnels located in the meteorological separation zone where natural wind is also obvious. Natural wind pressure influences the ventilation system significantly. This paper summarized theoretical derivation, numerical simulation and site measurement of natural wind pressure as well as the research route of effective utilization of natural wind was also determined. Practice proved that the effective utilization of natural wind could create a huge economic interest. For more and more tunnels located in the meteorological separation zone, it is suggested a monographic study of the natural wind pressure should be carried out.

1 INTRODUCTION

With the construction of highway tunnels at high altitude, an increasing number of tunnels located in the meteorological separation zone. Climate in the entrance and exit of this type of tunnel differs significantly, which generates natural wind. Considering the safety, the direction of natural wind and the direction of traffic are regarded, as the opposite directions in calculation of ventilation, namely, the natural wind pressure is considered as resistance. However,with further researches conducted, more and more scholars recognize that it's inappropriate and extremely uneconomical just to take the natural wind pressure as resistance. Based on this point, many scholars launch a study of the natural wind pressure, in order to take advantage of natural wind as dynamic pressure. At present, study on natural wind pressure mainly focuses on theoretical derivation, numerical simulation and site measurement and so on.

2 THEORETICAL DERIVATION

Shi (2011) first proposed the influencing factors of the natural wind pressure containing three major parts, environment factors outside the cave (ultra-static pressure difference caused by atmospheric horizontal partial pressure gradient between tunnel entrance and exit); environmental factors in tunnel portals (Structural Opening) (wind wall pressure difference caused when natural wind blowing to Structural Opening); environment factors inside the cave (thermal potential difference caused by the air temperature difference inside and outside the tunnel). He first established a formula used to calculate natural wind inside the tunnel based on ultra-static pressure difference, wind wall pressure difference and thermal potential difference.

And calculation method of natural wind in the presence or absence of vertical well was also discussed. According to Highway Tunnel Ventilation and Lighting Design Specifications, wind speed of the natural wind generated inside the tunnel can be calculated according to the ventilation resistance formula.

$$\Delta P = (\lambda_r \frac{L}{D_r} + \xi_\theta + 1)\frac{\rho v_n^2}{2} \tag{1}$$

$$\Delta P = \Delta P_1 + \Delta P_2 + \Delta P_3 \tag{2}$$

$$v_n = \sqrt{2\frac{P_1 - P_2 - 2\rho g H + 0.35[\rho_1(V_{\alpha 1}\cos\alpha_1)^2 \text{-} \rho_1(V_{\alpha 2}\cos\alpha_2)^2 + \frac{\rho_1+\rho_2}{2}\text{gH}]}{\rho(\lambda_r \frac{L}{D_r} + \xi_\theta + 1)}} \tag{3}$$

where P_1 = atmospheric pressure in the entrance (pa); P_2 = atmospheric pressure in the exit (pa); ρ = air density (kg/m^3); H = elevation difference between entrance and exit (m); $V_{\alpha 1}$ = natural wind speed (m/s); $V_{\alpha 2}$ = natural wind speed outside tunnel exit (m/s); ρ_1 = air density outside the entrance (kg/m^3); ρ_2 = air density outside the exit (kg/m^3); α_1 = inclined angle between natural wind and tunnel centerline outside the entrance (°); α_2 = inclined angle between natural wind and tunnel centerline outside the exit (°); λ_r = friction loss coefficient of tunnel wall; ξ_θ = loss coefficient in tunnel entrance; L = tunnel length (m); D_r = equivalent diameter of tunnel section (m).

You (2010) drew the Mechanism of action of the major influencing factors through independent analysis of the influencing factors inside and outside the long tunnel. He also deduced theoretical formula of the natural air flow by the action of ultra static pressure difference and thermal potential difference alone.

Zhou (1994) raised one-dimensional flow loop analysis used to calculate natural wind pressure of complicated constructions. Through linking research object and environmental fluid, external flow and internal flow, flow area and still area with closed loop, this method studyed the pressure change along the loop, established equilibrium equation of energy in the loop and solving the amount of natural wind pressure in the loop.

Lv (2006) from Chang'an University deduced calculation formula. of natural wind pressure in highway tunnels and pointed out natural wind resistance in tunnels shoud be determined by equivalent pressure difference of natural wind. In the absence of field observation data, natural wind resistance was assumed constant and the value was in the range from 10 to 30 Pa.

Zeng (2003) studied natural wind pressure of tunnels with shafts through theory of loop natural wind pressure and pointed out that the size and direction of the natural wind pressure in tunnels with shafts would change, which in return would affect the work of main fan of shaft. Therefore, more attentions should be paid to the design and operation of ventilation.

Li (2007) studied natural wind pressure analysis based on thermodynamics. From point of view of thermodynamics and considering the influence of thermal factor in tunnel on wind flow, the calculation method of natural wind was derived. But there was still some deficiency. The air flow needed to be in a relatively static state so that the pressure needed to be static pressure. The calculation method was also fussy.

Tan (2007) elaborated the cause and change law of natural wind pressure. The adverse effect on reverse air flow of the mine ventilation, wind pressure of the main fan, Ventilation network caused by the change of natural wind pressure was also analyzed. So was the preventive action.

Zhang (2009) elaborated the cause of natural wind pressure, the effect of natural wind pressure on ventilation system and the treatment method.

Gao (1998) proposed three manifestations to generate natural wind pressure in mine, namely natural thermal potential difference, horizontal hot press difference and natural wind, press difference and natural wind.

Analysis indicated that the major influencing factors included temperature and humidity inside and outside the cave, atmospheric pressure difference and elevation difference between tunnel entrance and exit, the size and direction of wind outside the cave and so on. The theoretical formula used included air flow balance, pressure balance and energy balance.

3 NUMERICAL SIMULATION

Wang & Guo (2011) used the software FLUENT to model three typical highway tunnels-the Fangdou Mountain Tunnel, the Tiefeng Mountain No. 2 Tunnel and the Erlang Mountain Tunnel that have been carried out field test. The measured external environmental factors after processing were inputted as the boundary condition and natural wind speed corresponding to the measured position was calculated. Compared with the measured value, the correctness of the method to calculate natural wind was verified. The calculation method that had been verified could be applied to the use of natural wind of unbuilt tunnels

Yu (2006) studied on winter natural ventilation of long and narrow drainage tunnel. Mathematical models of tunnels in the presence or absence of vertical wells were modeled to analyse air flow condition of the winter drainage tunnel. It turned out that the air flow condition of long and narrow tunnel in the absence of vertical wells was significantly affected by external wind pressure; hot press difference affected more significantly than wind pressure in tunnels with vertical wells and there was good airflow organization.

Wang & Yan (2011) used the fluid dynamics software FLUENT to model and calculate the natural wind speed of the Niba Mountain Tunnel in Ya'xi highway that had been completed. The calculated value matched well with the measured value.

In a word, it is effective to calculate the natural wind speed by modeling unbuilt tunnels. Meteorological parameters of tunnel entrance and exit as initial data inputted into computer were needed while modeling.

4 SITE MEASUREMENT

In order to use theoretical derivation calculation formula to get natural wind speed and the natural wind pressure further, site measurement of natural meteorological parameters in tunnel entrance and exit is needed. Therefore, meteorological stations with a long measuring period are needed. Meanwhile, in order to verify the correctness of theoretical calculation methods and numerical calculation methods of natural wind speed, site measurement of natural wind for completed tunnels is needed. Consequently, study on site measurement at home and abroad is divided into two parts, namely site measurement of meteorological parameters in tunnel entrance and exit for inbuilt tunnels and site measurement of natural wind for completed tunnels.

4.1 *Long-term site measurement of meteorological parameters for tunnels*

Wang & Guo (2011) established four long-term one-year-period (Actually testing 485 days) meteorological stations at Niba Mountain Tunnel, of which the testing parameters included wind direction, wind speed, atmospheric pressure, air temperature and humidity. A total of 465600 related data in 93120 groups were obtained for calculating natural wind. The installation and location of four automatic meteorological stations in tunnel entrance and exit and two inclined shafts of Niba Mountain Tunnel is shown in Figure 1.

Wang & Yan (2012) established two long-term one-year-period meteorological stations at Baron Mountain Tunnel in the Aba Prefectures of Sichuan. Eight kinds of meteorological parameters were automatically observed 24 hours a day, including temperature, precipitation, air pressure, earth temperature, humidity, wind direction, wind speed and radiation.

a) Weather station in entrance of Ya'an (1#)

b) Weather station in inclined shaft of Ya'an (2#)

c) Weather station in entrance of Lugu (3#)

d) Weather station in inclined shaft of Lugu (4#)

Figure 1. Automatic weather stations in entrance, exit and inclined shafts of Niba Mountain tunnel.

Analysis and evaluation of data from meteorological stations laid the foundation for further study of the natural wind. It could also provide helpful experiences for the establishment of meteorological stations in similar kinds of projects.

4.2 *Site measurement of natural wind in completed tunnels*

Wang & Guo (2011) verified the correctness and rationality of theoretical calculation methods and numerical calculation methods of natural wind speed, through site measurement of wind speed in three typical highway tunnels-the Fangdou Mountain Tunnel, the Tiefeng Mountain No. 2 Tunnel and the Erlang Mountain Tunnel.

Qiu (2004) conducted research on test of temperature in multi-shaft deep buried super long QinLing tunnel. Relying on long Qinling-Zhongnanshan tunnel, site measurement of wind speed and temperature inside and outside the tunnel was conducted. Research on temperature range and variation of natural wind in tunnel under the influence of formation temperature, changes of inside wind speed and changes of outside temperature of tunnel was conducted, which could provide evidence for design of ventilation and disaster prevention in multi-shaft deep buried super long tunnel. During test period of Highway tunnel, natural wind speed in tunnels is small. The maximum wind speed is 2.0 m/s and the average wind speed is 1.01 m/s.

Wang (2004) conducted research on grasping the law of natural wind and taking advantage of it after Yuanliangshan tunnel was completed. Five jet fans were closed in parallel guide-pit and main tunnel, only one single jet fan was reserved to adjust the air flow direction. The close of axial flow fan with a power of 110 kw reduced energy consumption, manpower, material resources and saved a lot of money.

In conclusion, field test data is the basis of calculation of natural wind speed. It can also verify the correctness of theoretical calculation and numerical calculation.

a) Weather station in tunnel entrance

b) Weather station in tunnel exit

Figure 2. Automatic weather station in entrance and exit of Balang Mountain tunnel.

5 RESEARCH ROUTE OF EFFECTIVE UTILIZATION OF NATURAL WIND

Study on natural wind pressure at home and abroad is mainly aimed at mine, while study on natural wind pressure of highway tunnel has just started. It is gradually recognized with the construction of high-altitude and mountain-crossing highway tunnels in the meteorological separation zone, so this is a relatively new research direction with significant economic value. It is perfectly possible to induce, control and utilize natural wind which had been completely thought as resistance along with the progress of science and technology.

Wang et al. (2010, 2011, 2012) conducted research on effective utilization of natural wind pressure and got technical research route based on Niba Mountain Tunnel in Yaxi Highway and Baron Mountain Tunnel in 303 Provincial Highway. Research route of energy saving and ventilation technology of Baron Mountain Tunnel based on effective utilization of natural wind is shown in Figure 3.

6 ECONOMIC VALUE OF STUDY ON UTILIZATION OF NATURAL WIND

Researches on utilization of natural wind for highway tunnel are not much, so are the research results. But its prospect is cheerful.

Liang (2005) measured natural ventilation rate of east mining area of Zhaokou Iron in Jinling Iron. The value of natural wind pressure and wind resistance was calculated, experimental research on natural ventilation in winter was carried out. Natural ventilation in winter not only ensures required wind rate of production process under shaft but also saves power expense by 39 000 Chinese dollars annually.

After research on Niba Mountain Tunnel in Yaxi Highway, Wang & Guo (2011) drew the conclusion: Using energy-saving ventilation when the directions of natural wind and driving are the same and natural wind speed is greater than design wind speed; if not, using natural ventilation.

Through this kind of energy-saving mode, Niba Mountain Tunnel could Energy saving up to 50.5% when real-time control is implemented and to 13.28% when controlled by time.

7 CONCLUSIONS AND SUGGESTIONS

1. Natural wind speed is the basis of natural wind pressure research, which could be obtained through theoretical derivation, numerical simulation and site measurement.
2. To achieve energy-saving ventilation and some economic value, research on natural wind pressure research is conducted.
3. Except necessary inducement and control equipment are needed, natural ventilation don't produce other fees, so energy saving effect is remarkable and application prospect

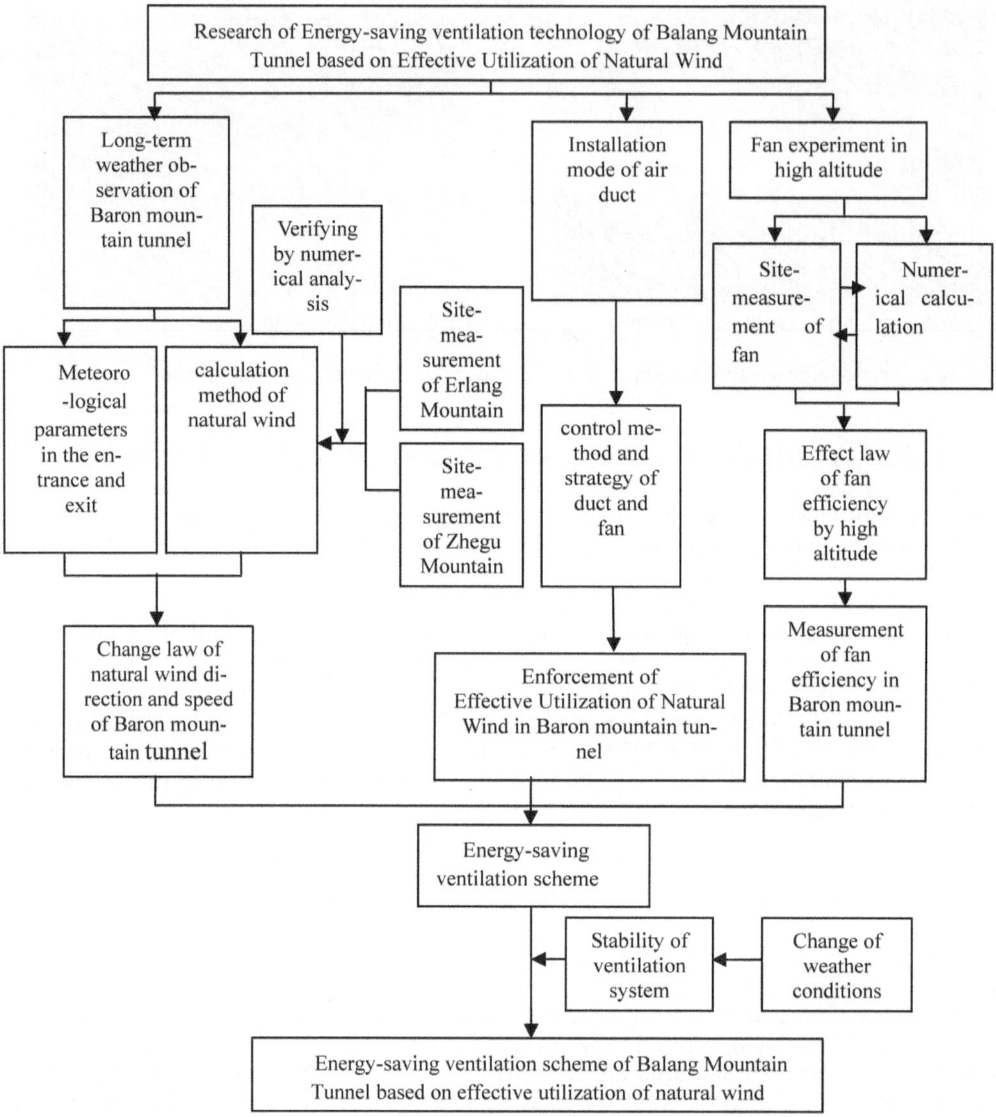

Figure 3. Research route of energy-saving ventilation technology of Balang Mountain tunnel based on effective utilization of natural wind.

is promising. With the construction of high-altitude and mountain-crossing highway tunnels, it is suggested special research to be conducted on natural wind pressure of tunnels in the meteorological separation zone.

The present research is still in its initial stage, so the research has certain practical value and innovation which should attract the attention of scholars at home and abroad.

REFERENCES

Gao, R.L. 1998. Probing into the concept of natural draft pressure. *Coal Engineer*. (4): 29–31.

Li, Y.C. 2007. Analysis method of natural wind pressure based on thermodynamics. *Express Information of Mining industry*. 456(4): 26–28.

Liang, X.C. & Guo, M.C. 2005. The application of natural wind pressure in mine ventilation. *Gold*. 26(8): 20–21.

Lv, K.C. 2006. Problems and their solutions in ventilation design for super long highway tunnels. *Modern Tunneling Technology*. 43(6): 35–39.

Qiu, Y.L. et al. 2004. Temperature measurement in the extra-long and deeply-embedded Qinling tunnel with many shafts. *Modern Tunneling Technology*. 41(3): 35–39.

Shi, Y.X. 2011. Study on calculation method of natural wind and energy saving ventilation technology on tunnel. Chengdu. *Southwest Jiaotong University doctor degree dissertation*.

Tan, H.W. 2007. Changing regulation and influence of natural wind pressure on ventilation system of mine. *Gold*. 28(11): 20–23.

Wang, G.H. & Li, Z.G. 2004. Ventilation of natural wind after Yuanliangshan tunnel was completed. *Tunnel Construction*. 24(5): 28–30.

Wang, M.N. & Guo, C. 2011. Research of key technology of deep buried Daxingling Niba Mountain tunnel. Southwest Jiaotong University.

Wang, M.N. & Yan, T. 2011. Research of ventilation and fire protection and rescue technology of single-tunnel-opposite-driving Balang Mountain tunnel. Southwest Jiaotong University.

Yan, T. 2012. Meteorological observation and analysis evaluation of Balang Mountain tunnel. *Technology Seminar of Alpine Highway Tunnel Construction*: 150–156.

YU, Y.L. et al. 2006. Natural ventilation in the long and narrow sewer tunnel in winter. *Chinese Journal of Underground Space and Engineering*. 2(3): 480–484.

You, H.B. 2010. Study on the effect factors and calculation method of natural wind in extra-long Tunnel. Chengdu. *Southwest Jiaotong University Master Degree Thesis*.

Zeng, Y.H. & Guan, B.S. 2003. Research on natural ventilation in tunnel with vertical shaft. *Underground Space*. 23(1): 69–71.

Zhang, J.Y. 2009. Qualitative analysis of the impact of ventilation system influenced by natural air pressure. *Modern mining*. 478(2): 117–119.

Zhou, M.R. 1996. Fluid Mechanics Pump and Fan. Beijing: China Architecture & Building Press.

Modeling and Computation in Engineering II – Xie (ed)
© 2013 Taylor & Francis Group, London, ISBN 978-1-138-00058-2

Stability analysis of a long-span double-arch-support steel structure

Xuchun Li & Lan Chen
Civil Engineering and Transportation Institute of South China University of Technology, Guangzhou, Guangdong, China

Wenbo Sun
Architectural Design and Research Institute of South China University of Technology, Guangzhou, Guangdong, China

ABSTRACT: A new large-span double-arch-support steel structure whose arch foots' span is 257 m is applied in the design of a large stadium, which is combined with the upper tube arch (large diameter) and lower space arch (triangle truss). The design concept and structure system is described firstly. And then, the stability analysis is calculated by Strand7 FEA software. Stability analysis steps: (1) according to engineering experience, the first overall bucking modes and linear buckling load is determined in the eigenvalue buckling analysis; (2) a consistent initial defect L/300 is imposed on the mode in the geometric nonlinear analysis; (3) the major factors (the slenderness ratio of upper arch and the prestress degree of limit cable) are taken into account in the effect of elastic stability. The main results of stability analysis are shown in detail.

1 ENGINEERING BACKGROUND

A new large-span double-arch-support steel structure is applied in the design of a large stadium, which is combined with the upper tube arch (large diameter) and lower space arch (triangle truss). The arch foots' span of the structure is 257 m, and the maximum span of the radial main beam is 50 m (the cantilever parts are not included). The shape of the steel roof structure is obtained approximately by tangent cylindrical surface combination. The roof structure is combined with tapered H-section steel beams, purling and central cross braces. Tapered H-section steel beams change gradually from the neighboring supports' section type (H600 × 1000 × 16 × 34) to the lower space arch' section type (H600 × 2000 × 16 × 34). Generally, the purling' section type is H300 × 500 × 8 × 16 with 3 m proportional spacing. The design sketch is shown in Fig. 1.

The upper tube arch is a steel pipe whose outer diameter is 2000 mm and thickness changes from 28 mm to 34 mm gradually. The lower space arch is combined with triangle arch trusses whose top chords' diameter is 800 mm and bottom chord's is 1100 mm. The upper tube arch is limited by the stable cables (Φ30) in order to ensure lateral stability. The suspender (Φ351 × 12) is set between the upper tube arch and the lower space arch to coordinate the deformation of two arches.

The ultimate capacity of a long-span structure usually depends on the stability of the structure so the stability analysis with Strand 7 must be considered in the design. Completely symmetrical on both sides of the structure, single sides is taken as model. The layout of double-arch-support steel structure is shown in Fig. 2.

Figure 1. The design sketch of a long-span double-arch-support steel structure.

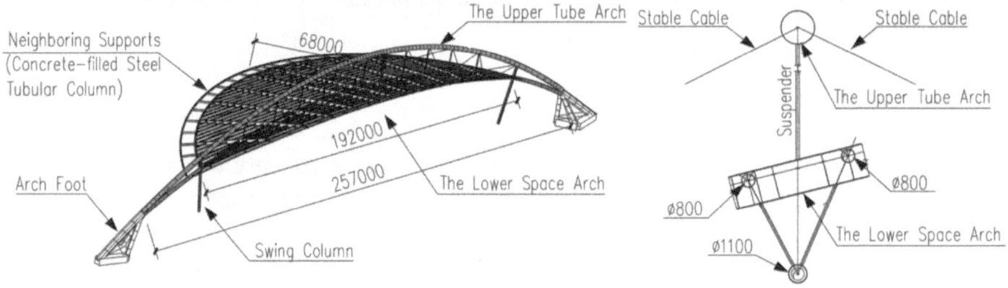

Figure 2. The layout of double-arch-support steel structure.

2 EIGENVALUE BUCKLING ANALYSIS

Eigenvalue buckling analysis is generally used to estimate the minimum critical buckling load based on structural linear stiffness that the effect of initial stress stiffness is considered. However, because actual initial defects and nonlinearity make structure not buckle in the elastic buckling strength, there is a fairly large gap between the stability bearing capacity estimated by eigenvalue buckling analysis and the actual force condition. So the eigenvalue buckling analysis which usually leads to nonconservative results can only provide an upper limit load of the design.

According to the clause 4.3.3 in "*Technical Specification for Space Frame Structures*", the uniform loads contains full-span uniform load (dead weight + dead load + full-span uniform live load) and half-span uniform load (dead weight + dead load + half-span uniform live load). The structural stability is analyzed and the top twenty order modes of overall buckling under two uniform loads are calculated respectively. The result shows that the buckling coefficient of the first overall buckling mode under full-span uniform load is 8.095 and the buckling coefficient under half-span uniform load is 8.844. The first overall buckling modes of the structure are shown in Fig. 3.

Due to the nonlinearity of the cable, the reference value of the eigenvalue buckling analysis with the elastic method is little. Stability bearing capacity of the structure is difficult to be reflected accurately and the impact of the initial defect is unable to be assessed in the eigenvalue buckling analysis, but linear buckling load can be regarded as the upper limit load of the nonlinear buckling analysis.

3 GEOMETRIC NONLINEARITY ANALYSIS

Geometric nonlinear analysis that a nonlinear technology gradually increasing the load (or gradually increasing the displacement) is used to obtain structural stability bearing capacity is more precise than eigenvalue buckling analysis. The characteristics such as large deformation, the initial defect, gap, plastic response can be considered with the geometric nonlinear analysis.

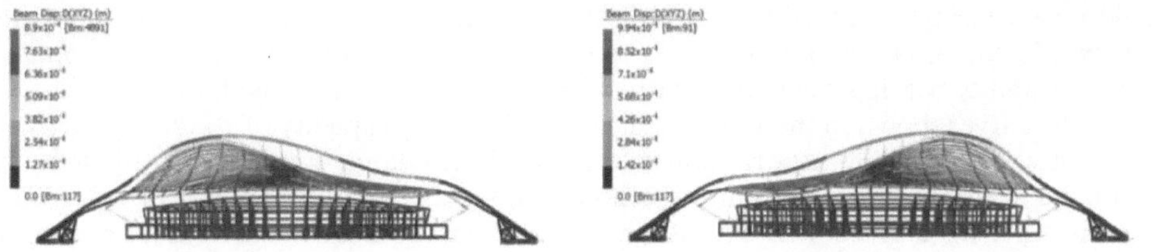

(a) Full-span uniform load (b) Half-span uniform load

Figure 3. The first overall buckling mode.

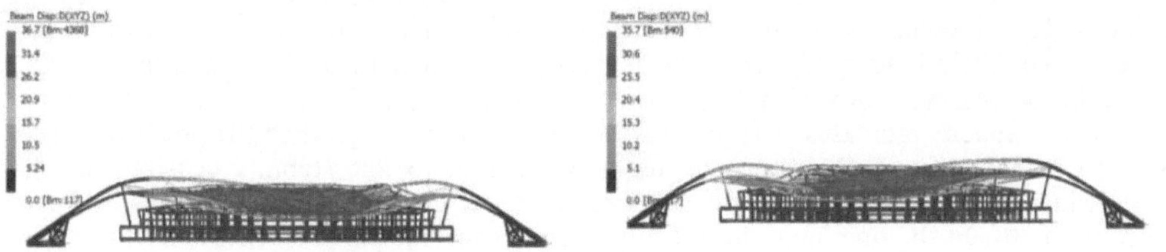

(a) Full-span uniform load (b) Half-span uniform load

Figure 4. The overall buckling mode of the structure with geometric nonlinear.

The consistent mode imperfections method is applied in geometric nonlinear analysis. According to the clause 4.3.3 in "*Technical Specification for Space Frame Structures*", through the top twenty order modes of overall buckling calculated by eigenvalue buckling analysis, imposing a consistent initial defect L/300 (the distance L between the swing columns is 192 m) on the modes in the geometric nonlinear analysis to obtain stability bearing capacity. The result shows that the coefficient of geometric nonlinear stability bearing capacity under full-span uniform load is 5.58 and the coefficient under half-span uniform load is 6.0. The overall buckling modes of the structure with geometric nonlinear are shown in Fig. 4.

The overall buckling modes of geometric nonlinear analysis are completely different with the modes of eigenvalue buckling analysis. Because eigenvalue buckling analysis ignores the impact of the initial defect, buckling modes are strictly ant symmetric; while the modes of geometric nonlinear analysis occur snap through buckling, which original upward arch suddenly changes sagging. The middle of the model deforms inwardly (downward), and both sides deform outwardly under uniform load. This deformation of the arch reflects the deformation compatibility.

4 PARAMETRIC ANALYSIS OF ELASTIC STABILITY

Elastic stability of a large-span double-arch-support steel structure is influenced by many factors, and the major factors (the slenderness ratio of upper arch and the prestress degree of limit cable) are taken into account in the effect of elastic stability.

4.1 *The slenderness ratio of upper arch*

Transformation law of elastic stability with the slenderness ratio of upper arch is studied, which should ensure the single tube arch satisfying local stability firstly. Five section types

(Φ2800 × 48, Φ2600 × 44, Φ2400 × 41, Φ2200 × 37 and Φ2000 × 34), which are numbered No. 1, No. 2, No. 3, No. 4 and No. 5 respectively, are chosen. The coefficient of geometric nonlinear stability bearing capacity λ_e with different slenderness ratio is shown in Table 1.

From the curve (shown in the Fig. 5), the stability bearing capacity of the structure plummets down with the slenderness ratio increasing, and the change trend under full-span uniform load and under half-span uniform load are identical. Because stiffness of upper arch and the integral structural stiffness decreases with the slenderness ratio increasing, the stability bearing capacity of the structure declines afterward.

4.2 *The prestress degree of limit cable*

Transformation law of elastic stability with the prestress degree of limit cable whose parametric ranges are "no cable", 0, 1.0, 5.0, 10, 20 and 25 is studied.

As seen in Fig. 6, the existence of cable is of great importance to the stability bearing capacity. The difference between the stability bearing capacity without cable and with cable under two uniform load is 27% and 32% respectively. Besides the stability bearing capacity with cable decreases slightly with the prestress degree increasing from 0 to 1; while the stability bearing capacity increases with the prestress degree increasing, when the prestress degree is more than 1. After the prestress degree is more than 15, the stability bearing capacity doesn't increase; while the stability bearing capacity decreases, when the prestress degree is more than 20. On the one hand, limit cable can improve out-plane stiffness and the stability bearing capacity; on the other hand, it's harmful for the stability that the axial force of upper arch increase with the prestress degree increasing. When the prestress degree and the axial force of upper arch are little, the effect of cable is mainly improving out-plane stiffness. When the prestress degree increases to a certain value, the influence of the prestress degree on the axial force of upper arch can't be ignored, and even exceed the influence of the prestress degree on out-plane stiffness, so the stability bearing capacity decreases with the prestress degree increasing.

The greater is initial prestress of cable, the greater is the breaking force of cable, so the larger section of cable must be chosen. If only the stability bearing capacity is considered, with the prestress degree less than 1 the bearing capacity of cable is big but prestress is little, which the section is small. But due to the demand of cable sag in the project, prestress

Table 1. The coefficient λ_e with different slenderness ratio.

Section types	No. 1	No. 2	No. 3	No. 4	No. 5
The slenderness ratio	40	43	47	51	57
λ_e under full-span uniform load	8.03	7.35	6.72	6.13	5.63
λ_e under half-span uniform load	8.77	8.08	7.47	6.86	6.05

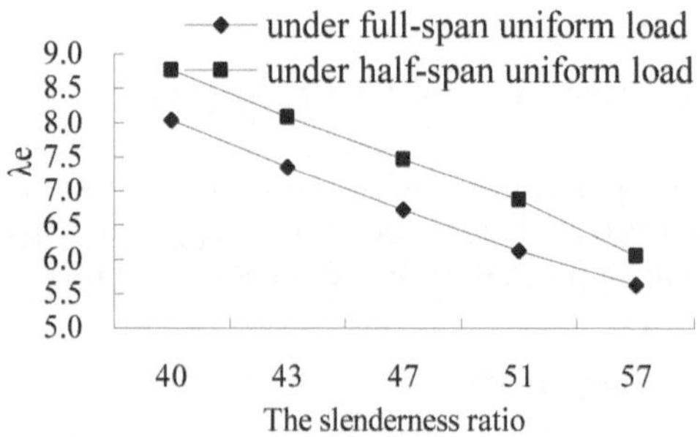

Figure 5. Influence of the slenderness ratio of upper arch.

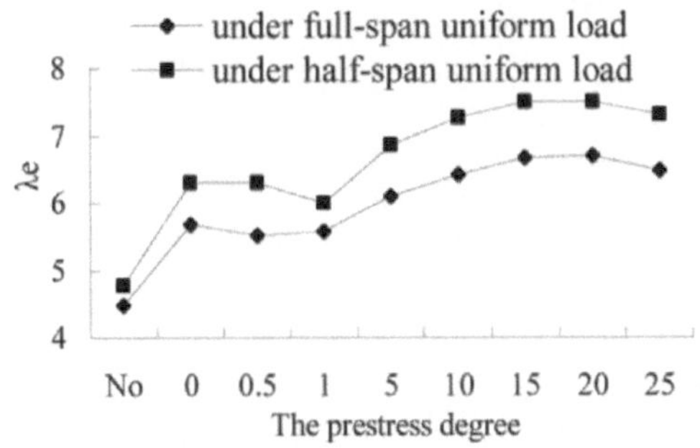

Figure 6. Influence of the prestress degree of limit cable.

can't be too little. If a cable of larger prestress is selected in pursuit of high bearing capacity, a larger section of the cable needs to be chosen and its economic impact needs to be considered. Therefore the section type Φ5 × 37 whose breaking force is 1213 kN is chosen in the project, which satisfies the economic requirements, the demand of cable sag and the minimum stability bearing capacity that satisfies the specification.

5 CONCLUSION

The stability analysis of a large-span double-arch-support steel structure is done by Strand 7 FEA software, and proposing for practical engineering, conclusion as follows:

1. With the nonlinearity of the cable, the buckling modes and buckling load of the double-arch-support structure are difficult to be reflected accurately in the eigenvalue buckling analysis which is based on linear theory, but linear buckling load can be regarded as the upper limit load of the nonlinear buckling analysis.
2. After a consistent initial defect L/300 being imposed on the overall buckling modes, the geometric nonlinear buckling modes is completely different with the eigenvalue buckling modes, and geometric nonlinear buckling mode is closer to the real buckling mode.
3. The stability bearing capacity of the structure decreases with the slenderness ratio increasing.
4. The existence of cable is of great importance to the stability bearing capacity. The stability bearing capacity increases with the prestress degree increasing, when the prestress degree is more than 1, while the stability bearing capacity decreases, when the prestress degree increases to a certain value.
5. The section Φ5 × 37 is chosen in the project, which satisfies the economic requirements, the demand of cable sag and the minimum stability bearing capacity that satisfies the specification.

REFERENCES

GB50017-2003. 2003. *Design code for the steel structures*. Beijing: China Planning Press. 2010. *Report of wind tunnel pressure measure experiment in stadium of Huaian sports center*, Guangzhou, China. Guangdong: Architectural Scientific Research Institute.

Guolun Peng. 2002. *Fortran95 program Design*, Beijing: China Electric Power Press.

JGJ7-2010. 2010. *Technical Specification for Space Frame Structures*. Beijing: China Architecture and Building Press.

Tongyan Lin. & Stotesbury, S.D. 1997. *Structural Concepts and Systems for Architects and Engineers* (Second Edition), Beijing: China Architecture and Building Press.

Modeling and Computation in Engineering II – Xie (ed)
© 2013 Taylor & Francis Group, London, ISBN 978-1-138-00058-2

A calculation method for longitudinal weld of rectangular box columns under buckling limit state

Chong Wu & Yu Wu
Tongji University, Shanghai, China

ABSTRACT: A method for calculating stress in longitudinal weld of rectangular box columns under buckling limit state is proposed. Shear stress of longitudinal weld at overall buckling state is given based on the column buckling theory and extra fiber yield criterion. Normal stress at local buckling state is calculated by assuming panels which consisting the cross-section are clamped at edges. A method to check combined stresses of the weld is put forward. The Calculation method is conservative for design.

1 INTRODUCTION

Steel pylon has been widely used in cable-stayed bridge and suspension bridge. For instance: Akashi-Kaikyo Bridge, Tatara Bridge in Japan, Golden Gate Bridge, Verrazano bridge and New San Francisco—Oakland Bay Bridge in America, the Second Bosporus Bridge in Turkey, the Third Nanjing Yangtze River Bridge and Taizhou Yangtze River Bridge in China, Wazirabad Bridge in India. Steel pylon is mostly under axial compression at serviceability limit states, column's bending deformation and plate's transverse distortion are both small. Consequently, the stress in longitudinal weld is very small. But due to the initial bending and residual stress, steel pylon's failure type is overall buckling or local buckling. The weld is in complex stress state with not only shear stress but also normal stress at ultimate state. The longitudinal welds in Wazirabad Bridge designed by SBP in Germany are full penetration welds or partial penetration welds with an effective throat thickness larger than plate thickness, which make the weld resistance no less than that of parent metal, namely, full strength criterion. At the same time, based on the elastic theory, Japanese use partial penetration welds or fillet welds with small throats according to allowable stress method. They recommend that the inner weld should use fillet weld with a leg $S \geq \sqrt{2t}$ while the outer weld shall apply U-groove with penetration $a \geq 2\sqrt{t}+3$ (t is the thinner plate connected). Considering the two aspects, Chinese use partial penetration welds with resistance a little less than parent metal when design longitudinal weld. For example, the total throat thickness of longitudinal weld in the Third Nanjing Yangtze River Bridge is 6 mm less than the plate thickness (Plate thickness is 42 mm), while that in Taizhou Yangtze River Bridge is about 60% of the plate thickness (Plate thickness is 60 mm, throat of outer longitudinal weld is 22 mm while that of inner is 12 mm). A weld with excessively small throat will fail before the plates while weld determined by equal-strength criterion will be difficult to fabrication, because they are always with large size which causes too much groove preparation, welding and large deformations, residual stresses. Consequently, it is necessary to research on the design philosophy and calculation method of longitudinal weld in steel pylon.

Considering the influence of overall buckling and local buckling to the stress of weld, the equation for calculating stress of longitudinal weld in rectangular box column at overall buckling or local buckling state has been derived, based on the design principle that the strength of longitudinal weld shall be no less than the resistance of the member.

2 SHEAR STRESS AT OVERALL BUCKLING STATE

As overall buckling mode for a column with box-section is bending buckling while torsional deformation is small. Only bending buckling is discussed and the pylon's torsion is ignored.

The deflection and bending moment of a simply supported column under overall buckling can be expressed as

$$y = y_0 \sin\frac{\pi z}{l};\ M = Ny = Ny_0 \sin\frac{\pi z}{l};\ Q = \frac{dM}{dz} = \frac{\pi N}{l} y_0 \cos\frac{\pi z}{l} \tag{1}$$

In which l is the column length. y_0 is the lateral deflection. Q and M are separately the shear force and bending moment.

Assuming the overall buckling resistance, N and the extra fiber stress at ultimate state could be expressed as:

$$N = \varphi A f_d \tag{2}$$

$$\frac{N}{A} + \frac{Ny_0}{W_x} = f_d \tag{3}$$

where A is area of the section; φ is stability factor; f_d is strength; W is the section modulus;

From Equations 2 and 3, the lateral deflection at overall buckling ultimate state could be expressed as

$$y_0 = \frac{W_x(1-\varphi)}{\varphi A} \tag{4}$$

Substituting Equation 2 and 4 into equation 1, the shear force was obtained:

$$Q_{max} = \frac{\pi W_x(1-\varphi) f_d}{l} \tag{5}$$

For a single cell box section, the shear flow at the weld under shear force Q_{max} is:

$$q = q_0 + q_1 \tag{6}$$

q_0 is the shear flow in the opened-section formed by making a slit at the closed-section.

$$q_0 = \frac{Q_{max} A_s y_s}{I_x} \tag{7}$$

In which q_1 is the shear flow at the slit:

$$q_1 = \oint_s \frac{q_0}{t} ds \Big/ \oint_s \frac{ds}{t} \tag{8}$$

A_s is the area from the slit to the point to be determined and y_s is the height of the centroid of this area above the point to be determined.

Shear stress of weld listed in table 1 can be given by

$$\tau = \frac{q}{\sum h_e} \tag{9}$$

$\sum h_e$ is the sum of throat thickness in table 1

Table 1. Sum of effective throat thickness for different welds.

	Full penetration weld	Partial penetration weld	Fillet weld
scheme	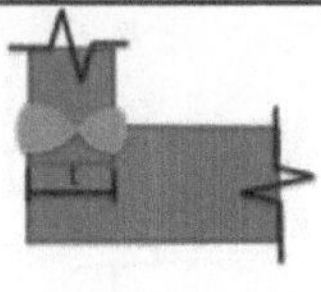	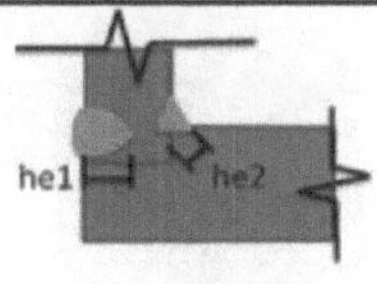	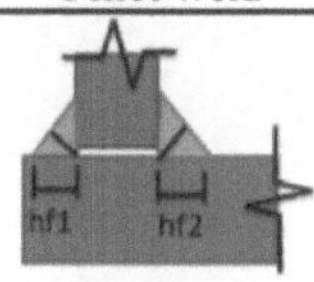
$\sum h_e$	t	$h_{e1}+h_{e2}$	$0.7(h_{f1}+h_{f2})$

3 NORMAL STRESS AT LOCAL BUCKLING STATE

In some cases the column cross-section is composed of more slender plate elements. As these slender-plate elements may buckle locally as shown in Figure 1, which cause stress out of the throat plane, a stiffened plate is selected from the four panels to calculate the normal stress in weld.

The deformation of stiffened plate at buckling will related to the slenderness of longitudinal stiffeners. As shown in Figure 2, for flexible stiffened plate, stiffeners bend with plate and for rigid stiffened plate, buckling will occur in sub-panels between stiffeners.

The differential equation of equilibrium of flexible stiffened plate can be written as

$$\left(D_x\frac{\partial^4 w}{\partial x^4}+2H\frac{\partial^4 w}{\partial x^2\partial y^2}+D_y\frac{\partial^4 w}{\partial y^4}\right)+N_x\frac{\partial^2 w}{\partial x^2}=0 \tag{10}$$

where D_x, D_y, are the equivalent bending stiffness parallel to the axis x and y. H is a parameter related to torsional stiffness.

For rigid stiffened plate, the differential equation of equilibrium of each sub-panel could be simplified as

$$D\left(\frac{\partial^4 w}{\partial x^4}+2\frac{\partial^4 w}{\partial x^2\partial y^2}+\frac{\partial^4 w}{\partial y^4}\right)+N_x\cdot \tag{11}$$

where D is the flexural rigidity unit width of the plate.

The edges of the panel at y = 0 and y = b (Figure 3) are neither clamped nor simply supported. In design, the boundary condition could be treated as rigid conservatively. (Scheme for calculating is shown in Figure 3). The longitudinal weld bears bending moments distributed along the clamped edges.

Rayleigh-Ritz idea can be used to solve differential equations (10, 11). While applies first order approximation and ignores high order terms, the deflection function in z direction is given by

$$w=C_m\sin\frac{m\pi x}{a}1-\cos\frac{2\pi y}{b_0} \tag{12}$$

where b_0 is b (for flexible stiffeners) or b_1 (for rigid stiffeners). C_m is the maximum deflection out of the plate.

Stress in the plate at point (x, y, z) can be expressed as

$$\sigma_x=\sigma_x^0-\frac{Ez}{1-\nu^2}\left(\frac{\partial^2 w}{\partial x^2}+\nu\frac{\partial^2 w}{\partial y^2}\right)$$

$$\sigma_y=\sigma_y^0-\frac{Ez}{1-\nu^2}\left(\frac{\partial^2 w}{\partial y^2}+\nu\frac{\partial^2 w}{\partial x^2}\right)$$

$$\tau_{xy}=\tau_{xy}^0-\frac{Ez}{1+\nu}\frac{\partial^2 w}{\partial x\partial y} \tag{14}$$

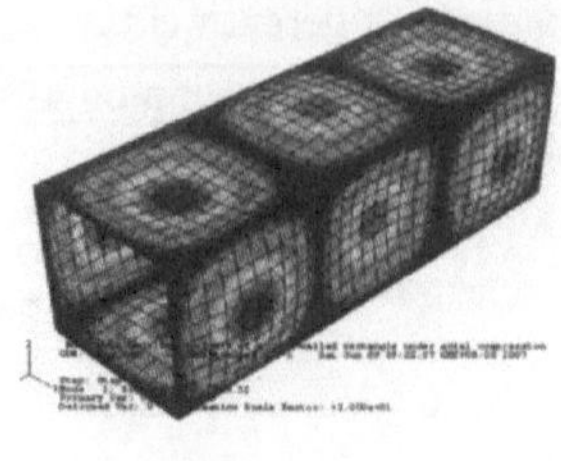

Figure 1. Local buckling.

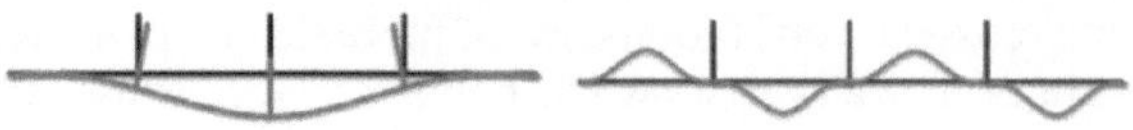

Figure 2. Buckling type for flexible stiffeners (the left) and rigid stiffeners (the right).

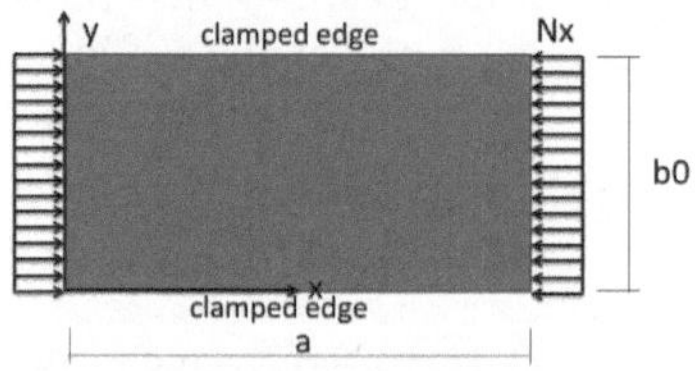

Figure 3. Scheme for a plate pined at $x = 0$, $x = 1$ while clamped at $y = 0$ and $y = b_0$.

where, σ_x^0, σ_y^0 and τ_{xy}^0 are the normal stress at x, y direction and shear stress in the mid plane. E is the Young's modulus of elasticity. ν is Poisson's ratio, for steel, $\nu = 1/3$.

For a plate under compression in x direction, the stress in ultimate state in the mid plane can be expressed as

$$\sigma_x^0 = -\sigma_{\lim};\ \sigma_y^0 = 0;\ \tau_{xy}^0 = 0$$
$$\sigma_{\lim} = \left(A_{eff}/A_c\right) f_d \tag{13}$$

where $\sigma_{\lim}$ is the average stress in ultimate state. A_{eff} is the effective area of the plate for plate buckling effects due to direct stresses at the ultimate limit state. A_c is the area of the whole section.

Evidently, the stress at position (a/2/m, b/2, −t/2) is most critical, submitting the coordinate (a/2/m, b/2, −t/2) and to Equation 12.

$$\sigma_x = -\sigma_{\lim} - \frac{3C_m E\pi^2 t(2a^2 + 3b^2m^2)}{8a^2b^2}$$
$$\sigma_y = -\frac{3C_m E\pi^2 t(6a^2 + b^2m^2)}{8a^2b^2}$$
$$\tau_{xy} = 0 \tag{15}$$

According to maximum distortion-energy theory

$$\sqrt{\sigma_x^2 + \sigma_y^2 - \sigma_x\sigma_y + 3\tau_{xy}^2} = f_d \tag{16}$$

Submitting Equation 15 to Equation 16, the maximum deflection C_m can be got.

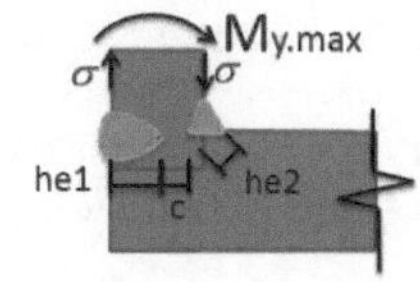

Figure 4. Relationship between stress and bending moments distributed along the clamped edges.

Bending moments distributed along the clamped edges is

$$M_y(x) = -D\left(\frac{\partial^2 w}{\partial y^2} + \nu\frac{\partial^2 w}{\partial x^2}\right) = \frac{Et^3\pi^2 C_m}{3b^2(1-\nu^2)}\sin\frac{m\pi x}{a} \tag{17}$$

Figure 4 shows the relationship between stress and bending moments distributed along the clamped edges.

The stress σ should be taken as:

$$\sigma = \frac{M_{y.max}}{W} \tag{18}$$

W is the section modulus per unit length about centroidal axis of weld which can be taken as same as beams.

4 COMBINED STRESSES IN LONGITUDINAL WELD

A uniform distribution of stress is assumed on the throat section of the weld, leading to the shear stresses parallel to the axis of the weld, normal stresses and shear stresses perpendicular to the throat. The design resistance of the fillet weld will be sufficient if the following are both satisfied:

$$[\sigma_\perp^2 + 3(\tau_\perp^2 + \tau_{//}^2)]^{0.5} \le f_u / \beta_w \gamma_{M2} \,\&\, \sigma_\perp \le \beta f_u / \gamma_{M2} \tag{19}$$

$\sigma_\perp$ is the normal stress perpendicular to the throat; $\tau_\perp$ is the shear stress (in the plane of the throat) perpendicular to the axis of the weld; $\tau_{//}$ is the shear stress (in the plane of the throat) parallel to the axis of the weld; f_u is the nominal ultimate tensile strength of the weaker part joined; β_w is the appropriate correlation factor relating the properties of a parent or base metal to the properties of a matching weld consumable; γ_{M2} is the Partial safety factors for joints; β is the correlation factor for material.

$\tau_{//}$ is the shear stress caused by overall buckling, while relationships between $\sigma_\perp, \tau_\perp$ to σ is:

$$\sigma_\perp = \sigma\cos\theta;\ \tau_\perp = \sigma\sin\theta \tag{20}$$

θ is the coangle to angle between throat and σ

The most dangerous stress state of the weld at ultimate state is not the sum of the stress calculated above because the most unfavorable stress caused by overall buckling and local buckling always occur at different points. It is conservative to assume the three stress components acting on same point.

5 SUMMARY

This paper provides a design philosophy of longitudinal weld based on ultimate state by analyzing the load acting on weld at different ultimate states; set up mechanical models of the weld and derive equations for calculating stress in it. But there are still problems for further research, including:

Firstly, all the results are only fit for columns under axial compression with no bending.

Secondly, the boundary condition of the panel is clamped, distortion of adjacent panels and stiffeners are ignored.

The next, deflection function is only first order approximation when solving the differential equation for panel.

Last, no cooperation has been done with engineering practice before. The result remains to be verified.

REFERENCES

Beg, D., Kuhlmann, U., Davaine, L. et al. (ed.) 2010. *ECCS Eurocode Design Manual Design of Plated Structures*. Paris.

Cui, B., Meng, F.C., Feng, L.P. et al. 2005. Design and Fabrication of the Tower Column of Nanjing No.3 Yangtze River Bridge. *China Railway Science. 26(3)*:42–47.

Dai, Y.N. (ed.) 2005. *Technique for the Tower Column of Nanjing No.3 Yangtze River Bridge*. Beijin: China Communications Press.

EUROPEAN COMMITTEE FOR STANDARDIZATION. EN 1993-1-5, (2006). Eurocode 3: Design of steel structures—Part 1–5: Plated structural elements.

Hua, X., Zheng, X.D., Zhou, Y.F. et al. 2009. Design of Middle Steel Tower on Triple-Tower Suspension Bridge of Taizhou Yangtze River Bridge. *Highway. No. 7:* 68–74.

Konishi, I (ed.) 1980. *Steel Bridge*. Beijin: People's Railway Press. (in Chinese)

Li, H.F., Cao, P.Z., Wei, F.F. et al. 2008. Effect of ultimate stability capacity of steel boxing column with residual stress and induced bending. *Sichuan Building Science. 34(3):* 30–33.

Manzanarez.R., Nader, M., Abbas, S. et al. 2000. Design of the new San Francisco—Oakland Bay Bridge. *Structures Congress 2000: Advanced Technology in Structural Engineering. Philadelphia, PA, United states*.

Qu, Q.Z (ed.) 2000. *Elastic Plate Theory*. Beijin: China Communications Press.

Schlaich Bergermann und Partner. *Consultancy service for detailed structural design and drawings of proposed signature bridge at Wazirabad, Delhi 2007*.Berlin.

Shao, Y.B. & Peng, L. 2008. Finite Element Analysis of Local Buckling of Thin-Walled Square Hollow Section Tube under Axial Compression. *Journal of Yantai University (Natural Science and Engineering Edition). 21(4):* 304–309.

Wang, S., Wang, D., Wang, X. et al. 2010. Finite Element Analysis of Connection for Steel Pylon and Concrete Foundation. *Journal of Architecture and Civil Engineering. 27(5):* 89–95.

Wang, X., Wang, C.S., Yu, X. et al. 2008. Research on Structure Design of Steel Bridge Towers. *Highway. No. 5:* 46–49.

Wu, C. (ed.) 2006. *Mordern Steel Bridge*. Beijin: China Communications Press.

Zhu, Y.P. 2006. Study on the Overall Stability of Axially Compressed Built-Up Steel Columns with Batten Plate. Chengdu: Southwest JiaoTong University.

Modeling and Computation in Engineering II – Xie (ed)
© 2013 Taylor & Francis Group, London, ISBN 978-1-138-00058-2

Analysis on formation mechanism and feature of Yanjiagou rockfalls in Pingchang of Sichuan

P. Su & W. Chen
Southwest Petroleum University, Chengdu, China

ABSTRACT: The security of local people and normal operation of highway are directly threatened by Yanjiagou developed rockfalls. Based on the scene investigation of Yanjiagou 56 rockfalls, the paper statistically analyzed formation mechanism and feature of those rockfalls, found the features of rockfalls included monomer rockfall with colonial development, local chain laws, massive structure, smaller scale, multipoint and wide area, leaded to the formation of raindrops style disasters, realized that the discrepancy of formation mechanism and shape of rock cell, developmental feature of rockfalls were caused by construction and combinations of rock, classified rock cell into two types as smooth rock cell and rough rock cell, studied the major controlling structural plane consisted of ledger rock and high-angle planar X-type conjugate shear joints, finally concluded the formation of rockfalls mainly influenced by four factors as topography, interbedded sandstone and mudstone, surface water, plant roots split role, and failure mechanism involved slip-tensile, bending-tension.

1 INTRODUCTION

The consequences would be hardly predicted as the sudden and irresistible of rockfall. Rockfalls are the primary geological disaster in Yanjiagou, which seriously threatened the safety of local people and the normal operation of highway. Rockfalls in Yanjiagou which is located in the hilly area, with the characteristics of smaller relative slope elevation and vegetation, possess the nature of concealment and easily to be overlooked. Formation mechanism and developed characteristics of these rockfalls have similar laws with the major emerged interbedded sandstone and mudstone, such as rockfall of Simian mountain in jiangjin, rockfall of the Giant Stone Buddha at Leshan etc. (W. Chu 2009, H. Siming 2009, W. Lei 2007, L. Changhai 2010, Y. Siqiao 2005, L. Ming 2008, L. Feng 2009). According to the fieldwork, this article takes the region rockfalls of Yanjiagou in Pingchang as research subjects, counts and analyzes the characteristics of rockfalls along the ditch line, thus systematically analyzes and investigates the formation mechanism and feature of local rockfalls, therefore provides a basis theoretical for prevention of rockfalls.

2 GEOLOGICAL ENVIRONMENT

2.1 *Topography*

Pingchang is located in the north of Sichuan basin, 150 km from the Daba Mountain, belonged to hilly landscape, mainly influenced by tectonic denude and water erosion and incision role. Terrain of Yanjiagou which is situated in the north of Pingchang, is 390~800 m altitude and 50~200 m relative elevation, high in the north and lower in the south, presents as ladder square mountain. Each step is 50~200 m height and gradient of 60~80°. Transitionary gentle slope zone between two steps is 20~100 m width and gradient of 10~30°. U-type valley is developed in this region, which is 10~50 m width and gradient of 40~70°, and the maximum cutting depth of 100 m.

2.2 *Formation lithology*

Only Penglaizhen red and purple sandstone and mudstone of early Jurassic (j3p) can be seen in Yanjiagou, which demonstrated a large interlaced and parallel bedding structure distribution. Sandstone principally consists of quartz and feldspar, includes a smidgen of dark minerals, glauberite geode and stripes, presents middle-thick bedding structure, belongs to brittle rock with silvery percussive sound and shock, which is cemented by clay and calcium. Strong weathering mudstone with obvious alteration of mineral which lead to irregular erosion holes which diameter is 1~3 cm width, presents thin-middle bedding structure intercalated thin-bedded siltstone, belongs to plastic rock with developed weathering crevasse which impregnates a large quantity oxides of black iron and manganese, easily disintegration with water, liable collapse by rap and weak capability of weather resistant, which is filled and half-filled by clay and calcium.

2.3 *Geological structure*

Upon a regional point of view, Pingchang lied in Yangzi paraplatform, shaped into lotus-form structure with Yilong and Bazhong, located between Tanxikou anticline and Shuijiacao anticline. Yanjiagou located in northern flank of Pingchang syncline, 175 km away from the wastern Longmeng mountain riftzone, passed no large riftzone and faultage through this area, which is slightly impacted by tectogenesis since the Jurassic with 5~10° strata inclination. Sedimentary structure plane is the major original structure plane. Based on the nature of region tectonic joint, tectonic structural plane can be divided into four types, such as high-angle plane X-type conjugate shear joints, low-angle cross-section X-type conjugate shear joints, vertical-tensional joints, horizontal-tensional joints (F. Jingyu 2009), which have formed in three period of Phase of Yanshan, medium-term and later Himalayan Epoch. High-angle plane X-type conjugate shear joints are the major tectonic joints, which dominant orientations are 110°∠72°, 8°∠76°. Epigenetic structure plane in this district includes unloading fissuring, weathering fissuring, gravitational spreading deformation fissuring.

2.4 *Hydrogeological condition*

Pingchang belongs to subtropics monsoon climate region of basin with clearly seasons and mild situation. Local water system developed as arborization, which belongs to Qujiang water system and includes 11 major rivers, such as Bahe, Tonghe. Two or three flood disasters were happened between July and August each year. The total rainfall of 2010 is 1200 mm. Surface water is primarily supplied by precipitation and river which is the tributary of Bahe that trickled from northeast to southwest in front of Yanjiagou. Groundwater is consisted of perched water in loose accumulation body of quaternary system, bedrock fracture water and void water. 31 gulches are passed by the line of investigation, which involve 8 gulches with running water throughout the entire year and other 12 gulches with seasonal water.

3 YANJIAGOU ROCKFALLS' FEATURE

3.1 *Space geometry and distribution feature of rockfalls*

Z-type crossing method is used in this investigation to status and survey rockfalls along the line and collapsing accumulation area. 56 rockfalls and 16 collapsing accumulation area are observed in the 10 km distance, which concentrated distribution, multipoint and wide area. Yanjiagou rockfalls' developmental distribution is shown in Fig. 1.

The primary feature of Yanjiagou rockfalls involve monomer rockfall with colonial development, local chain laws (C. Hongkai 2009), smaller scale of individual rockfall whose major volume is between 20~100 m^3 that accounting for two-thirds of this survey. Form of rockfalls includes plate-like, columnar, lump, shell-like (Fig. 2). Each step's exposure face is densely covered with rockfalls along the lateral line of Bahe, which gradient of 50~70° that can be divided into two stairs from top to bottom-500~550 m altitude for the first stair, 440~460 m altitude for the second stair.

3.2 *Rock cells' feature*

Rock cell is widespread in the interbedded region of sandstone and mudstone. Formative factor and shape of rock cells, as well as the developmental situation of rockfalls would be affected by the difference of lithology, structure and compound mode of rock. Depended on the smooth degree of rock cells surface, rock cells can be divided into two categories, smooth rock cells and rough rock cells.

Lithology in smooth rock cells includes sandstone or argillaceous sandstone with thin-middle bedding structure. Upper rockfalls are consisted of sandstone with middle-thick bedding structure. Conformation of rock cells are controlled by the thickness of strata and shear strength of rock in the rock cells which present as inverted ladder with 2~4 stairs about 200 mm width. Rock cells with smooth surface and micro-weather create the chain development of rockfalls which shown as plate and shell.

Lithology in rough rock cell involves mudstone or sandy mudstone. Upper rockfalls are consisted of sandstone or argillaceous sandstone with middle-thick bedding structure

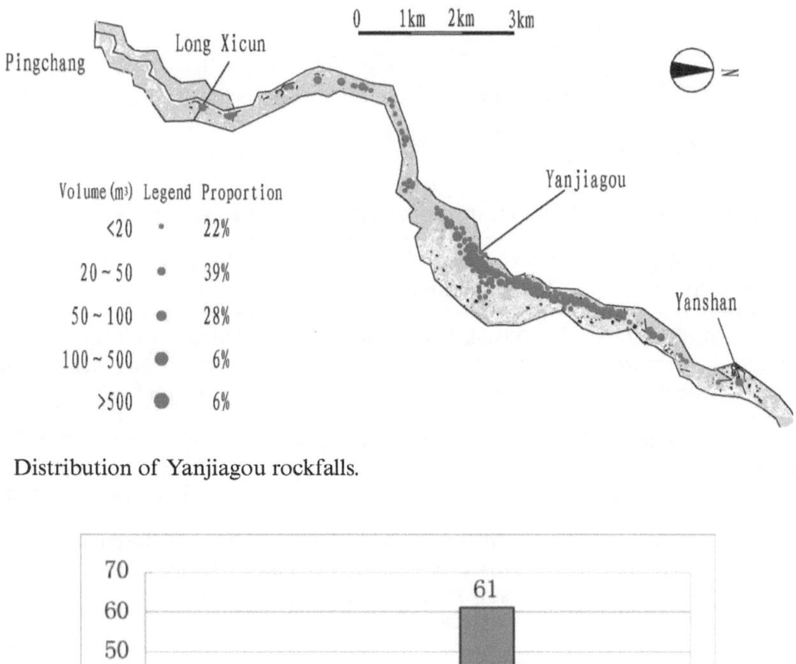

Figure 1. Distribution of Yanjiagou rockfalls.

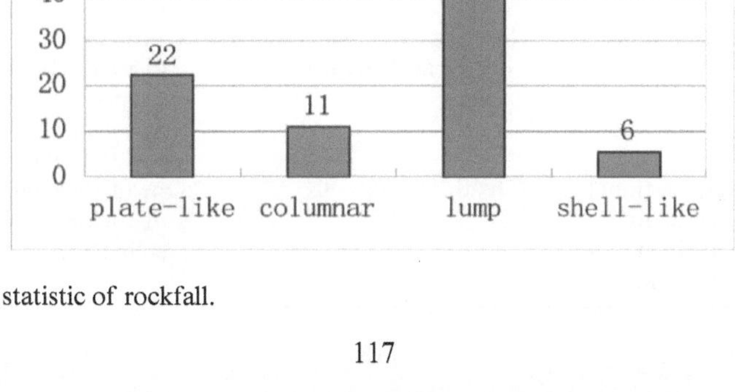

Figure 2. Form statistic of rockfall.

(Fig. 3-b). Conformation of rock cells which present as ⊂-type, is controlled by the diversity weather speed of sandstone and mudstone. Rock cell with rough surface, high degree of weather and weak capability of weather resistant bring about the monomer development of rockfalls which shown as lump and pillar.

3.3 *Feature of structural plane*

Formation and destruction of rockfalls are controlled by structural plane as it is a weak side in rock. Conformation and feature of controlling structural plane of Yanjiagou rockfalls are shown in Tab.1.

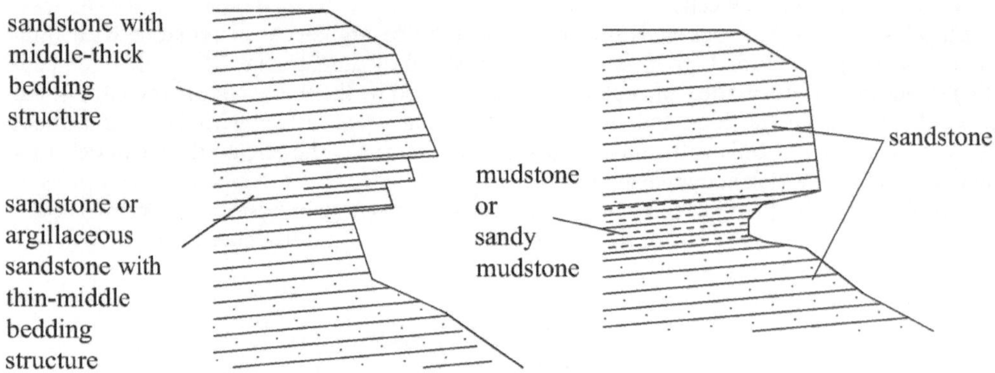

A Smooth rock cell B Rough rock cell

Figure 3. Feature of rock cell.

Table 1. Analysis of conformation and feature of structural plane of Yanjiagou rockfalls.

Conformation of structural plane	Geologic structure	Distribution	Evaluation of geological feature
Original structure plane	Sedimentary structure plane	Visible in each lithology	Major conformation factor of chain laws, smooth rock cells and rockfalls
Tectonic structural plane	High-angle plane x-type conjugate shear joints	Visible in each lithology	Major conformation factor of rockfalls, approximately orthogonal to the strata plane
Epigenetic structure plane	Unloading fissuring	Commonly visible in cliffy slope, parallel with trend of stream valley and sloping surfaces	One of the conformation factor of rockfalls, unloading is generally developed in this area with the influence of topography, which extended along the tectonic structural plane
	Weathering fissuring	Developed in mudstone and original structure plane	One of the conformation factor of rockfalls, reduce engineering properties of structural plane, major reason for destruction of rockfalls with rough rock cell
	Gravitational spreading deformation fissuring	Depth of rock cells	One of the conformation factor of rockfalls, which control the stability of rockfalls

4 FORMATION MECHANISM OF ROCKFALLS

4.1 *Analysis of mechanism*

Conformation of Yanjiagou rockfalls are primarily controlled by the following four aspects:

1. Steep terrain: The angles of slope are greater than 45°. Complete exposure of bedrock is directly weathered and denuded with developed tectonic joints which are propitious to the extending and transfixion of epigenetic structure plane.
2. Alternated distributed hard and soft stone: Diversity weather of rock is caused by the discrepancy of lithological nature, which can take shape of rock cell bellowed hard rock. Thus stress in slope is redistributed, and concentrated in some weak district such as controlling structural plane and exposure facial corner. Finally rockfalls are destroyed in the gravitational stress.
3. Developed surface water: Stress of fracture water would be aggrandized by enrichment of water. Meanwhile, strength of rock would be reduced by the immersion of water which can soften stone.
4. Plant roots split role: Over 80~95% area is covered by plants in the whole county. With developed hole in sandstone which has the inartificial force storage capacity of sandstone, together with the compactness and impermeability of mudstone, plants can be provided sufficient water, and even grow up in the region where bedrock is well exposed, especially for xylophyta whose rootstalk is grown fast. In the extending and expanding process of rootstalk growth, cranny is continuatively developed until the tensile failure of rock, which is formed by the hoop stress in rock.

4.2 *Destruction mechanism*

Two factors are involved for the destruction mechanism of Yanjiagou rockfalls: slip-tensile, bending-tension:

1. Slip-tensile: Rockfalls are slipped from the plane of strata or latter high-angle joints' planar to the exposure surface under the influence of gravity which engenders the ripped wreck of rock into two instable modes -slump and falling. Among these rockfalls, falling destruction is commonly seen in the smooth rock cell with plate-like and shell-like, which developed with chain role. 19 rockfalls with this destruction mechanism are counted in this investigation, accounted for 33.9% compared with the whole rockfalls.
2. Bending-tension: Under the flexural moment effect of gravity, rock bends like socle beam from the force into exposure face. A high-angle structural plane which causes topple and falling types destruction of rockfalls, is formed from top to bottom after the appearance of ripped wreck in the after skirt of rockfalls. Rockfalls with lump and pillar formation can be seen both in these two species of rock cell. 37 rockfalls with this destruction mechanism are counted in this paper, accounted for 66.1% compared with the whole rockfalls.

5 CONCLUSIONS

Yanjiagou in Pingchang with it steady regional structure, belongs to typical ladder hilly district, whose formation lithology is consisted of sandstone and mudstone.

1. Features of rockfalls in this region include smaller scale, monomer rockfall with colonial development, local chain roles, massive structure, dense coverage, thus result the occurrence of raindrop type disaster.
2. The difference of conformation mechanism and shape of rock cell, as well as developmental characteristics of rockfalls were brought about by the situation of lithology, construction and combinations of rock, classified rock cell into two types as smooth rock cell and rough rock cell.

3. Conformation of controlling structural plane of Yanjiagou rockfalls is mainly influenced by strata plane and high-angle planar X-type conjugate shear joints, and is secondarily impacted by unloading fissuring, weathering fissuring and gravitational spreading deformation fissuring.
4. Formation mechanism of Yanjiagou rockfalls involves Steep terrain, Alternated distributed hard and soft stone, developed surface water and plant roots split role. Failure mechanism of them is slip-tensile and bending-tension.

REFERENCES

Chen Hongkai, Tang Hongmei & Xian Xuefu. Chained Developing Pattern For Roek Slopes With Gentle Dip Strata. Journal of Lanzhou University (Naturalseienees), 2009, 45(1):20–25.

Chen Hongkai, Tang Hongmei, Ye Siqiao et al. Prevention Theory of Rockfall. Beijing: Earthquake Press, 2006.

Fan Jingyu, Qin Qirong & Su Peidong. Fracture Genesis in Tongnanba Structural Belt. Special Oil and Gas Reservoirs, 2009, 16(1):38–41.

He Siming, Zhang Xiaogang, Shen Jun et al. Analysis and Prevention of the Unstable Rock near Jingyun Pavilion In Leshan Giant Buddha Scenic Area. Journal of Catastrophology, 2009, 24(3):39–44.

Liang Feng, Li Bao, Li Jianfeng et al. Analysis on Formation and Development Mechanism of Simianshan Toudaohe Dangerous Rock Belt in Jiangjin County of Chongqing. The Chinese Journal of Geological Hazard and Control, 2009, 20(1):55–59.

Li Changhai, Dun Hui, Zhang Ming et al. Stability Evaluation and Prevention Measures of Potential Unstable Rock Mass along the Slope of Qingshiyan Section of Yu-Li Railway. Journal of Geological Hazards and Environment Preservation, 2010, 21(1):63–68.

Li Ming, Chen Hongkai, Ye Siqiao et al. Failune Mechanism of Hongyadong Dangerous Rockmass, Chongqing City. The Chinese Journal of Geological Hazard and Control, 2008, 19(2):1–6.

Wang Lei & Yang Huijian. Characteristic and Formation Mechanism of Zhangjiagou Area Rockfalls for Zhongwu Gas Pipe Line. China Water Transport, 2007, 5(4):62–63.

Wu Chu & Chen Hongkai. Risk Zoning Assessment of Geological Disaster in Simian Mountain National Park. The Chinese Journal of Geological Hazard and Control, 2009, 20(3):76–79.

Ye Siqiao, Tang Liangqin, Chen Hongkai et al. Study on Mozilling Unstable Rock and It and Control in Yunyang County, Chongqing City. Journal of Geological Hazards and Environment Preservation, 2005, 16(1):17–22.

Zhang Zuoyuan, Wang Shitian, Wang Lanshen et al. Analysis Theory of Engineering Geology. Beijing: Geology Press, 2006.

Modeling and Computation in Engineering II – Xie (ed)
© 2013 Taylor & Francis Group, London, ISBN 978-1-138-00058-2

Relationship research between fault rotation and surface deformation based on the disclination theory

Y. Zhang, Y.Z. Zhang, Y.J. Wu & W. Qu
Institute of Geological Engineering and Gomatics, Chang'an University, Xi'an, China

ABSTRACT: Fault rotation is one of the key factors that cannot be ignored which cause the surface deformation. This paper established the model between fault rotation and surface displacement in the infinite elastic medium space based on the disclination theory. We took Longmen mountain as study area, which is a strong faults activities area in China, then used the above model to study the horizontal and vertical displacements spatial distribution characteristics induced by Longmen Mountain fault aseismic rotation. Then it is comparised the simulated results with GPS measured coseismic results. The model simulation results show that (1) the Longmen mountain fault has significant nature of thrust fault; (2) The closer to the fault, the larger displacements the land surface have, the maximum value located in Beichuan area; (3) At the two end points of Longmen mountain fault the simulated displacements present a messy state. The Simulated results have better consistent with GPS measured coseismic results, and there are certain differences in the displacements magnitude.

1 INTRODUCTION

Cracks activities of odd change sources in crustal (fault) is one of the main incentives for local deformation. The relationship between fault translation and surface deformation field, strain field and strain gradient can be described by the rectangular dislocation theory in semi infinite elastic medium space (Xu, K.L. (1989)). Since the dislocation theory first introduced by Steketee (1958) to study the problem about deformation of the earth related to fault activitied, many mathematical models describe the relationship between dislocation and ground deformation in the uniform infinite space have been produced (Iwasaki, T. & R, Sato. (1979); Takemoto, S 1981. (1981); Segall, P. & D.F. McTigue. (1984); Sato, R. 1979. (1979); Heki, K. & S. Miyazaki. (1997); Shen, C.Y. & Wang, Q. (2002)) after decades of effort of the domestic and foreign scholars, as well as with the increase and improvement of the types of "source". Most of the above-mentioned studies were carried out based on the dislocation theory, which describes the relationship between the fault activities and surface deformation. However, in addition to the strike-slip, dip-slip and rifting activities, the fault also has rotating activities, and the rotation of the fault is one of the key factors that can not be ignored which causes surface deformation. Therefore, a new theory which describes the relationship between the fault rotation and land surface deformation in elastic medium has been put forward (Yang, S.H. (1998); B.A. Лихачев (1989)). In this paper the author has been deduced and established the fault rotation in deep crustal and land surface displacement model based on the disclination theory, as well as taken Longmen mountain as study area, which is a strong faults activities area in China, then used the above model to study the horizontal and vertical displacements spatial distribution characteristics induced by Longmen Mountain fault aseismic rotation. Then we comparised the simulated results with GPS measured coseismic results.

2 THE THEORETICAL RELATIONSHIP BETWEEN FAULT PLANES ROTATION AND LAND SURFACE DEFORMATION

2.1 *The relationships between the odd change source and any point displacement in elastic medium*

Assumed that a fault plane (not consecutive) exist in elastic media space, the discontinuity of the rotation vector and translation vector that the hanging wall relative to the footwall can be expressed as (Yang, S.H. (1998)):

$$\Omega^+ - \Omega^- = \omega$$

$$u^+ - u^- = b + \omega \times r \tag{1}$$

where in, superscript's positive and negative signal represent the upper and lower fault, displacement discontinuity formed by the translation vector b and mutual rotation angle ω. In order to study the propagation of odd variable source (dislocation, disclination) in an elastic medium, it is considered a homogeneous, isotropic, infinite free elastomer (physical zero). First, given a plastic (intrinsic) strain e_{kl}^p, it is generally the function of the spatial location. The total strain field of the medium at any point e_{kl}^T is the sum of elastic strain and the plastic strain:

$$e_{kl}^T = e_{kl} + e_{kl}^p \tag{2}$$

According to the elastic mechanics geometry equation, the total strain displacement using the displacement to coordinate derivative is represented as:

$$e_{kl}^T = u_{k,l}^T = \frac{1}{2}(u_{l,k} + u_{k,l}) \tag{3}$$

According to the elastic medium constitutive relation, the intrinsic stress:

$$\sigma_{ij} = c_{ijkl}(e_{kl}^T - e_{kl}^P) = c_{ijkl}(u_{k,l}^T - e_{kl}^P) \tag{4}$$

Using (4) in stress equilibrium equations $\sigma_{ij,i} = 0$ makes:

$$c_{ijkl}\, u_{l,ki}^T = c_{ijkl}\, e_{kl,i}^p \tag{5}$$

Solve the equation (5) we can get Green function tensor representation of general integral formula for displacement field:

$$u_n^T(R) = -\int c_{ijkl}\, G_{jn,i}(x - x')\, e_{kl}^p(x')\, dV' \tag{6}$$

where in, $c_{ijkl} = \dfrac{2\mu\nu}{1-2\nu}\delta_{ij}\delta_{kl} + \mu(\delta_{ik}\delta_{jl} + \delta_{il}\delta_{jk})$,

Green tensor function $G_{jn}(R) = \dfrac{1}{8\pi\mu}\left(\delta_{jn} R_{,qq} - \dfrac{1}{2(1-\nu)} R_{,jn}\right)$, δ_{ij} is symbol for Kronecker. The formula (6) shows the displacement at arbitrary point P generated by the plastic strain in the region of V' in the infinite homogeneous elastic space. Using $e_{kl}^P = -\delta_k(S)[b_l + \varepsilon_{lqr}\omega_q(x - x')]$ in (6) makes (Xu, K.L. (1989)):

$$\begin{aligned} u^T(R) &= \int c_{ijkl}\, G_{jn,i}(R)\delta_k(s')[b_l + \varepsilon_{lqr}\,\omega_q(x - x')]dV' \\ &= \int c_{ijkl}\, G_{jn,i}(R)[b_l + \varepsilon_{lqr}\,\omega_q(x - x')]ds' \end{aligned} \tag{7}$$

In the formula (7), ε_{lqr} is Einstein convention symbol, where b_l is zero, representing the displacement of the point P whose rotation angle is ω from X point with respect to X' point (disclination), where ω is zero, representing displacement caused by translational.

2.2 *Relationship between rectangular fault plane rotation and surface deformation*

Where b_l is zero, Equation (7) describes the displacement in the medium at arbitrary point generated by the translational and rotational of dots of a particular region in the infinite homogeneous elastic medium with respect to a fixed shaft. In the actual fault study, fault surface can be approximately described by a rectangular plane in order to describe the quasi-static deformation field of the surface generated by the rotation of the hanging wall relative to the footwall by the disclination theory. As shown in Figure 1, being taken a corner point of the rectangular fault as axis of rotation, rotation angle of hanging wall relative to the footwall is ω, if on the ground in parallel to the strike of the fault as the X axis, perpendicular to the strike of the fault as Y axis, and Z axis perpendicular to the ground as well as through the intersection point of the axes of rotation and fault plane downward to establish local space rectangular coordinates system. What's different with the dislocation theory is that in the process of mutual rotation of fault planes, the strike-slip displacement U_1 and dip-slip displacement U_2 caused by rotation on the fault plane is no longer constant, but a function of the rotation as independent variable parameter, whose magnitude related to the rotating shaft position.

The relationship between strike-slip component U_1 and dip-slip component U_2 at any point Q on the fault plane and the rotation parameter ω can be expressed as:

$$\begin{cases} u_1 = r\,(1-\cos\omega) \\ u_2 = r\sin\omega \end{cases} \tag{8}$$

where r is the distance from an arbitrary point Q on the fault plane to the axis of rotation. After the determination of the displacement components in the strike slip and dip slip direction on the fault plane, point source of a semi-infinite elastic space dislocation theory model and the relationship between any point on the fault plane and ground displacement can be expressed by the sliding displacement on the fault plane as (Okada, Y. (1982)):

$$u_i = \frac{1}{F}\iint_\Sigma U_j\left[\lambda\delta_{ik}\frac{\partial u_i^n}{\partial \xi_n} + \mu\left(\frac{\partial u_i^j}{\partial \xi_k} + \frac{\partial u_i^k}{\partial \xi_j}\right)\right]\upsilon_k d\Sigma \tag{9}$$

Equation (9) represents the displacement field $u_i(x_1, x_2, x_3)$ on the ground point P generated by the source dislocation at an arbitrary point Q of the dislocation surface, where, λ, μ are

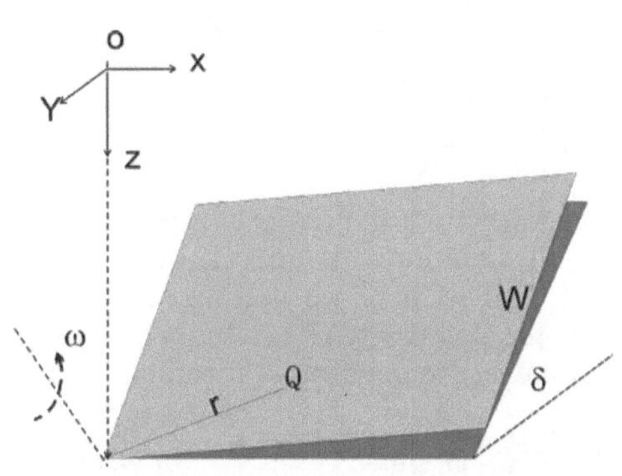

Figure 1. The relationship between rectangular fault rotation and local coordinate.

Lame constants, v_k is the direction cosine between the fault plane element $d\Sigma$ and vertical; the brace is the Green function in equation (7), u_i^j is the displacement components at the i direction produced by force component F_j at j direction on the dislocation point (ξ_1,ξ_2,ξ_3) (Okada, Y. (1982; 1992)).

For ease of calculation, it is assumed that the axis of the rotary shaft is at point $(0,0,d)$. U_1, U_2 express the hanging wall's strike-slip component, dip-slip component respectively relative to the footwall on the rectangular fault plane whose angle is δ, under depth is d (km), length is L (km), width is W (km).

Using the strike-slip component and dip-slip component in equation (9), we can get the displacement field of arbitrary point source on the ground at the disclination surface respectively.

At the strike-slip direction:

$$\frac{1}{F}\mu U_1 \Delta\Sigma\left[-\left(\frac{\partial u_i^1}{\partial \xi_2}+\frac{\partial u_i^2}{\partial \xi_1}\right)\sin\delta+\left(\frac{\partial u_i^1}{\partial \xi_3}+\frac{\partial u_i^3}{\partial \xi_1}\right)\cos\delta\right]$$

At the dip-slip direction:

$$\frac{1}{F}\mu U_2 \Delta\Sigma\left[-\left(\frac{\partial u_i^2}{\partial \xi_3}+\frac{\partial u_i^3}{\partial \xi_2}\right)\cos 2\delta+\left(\frac{\partial u_i^3}{\partial \xi_3}+\frac{\partial u_i^2}{\partial \xi_2}\right)\sin 2\delta\right] \tag{10}$$

Assuming $\xi_1 = \xi_2 = 0$, $\xi_3 = d$, using u_i^j in equation (10), the displacement field by the point source at the point $(0,0,d)$ on the ground can be obtained. The final result using $(x,y,0)$ instead of (x_1, x_2, x_3), strike-slip component produces a three-dimensional displacement field in the ground (B.A. Лихачев. 1989):

$$\begin{cases} u_x{}^0 = -\dfrac{U_1}{2\pi}\left[\dfrac{3x^2q}{R^5}+I_1{}^0\sin\delta\right]\Delta\Sigma \\ u_y{}^0 = -\dfrac{U_1}{2\pi}\left[\dfrac{3xyq}{R^5}+I_2{}^0\sin\delta\right]\Delta\Sigma \\ u_z{}^0 = -\dfrac{U_1}{2\pi}\left[\dfrac{3xdq}{R^5}+I_4{}^0\sin\delta\right]\Delta\Sigma \end{cases} \tag{11}$$

Dip-slip component produces a three-dimensional displacement field in the ground

$$\begin{cases} u_x{}^0 = -\dfrac{U_2}{2\pi}\left[\dfrac{3xpq}{R^5}-I_3{}^0\sin\delta\cos\delta\right]\Delta\Sigma \\ u_y{}^0 = -\dfrac{U_2}{2\pi}\left[\dfrac{3ypq}{R^5}-I_1{}^0\sin\delta\cos\delta\right]\Delta\Sigma \\ u_z{}^0 = -\dfrac{U_2}{2\pi}\left[\dfrac{3dpq}{R^5}-I_5{}^0\sin\delta\cos\delta\right]\Delta\Sigma \end{cases} \tag{12}$$

From (B.A. Лихачев, 1989) we know the surface displacement caused by fault rotation can be obtained by integral of equation (11) and (12) on the whole fault plane. For the complexity of the integral formulae, here we use the method of numerical integration. Considering the precision and the singularity and stability of the numerical integral value, we used two-dimensional Gauss–Legendre quadrature formula:

$$\int_{-1}^{1}\int_{-1}^{1} f(x,y)dxdy \approx \sum_{k=0}^{n}\sum_{i=0}^{n} A_k A_i f(x_k, y_i) \tag{13}$$

We finally get the displacement of U_1 and U_2 at x, y, z directions on the ground surface respectively are (u_x^1,u_y^1,u_z^1) and (u_x^2,u_y^2,u_z^2) by the numerical calculation. A final displacement generated by relative rotation of hanging wall of fault to footwall in ground surface is:

$$\begin{cases} u_x = u_x^1 + u_x^2 \\ u_y = u_y^1 + u_y^2 \\ u_z = u_z^1 + u_z^2 \end{cases} \tag{14}$$

Ground surface displacement generated by multiple faults is:

$$u_x = \sum_{i=1}^{n} u_x^i,\ u_y = \sum_{i=1}^{n} u_y^i,\ u_z = \sum_{i=1}^{n} u_z^i \tag{15}$$

3 EXAMPLE ANALYSIS

Longmen Moutain fault is located in Sichuan province, China, stretching about 500 km from northeast to southwest along the edge of the Sichuan Basin, 70 km wide. Cutting through along the bottom of the northwestern margin of the Sichuan Basin, the thickness of the crust suddenly changes, 60–70 km in the west and of less than 50 km to the east. During the Himalayan Orogeny process in the past one hundred million years ago, the Indian plate moves northward, squeezing the Eurasian plate, causing the uplift of the Qinghai-Tibet Plateau. While the uplifting, the Plateau also moves eastward, squeezing the Sichuan Basin. Longmen Moutain fault zone is an earthquake-prone area. On May 12, 2008, the Richter 8.0 earthquake took place. Longmen Moutain fault has significant nature of thrust fault. Then this paper use above established model to study the Longmen Moutain area land surface displacements distribution characteristics caused by Longmen moutain fault rotation.

The ground horizontal displacements component of 20 sub-faults were calculated by formula (15) are superimposed. The horizontal displacements generated on the land surface when Longmen moutain fault rotating with 5 seconds (shown in Fig. 2). Figure 2 shows the surface horizontal displacement caused by fault rotating.

As can be seen from Figure 2, the surface horizontal displacements caused by the fault rotation, revealing that the Longmen moutain fault has significant nature of thrust fault. The

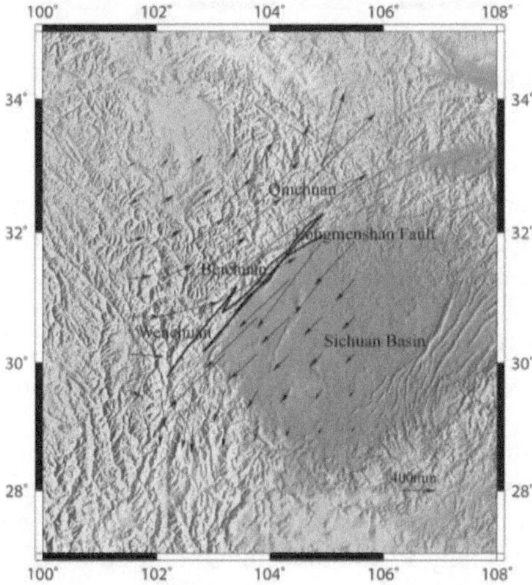

Figure 2. The horizontal displacements simulated of the model.

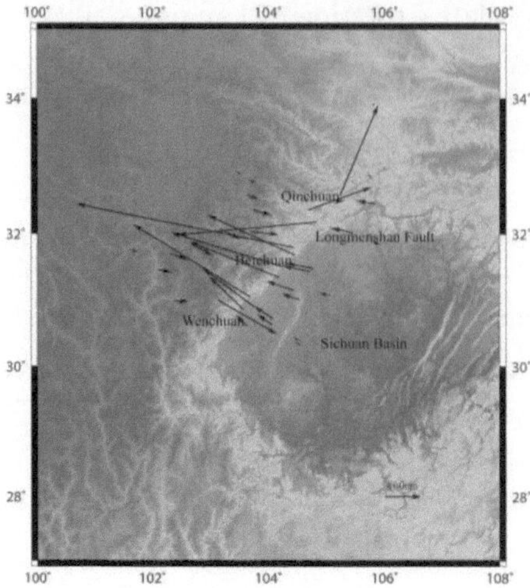

Figure 3. GPS obsercation results.

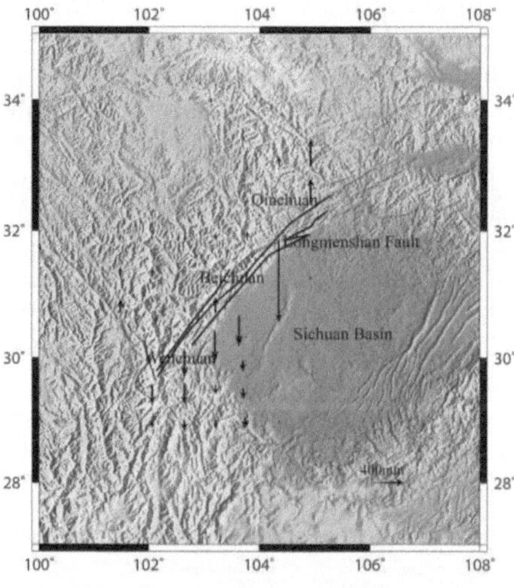

Figure 4. The vertical displacements simulated of the model.

value of horizontal displacements in Beichuan ares reached the maximum magnitude. At the two end points of Longmen mountain fault the simulated horizontal displacements present a messy state, with no obvious movement trend (Qinchuan area). The results are similar with the surface horizontal displacements characteristics of the Longmen Moutain fault zone revealed by the coseismic GPS obsercation showing in Figure 3. But there are some differences in magnitudes. The coseismic horizontal displacements magnitude is larger than that of the simulation results.

As can be seen from Figure 4 that, the vertical displacements distribution of the Longmen moutain fault has the following characteristics: (1) the largest value of vertical displacements located in Beichuan area in the central of the study area; (2) the endpoint of northern fault is showing a certain rise especially in the area of Qinchuan area. The characteristics of the model simulated results has better consistent with GPS measured coseismic results showing in Figure 5.

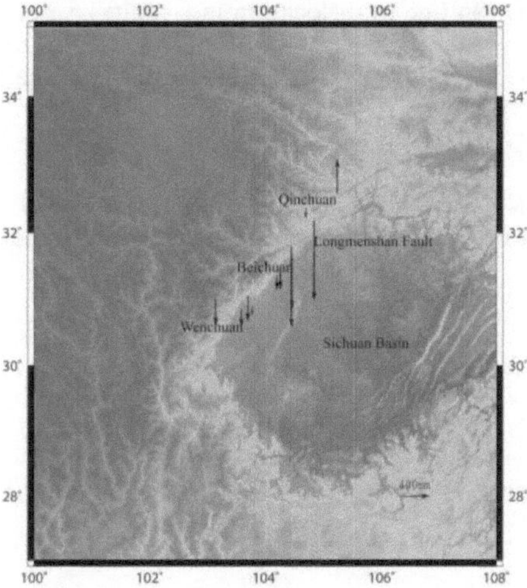

Figure 5. GPS obsercation results.

4 CONCLUSIONS AND DISCUSSION

In this paper the model between the fault rotation in deep crustal and land surface deformation is established based on the disclination theory. Taking Longmen mountain fault as study area, the results show that this model can be used to better describe the spatial distribution of surface deformation caused by the rotation of the fault. The research results will be used as a powerful complement of the Okada model, thus having important theoretical significance for using the GPS data to explore the nature of the faults rotation in deep crustal and the resulting seismic activity.

ACKNOWLEDGEMENTS

This study is supported by Nature Science Fund of China (NSFC) (project No: 41202189, 41274005), and a Key Project of the Ministry of Land & Resources, China (project No: 1212010914015), and the Fundamental Research Funds for the Central Universities (project No: CHD2012 JC076).

REFERENCES

B.A. Лихачев. 1989. Introductory Theory of Disclination. *Wuhan: Wuhan Press, 194.*

Heki, K. & S. Miyazaki. 1997. Silent fault slip following an interpolate earthquake at the earthquake at the Japan Trench. *Nature, 386, 595–597.*

Iwasaki, T. & R, Sato. 1979. Strain field in a semi-infinite medium due to an inclined rectangular fault. *J. Phys. Earth, 27,285–314.*

Okada,Y. 1982. Surface Deformation due to shear and tensile faults in a half-space. *Bull. Seismol. Soc. Am., 1018–1040,1985.*

Okada,Y. 1992. Internal Deformation due to shear and tensile faults in a half-space, *Bull. Seismol. Soc. Am.,75,1135–1154.*

Steketee, J.A. 1958. On Volterra's dislocation in a semi-infiinte elastic medium. *Can. J. Phys, 36, 192–205.*

Segall, P. & D.F. McTigue. 1984. Vertical displacements from a dip slip fault beneath surface topography,Abstract Chapman Conference on Vertical crustal Motion: Measurement and modeling. *AGU, Washington D.C.*

Sato, R. 1979. Crustal deformation due to dislocation in a multi-layered medium. *J. Phys. Earth, 19, 31–46.*
Shen, C.Y. & Wang, Q. 2002. GPS data inversion of kinematic model of main margin in Sichuan-Yunnan block. *Chinese J. Geophys. (in Chinese), 45(3):352–361.*
Takemoto, S. 1981. Effects of local inhomogeneities on tidal strain measurents. Bull. Disis. Pre. Res. Inst. Kyoto Univ. 33.15–46.
Xu, K.L. 1989. Structural geology (the second edition). *Beijing: Geology Publishing.*
Yang, S.H. 1998. Basic Theory of Crystal Dislocation (the second volume), *Beijing: Science Press, 546.*

Modeling and Computation in Engineering II – Xie (ed)
© 2013 Taylor & Francis Group, London, ISBN 978-1-138-00058-2

Resonant column test on dynamic shear modulus and damping ratio of the remolded red clay and its law analysis

Jian Li, Shang-Xiong Chen & Shu-Dan Xiong
State Key Laboratory of Geomechanics and Geotechnical Engineering, Institute of Rock and Soil Mechanics, Chinese Academy of Sciences, Wuhan, P.R. China

ABSTRACT: The remoulded red clay retains the beneficial engineering characteristics such as high strength and low compressibility of the original red clay, and eliminates the physical and mechanical properties such as high aquosity and cranny property which are adverse to the high-speed railway engineering. Therefore, the research on the dynamic shear modulus and damping ratio of the remoulded red clay under small strain amplitude can be one of the discriminating standards to decide whether the remoulded red clay can be used as the foundation of the high railway. On the basis of the indoor experimental such as grain analysis, liquid-plastic limit analysis and hit-solid tests, the paper does study on the standard red clay specimens under different confining pressures by using the resonant column test instruments. After sorting out the relevant test data, the paper analyzes the correlation between the dynamic shear modulus, dynamic damping and dynamic shear strain and its law. Then the paper does comparative research on the dynamic property between the remolded red clay and other different kinds of soil, and analyzes their differences and reasons causing the differences. Therefore, the unique dynamic properties are found. The research results can provide basic data and analysis reference for the relevant high-speed railway subgrade.

1 INTRODUCTION

With the development of high-speed railway, the train speed has been significantly improved. But more and more disease problems are highlighted with the accumulation of service time (Chen etc. 2010). To ensure safe and comfortable operation with high speed and to reduce maintenance as possible, subgrade settlement of the high-speed railway should be strictly controlled. Under the premise of the quality of subgrade filler and construction, the compaction deformation value of subgrade under gravity is limited, thus on this basis and under the condition of small strain amplitude, dynamic response analysis of foundation becomes the key of subgrade deformation. The dynamic shear modulus and dynamic damping ratio are two important parameters in the dynamic analysis. Dynamic shear modulus and dynamic damping ratio reflect the relationship of dynamic shear stress-strain and hysteresis quality of stress-stain respectively (Xie 1988).

At present, under small strain amplitude, researches about dynamic shear modulus and dynamic damping ratio have been focused on sand and regional conventional soil. For example, Quan-ming Wang etc. made a series of resonant column test for medium-coarse sand in Xiamen and obtained variation law of dynamic shear modulus and damping ratio in medium-coarse sand with mud (Wang etc. 2005). Estelle Delfosse-Ribay etc. used resonant column test to study the dynamic response of the grouted sand under different gradation (Estelle etc. 2004). I.A. Pantazopoulos etc. studied the dynamic characteristics of microfine cement grouted sand under the conditions of small strain amplitude (Pantazopoulos etc. 2012). Guoxing Chen and Xuezhu Liu etc. used free vibration column test to study the dynamic properties of recently deposited soil in Nanjing and got test curves of $G/G_{max} \sim \gamma$ and $D \sim \gamma$ (Chen etc. 2004, Liu 2003). Yajun Zhang etc. conducted a statistical analysis of the dynamic characteristics of the various types of soil in Shanghai coastal plain and gave the resonant column test results (Zhang etc. 2010). Xiaoming Yuan etc. conducted free vibration column test to do dynamic property test for vavious types of conventional soil in China under the conditions of small strain amplitude and given their dynamic characteristics regularity (Yuan 2000).

Under the condition of small strain amplitude, relatively little researches were done on the dynamic characteristics of the special soil. Yimin Xu etc. used resonant column test to study the relationship between the maximum shear modulus and confining pressure on sludgy loam (Xu 1991). Pengfei Jia studied the dynamic characteristics of Jingmen expansive soil under the conditions of small strain amplitude (Jia 2012). Zhi-jie Wang studied the dynamic characteristics of the undisturbed loess in different areas, as well as the factors that affect the dynamic characteristics (Wang 2010). Xiaohong Liu used free vibration column test instrument to study the dynamic characteristics of the undisturbed red clay (Liu 2011).

Red clay is a special soil with high strength and low compression, generally considered to be a natural foundation. However, due to its adverse geologic properties such as cranny property and uneven distribution and so on, security risks are given to the high-speed rail project and it often needs to be processed. Currently there are extremely little researches on dynamic characteristics of the remolded red clay under the conditions of small shear strain amplitude. In this paper, the resonant column test is used to study on dynamic shear modulus and damping ratio of the remolded red clay under small shear strain amplitude condition, and analyze its law, which can provide reference for the high-speed railway construction engineering.

2 RESONANT COLUMN TEST

2.1 *Survey of resonant column test*

Resonant column test is an experimental method to conduct torsion or bending vibration on the soil sample with certain condition of humidity, density and stress to obtain the resonant frequency through stepwise changing the driving frequency and then cut down the power to obtain the decay curves. According to the resonant frequency and the geometry and end condition of the soil sample, the dynamic shear modulus can be calculated, and according to the decay curve, the damping ratio D can be calculated. In order to reduce the test error, the resonant test demands the diameter of sample more than 2 times the length.

The resonant column test instrument uses GDS resonant column test system of Wuhan Institute of Rock and Soil Mechanics, Chinese Academy of Sciences, State Key Laboratory. The system is mainly composed of drainage system, monitoring system, driving system and pressure chamber. The strain is got by the accelerometer on the drive system, and its accuracy is up to 10^{-10}. The driving force of the drive system provides by the electromagnetic coil, the range of on-load voltage is 0.0001V~1V, and the range of dynamic strain is 1×10^{-6} and above.

2.2 *Test soil sample and its basic physical parameter*

Test soil samples selected from the test section of Wuhan to Xianning intercity railway, and it is remoulded. To study the engineering properties of the red clay, as well as to determine the test parameters, particle analysis, liquid and plastic limit, compaction and other physical and mechanical characteristics test were conducted.

2.2.1 *Particle analysis*

Particle analysis results: the particle size of the red clay studied was all less than 0.075 mm, which belong to fine-grained soil. The test results are shown in Table 1.

2.2.2 *Liquid and plasitcity limit test*

Liquid and plasticity limit test results: the liquid limit is 45%, plasticity limit is 20.9%, and the plasticity index is 23.1. From liquid and plastic limit analysis, it can be known that the red

Table 1. Test result of particle analysis.

Particle size/mm	0.5~0.25	0.25~0.075	0.075~0.005	<0.005
Content/%			57.4	42.6

Note: <0.05 mm content 90.7%, <0.01 mm content 52.7%, <0.002 mm content 29.6%.

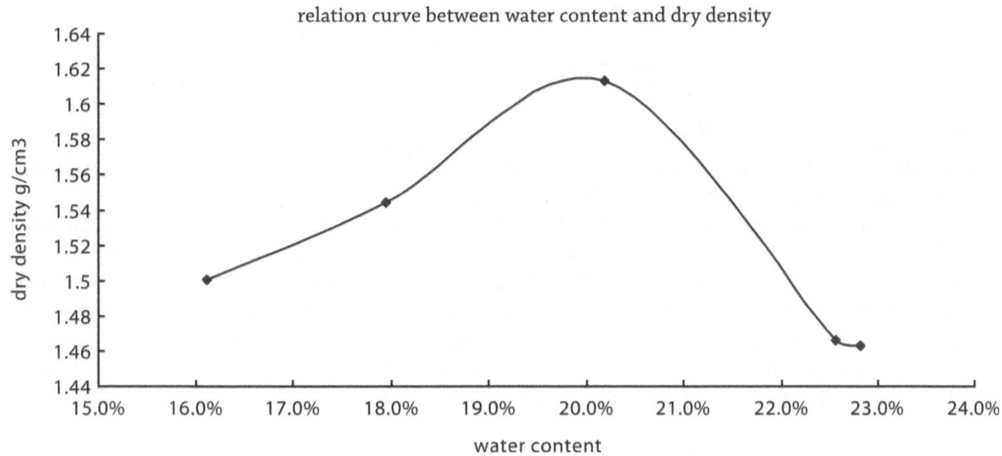

Figure 1. The compaction test result.

clay has a relatively higher plasticity index meaning the particales of this red clay is fine and the clay particle content is high, which correspond to the particle analysis result. According to the <Code for Soil Test of Railway Engineering>, it belongs to high liquid limit clay (The Ministry of Railways of the people's Republic of China 2010).

2.2.3 *Compaction test*

Compaction test results: the optimum water content of the test soil is 20% and the maximum dry density is 1.616 g/cm^3. Figure 1 shows the test results.

2.3 *Test scheme*

The basic parameters of the sample in test are determined according to the aforementioned physical and mechanical property test. The water content takes the optimum moisture content, 20%. The density is based on requirements of <Code for Soil Test of Railway Engineering>: Compactnes of the railway subgrade filler need to achieve more than 95%, taking $\lambda c = 0.95$, therefore, the density of sample is 1.84 g/cm^3. To make a comprehensive study on the dynamic characteristics of the remolded red clay, the test also produced soil samples with compactness 0.9, 0.85 and 0.8 to do experimental study, and their densities were respectively 1.74 g/cm^3, 1.64 g/cm^3, 1.54 g/cm^3.

Samples were produced by hydraulic jacks. Test soil sample geometry was $\Phi 50$ mm $\times$ 100 mm. In order to ensure the accuracy and comparability of the test datum, parallel tests of three groups of soil sample with compactness 0.95 were done. Each group includes four soil samples which were respectively conducted resonant column test under confining pressure 50 kPa, 100 kPa, 150 kPa and 200 kPa. For the convenience of datum processing, identifiers were given to the three groups of soil samples with compactness 0.95. The three groups were respectively identified by TS1, TS2 and TS3.

At the same time, according to previous references (Xie 1988, Zhang 2010, Yuan 2000, Xu 1991, Jia 2012), when the shear strain belongs to the range of 1×10^{-4}~1×10^{-2}, the modulus

Table 2. Resonant column test program.

Test number	Compactness				Consolidation ratio	Moisture content
	0.8	0.85	0.9	0.95		
Confining presure/kPa						
50	1	1	1	3	1.0	20%
100	1	1	1	3		
150	1	1	1	3		
200	1	1	1	3		

would change significantly. Therefore, relatively larger shear strain resonant column test was conducted to TS1 in order to analyze the variation regularity of the dynamic shear modulus and damping ratio under large shear strain condition.

Specific test programs are as follows in Table 2. Consolidation in the test was homonymous drainage consolidation. The completion criteria conform to <Code for Soil Test of Railway Engineering>: the change of consolidation displacement is not greater than 0.1 m^3 in 1 hour, or the axial deformation in 5 minutes is not greater than 0.005 mm. But in the test process, it was found that the above conditions were easily met, therefore, the consolidation time was unified by 1 hour. And the consolidation could make a better close contact between the instrument and the soil sample to ensure the test results.

3 TEST RESULTS AND ANALYSIS

3.1 *Test results comparision under different compactness*

Figures 2–4 show the test results under different compactness conditions of the four confining pressure.

In can be seen from Figure 2 that the test datum of the three groups of sample in different confining pressure show good consistency and the difference between the datum is small.

Meanwhile, it can be known that the relation cureve of $1/G \sim \gamma$ is linear, which accords with the hyperbolic model proposed by Duncan and Chang (Duncan & Chang 1970). Moreover, when the shear strain is small, the relation curve of $D \sim \gamma$ meets logarithmic relationship, but when the shear strain is over 1×10^{-4}, the relation meets linear which does not accord with the formula about damping ratio proposed by Hardin and Denevich (Hardin & Drnevich 1972a, 1972b).

3.1.1 *Variation regularity of dynamic shear modulus*

From Figure 3, it can be seen that along with the development of shear strain, the shear modulus of the remolded red clay decrease gradually. Under the same shear strain condition, along with the increase of compactness, the shear modulus is improved continually. Especially in the lower compactness case, through increasing compactness, the effect of improving dynamic shear modulus is notable, but it is not big when the compactness reaches 0.9, which is shown in Figure 3 that the dynamic shear moduli of the samples with compactness 0.9 and 0.95 are with little difference.

This means that along with increase of the compactness, soil's capacity of resisting deformation is improved constantly. Especially under the low compactness condition, the effect of improving soil's capacity of resisting deformation is notable. Therefore, for the high speed railway which strictly controls the post construction settlement, compactness is a technical index can not be igored in the foundation construction.

3.1.2 *Variation regularity of damping ratio D*

It can be seen from Figure 4 that along with the development of shear strain, the damping ratio of soil increase continually. Along with the improvement of the compactness, the

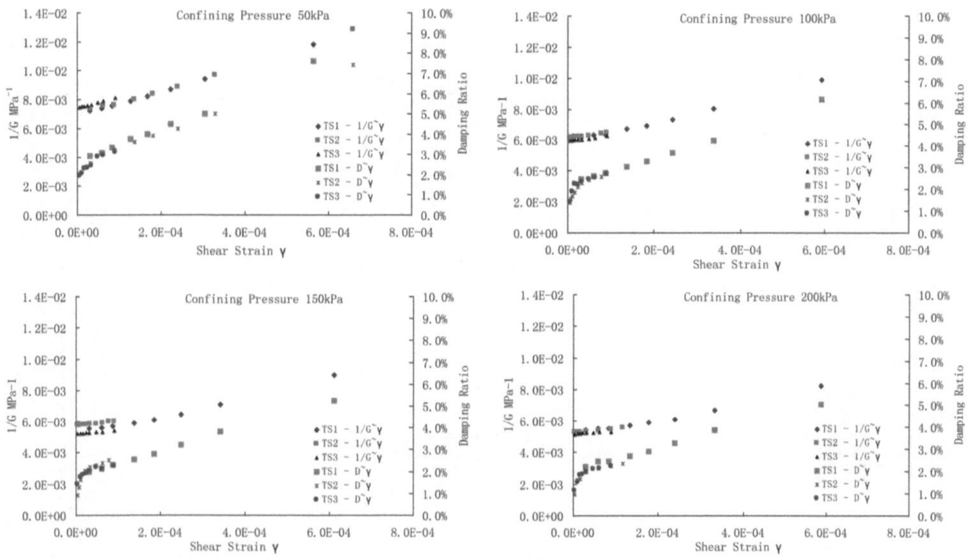

Figure 2. D~γ and 1/G~γ test curves under different confining pressure of the three groups of soil samples with compactness 0.95.

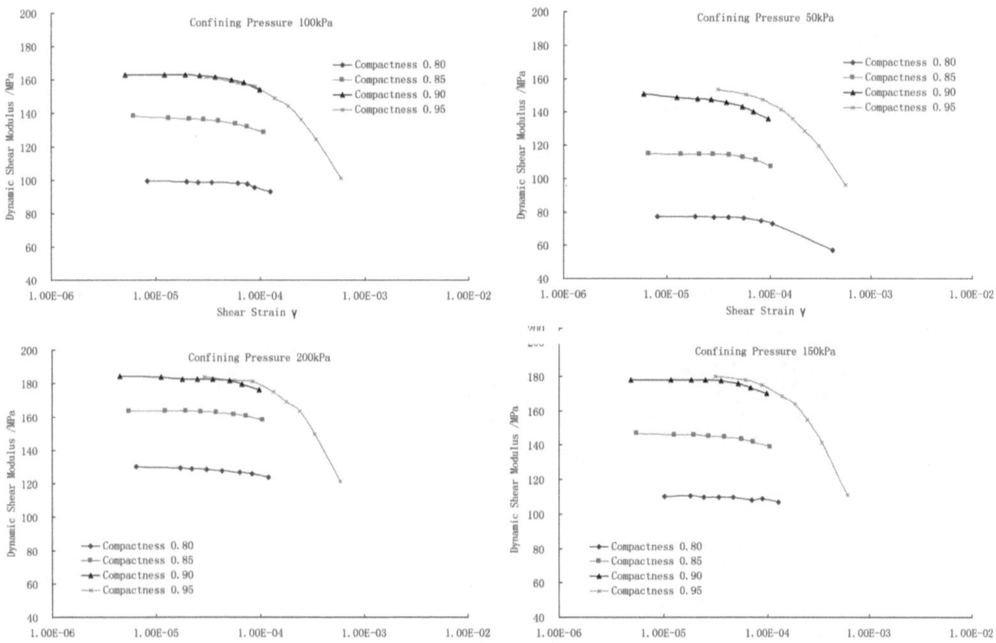

Figure 3. G~γ test curves of the soil samples with different compactness under same confining pressure.

damping ratios slightly reduce. But the regularity is not obvious, and the damping ratios between samples with different compactness are small in difference. This means that when the compactness of soil reaches a certain value, the stress wave can propagate into the deeper area of the remolded red clay foundation with less energy attenuation, which is conducive to mobilize deep soil to collectively resist the deformation of external loads.

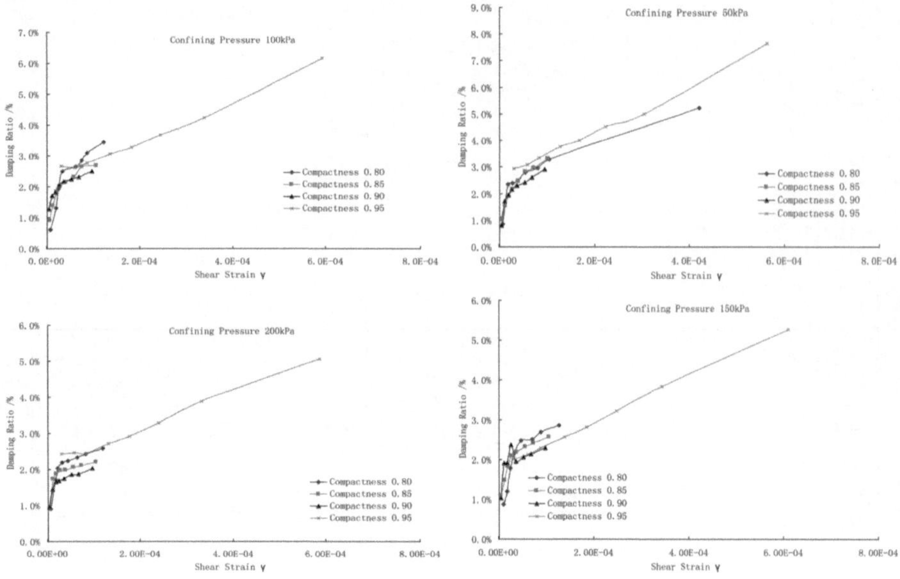

Figure 4. D~γ test curves of the soil samples with different compactness under same confining pressure.

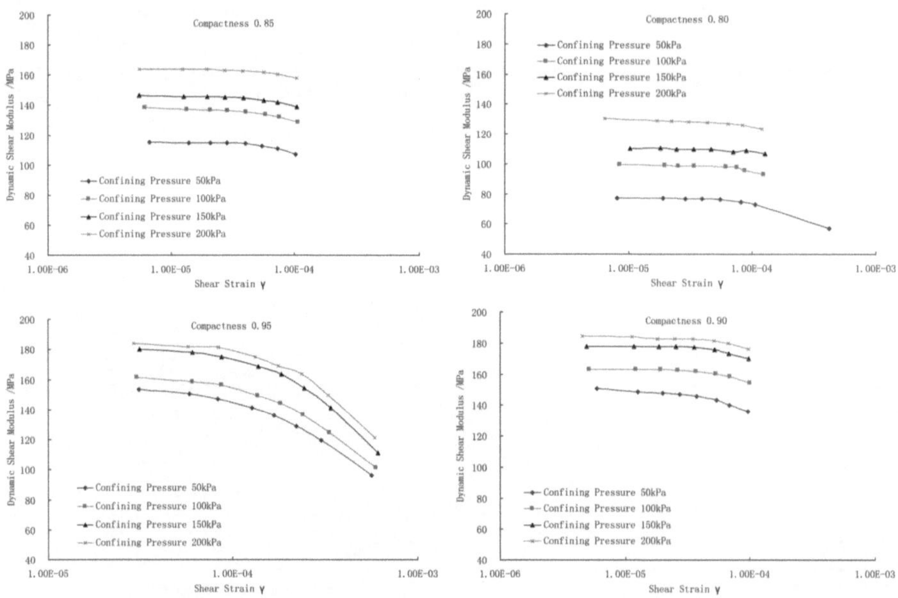

Figure 5. G~γ test curves of the soil samples with different confining pressure under same compactness.

3.2 *Test results of one group of samples comparision under different compactness*

Figures 5 and 6 respectively show the resonant column test results with different confining pressure under condition of same compactness.

3.2.1 *Variation regularity of dynamic shear modulus G*

Figure 5 shows the G~log (γ) test results of same group of samples under different confining pressure condition.

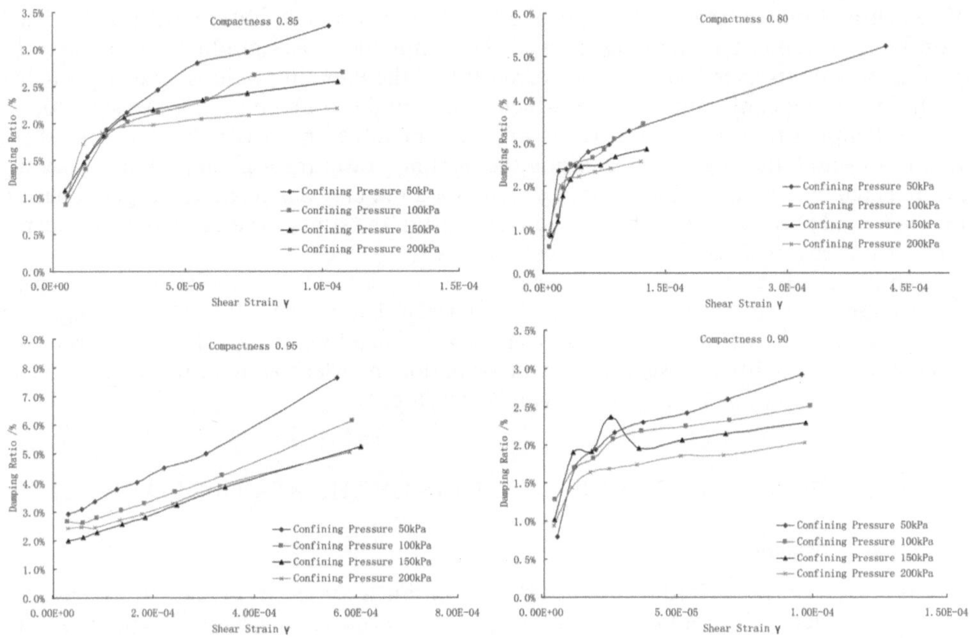

Figure 6. D~γ test curves of the soil samples with different confining pressure under same compactness.

Through analyzing Figure 5, it can be known that under certain confining pressure condition, with the increase of strain, shear modulus G decreases. But under small strain condition ($\gamma < 1 \times 10^{-4}$), their relationship is linear and the rule of remolded red clay is all the same to sand (Senetakis & Anastasiadis etc. 2011, 2012). When the shear strain is over 1×10^{-4}, the decreasing rate of dynamic shear modulus is faster. Thus it can be supposed that under small strain amplitude condition, the soil sample is in elastic range and the shear modulus changes small. Once the plastic deformation appears, the structure of soil sample is damaged and the void ratio increases, which cause the shear modulus drop faster. Therefore, for the high-speed rail project, it should be as far as possible to improve soil strength so as to enhance its ability to resist deformation, and make sure that the soil is in elastic deformation range as much as possible in order to effectively reduce the settlement.

Under certain strain condition, with the increase of confining pressure, shear modulus G increases gradually. As can be seen from Figure 5, the increment of the shear modulus is proportional to that of the confining pressure, and their relationship is linear. This is mainly because of the increase of confining pressure that compact the soil sample and make the sample become much denser. But from the pictures is can be seen that when the compactness is over 0.9, the effect of increasing shear modulus through increasing confining pressure is not significant. It is indicated that the soil samples themslves are much dense, thus the effect is not significant, which is different to the sand and silt (Liu 2003). All the test results indicate that the increase of compactness is significant to the effect of enhancing the ability of resisting deformation of the soil. Therefore the compactness standard must be strictly controlled.

3.2.2 *Variation regularity of damping ratio D*

Figure 6 shows the D~log (γ) test results of the same group of samples under different confining pressure condition.

As can be seen from Figure 6, when the shear strain is small ($\gamma < 1 \times 10^{-4}$), under certain confining pressure condition, the damping ratio increases with the increase of the shear strain. When the shear strain is less than 5×10^{-5}, the relationship between the damping ratio and shear strain is exponential; when the shear strain is over 5×10^{-5}, the relationship is linear.

When shear strain is small ($\gamma < 1 \times 10^{-4}$), under certain strain condition, with the increase of confining pressure, the damping ratio of the sample decreases gradually. But when the confining pressure is over 100 kPa, the decrement of the damping ratio is relatively smaller through increasing confining pressure, which means under high confining pressure condition, the damping ratio of the soil tends to a certain value and it is only related to shear strain. But overall, the effect of decreasing the damping ratio through improving confining pressure is relative smaller. That means the remolded red clay has relatively larger cohesion between the soil particles and higher strehgh, as a result, the stress wave can propagate deeper due to the less energy decay in the propagation process.

As can be seen from Figures 6, under small strain ($\gamma < 1 \times 10^{-4}$) condition, the damping ratio increases at a slower rate. When the shear strain is over 1×10^{-4}, the damping ratio increases rapidly with the increase of the shear strain. Therefore, it should be as far as possible to improve the ability of resisting shear deformation in order to reduce the damping ratio and the impact of the soil in deeper area can be developed.

4 THE MODEL ANALYSIS OF TEST RESULTS OF THE REMOLDED RED CLAY

4.1 *Davidenkov model*

On basis of experiments, Hardin and the others (Hardin & Drnevich 1972a, 1972b) have got a result that under cyclic loading, the stress-strain backbone curve of soil is hyperbola. The equation is as follows:

$$\tau_d = \frac{\gamma_d}{1/G_{\max} + \gamma_d/\tau_y} \tag{1}$$

In this equation: G_{max} is the initial maximum shear modulus; τ_y is the maximum dynamic shear stress. If called as reference strain γ_{ref} which is the cross point between G_{max} grade line and τ_y level line, then γ_{ref} is equal to τ_y/G_{max}, so the Hardin model is got as follows:

$$G_d = G_{\max}(1 - f(\gamma)) \tag{2}$$

In this equation,

$$f(\gamma) = \frac{\gamma_d/\gamma_{ref}}{1 + \gamma_d/\gamma_{ref}} \tag{3}$$

The equation of Damping ratio is:

$$D = D_{\max}\left(1 - \frac{G}{G_{\max}}\right) \tag{4}$$

Combining equation (2), type (4) into:

$$D = D_{\max}(f(\gamma))^n \tag{5}$$

Based on Hardin model, Martin and his groups (Martin & Seed 1982) improved equation (3) and (5), and then put forward the Davidenkov model with three parameters and the fitting formula of damping ratio with power form, which is as follows:

$$f(\gamma) = \left(\frac{(\gamma_d/\gamma_0)^{2B}}{1 + (\gamma_d/\gamma_0)^{2B}}\right)^A \tag{6}$$

$$D = D_{max}(f(\gamma))^n \tag{7}$$

It should be point out that in equation (6), γ_0 isn't a parameter with clear physical meanings any more, but just is a fitting parameter.

Based on Hardin model, the three parameter Davidenkov model has more three parameters, whose advantage is that through adjusting parameters, the model can fit the test dates better, and thus the dynamic shear modulus can be better predicted. Whereas the disadvantages are that firstly, the referential shearing strain γ_{ref} with clear physical meanings in the Hardin model is replaced by the fitting parameter γ_0 with no real physical meaning; secondly, due to the relative more parameters and no standard for the evaluation of the parameters, the application of the model is complex.

In order to decrease fitting parameters as much as possible and get more parameters from tests, the γ_0 in the Davidenkov model is still used γ_{ref}.

4.2 *Parameters determination of the Dvidenkov model*

4.2.1 *Determination of τ_y*

In hyperbolic model, the maximum dynamic shear stress τ_y means the maximum shear stress in backbone curve. In general, it can be got by two methods: firstly, the τ_y can be obtained by the backbone curve which is the peak point ligature of the hysteretic cureve gained by the step load test; secondly, the backbone curve can be regarded as the stress-strain relation curve of the static fast shear test, then the τ_y can be obtained. However, according to the research of Vueetic and his group (Vucetic 1990), the initial backbone curve got by the second method can better reflect the mechanical properties of the clay. The errors can be avoided, which is caused by the strain rate decrease of the peak points of the hysteresis curve. Therefore, the static fast shear three axial tests of non consolidation and non drain are done with a shear velocity of 0.4 mm/min.

Table 3 shows the test results with different confining pressure and compactness. It can be known that the fast shear strength of the remolded red clay has a logarithmic relationship with the confining pressure and has an exponential relationship with the compactness, thus their relationship can be assumed:

$$\tau_y = a \times \ln(\sigma_3 + b) \times e^{(c \times \lambda_c)} \tag{8}$$

By nonlinear fitting, the parameters in formula (8) can be obtained, as showing in Table 4. The fitting effect is preferable and $R^2 = 0.99$.

Table 3. The quick shear strength under different confining pressure and compactness.

Confining pressure/kPa	50				100	150	200
Compactness	0.80	0.85	0.90	0.95	0.80	0.80	0.80
Fast shear strengh/kPa	110	150	210	290	240	285	310

Table 4. The fitting results between the biggest shear stress and confining pressure and compactness.

Fitting parameter	*a*	*b*	*c*
Fitting results	0.33	−43.99	6.53

4.2.2 *Determination of G_{max}*

Table 5 shows the maximum dynamic shear modulus under conditions of different confining pressure and compactness. It can be seen that the maximum dynamic shear modulus is related with the confining pressure and compactness. From Table 5, it can be known that the maximum dynamic shear modulus has an approximate linear relationship with confining pressure and has a logarithmic relationship with compactness, therefore their function relationship can be assumed as follows:

$$G_{\max} = k \times Pa\left(\frac{\sigma_3}{Pa}\right)^n \times (\ln(\lambda_c + e))^f \tag{9}$$

In this formula, σ_3 is the confining pressure; λ_c is the compactness; Pa is the standard atmospheric pressure with a value of 101.4 kPa.

By nonlinear fitting, the parameters in equation (9) can be obtained, showing in Table 6. The fitting effect is pretty good and the R^2 is higher than 0.95.

4.2.3 *Determination of A and B*

Table 7 shows the values of A and B in the Davidenkov model, obtained by nonlinear fitting based on the aforementioned test results.

From Table 7, it can be known that the parameter A and B have a good linear relationship with confining pressure and compactness, which can be expressed by:

$$A/B = r \times \sigma_3^s \times \lambda_c^t \tag{10}$$

By nonlinear fitting, the parameters in equation (10) can be got, showing in Table 8. The fitting effect is preferable and the fitting degree is relative higher and the R^2 are respectively 0.98 and 0.91.

Table 5. The fitting results between the biggest shear modulus and confining pressure and compactness.

Confining Pressure/kPa	50				100				150				200			
Compactness	0.80	0.85	0.90	0.95	0.80	0.85	0.90	0.95	0.80	0.85	0.90	0.95	0.80	0.85	0.90	0.95
G_{max}/MPa	78	114	151	155	100	139	164	169	111	147	178	179	131	164	194	195

Table 6. The fitting results between the biggest shear modulus and confining pressure and compactness.

Fitting parameter	*k*	*n*	*e*	*f*
Fitting result	2.65	0.21	0.21	0.23

Table 7. A and B of the Davidenkov model.

Confining pressure/kPa	50				100	150	200
Compactness	0.80	0.85	0.90	0.95	0.95	0.95	0.95
A	0.11	0.08	0.053	0.049	0.32	0.26	0.22
B	5.1	7.2	8.1	8.5	9.2	12.6	13.7

Table 8. A and B of the Davidenkov model.

	A		B	
Fitting parameter	s	t	s	t
Fitting result	5.1	7.2	8.1	8.5

5 CONCLUSION

Through analyzing the resonant column test results of the remolded red clay and comparing with other types of soil, the main conclusion obtained are as follows:

1. Compactness is an important technical index to assess the foudation construction quality of the high speed railway. Along with the compactness and confining pressure increase, the dynamic shear modulus increase and the damping ratio reduce. After the compactness reaches a certain extent, the effect of improving shear modulus through improving compactness or confining pressure is not obvious.
2. The dynamic properties of soil relate to factors such as soil type, moisture content, compactness (void ratio), consolidation pressure and strength and so on. Compared with other general undisturbed soil, the remolded red clay has a lower moisture content, higher compactness. Moreover, the remolded red clay does not include crackes and other adverse geological factors and its strength is higher, thus the dynamic shear modulus is higher. As to be high speed railway foundation, it can bear relatively larger dynamic loads and the settlement of the whole foundation is relatively smaller.
3. The particles of the remolded red clay with high compactness has more contact point and larger contact area, which make the stress wave propagate faster and deeper with less energy attenuation, thus the damping ratio is smaller than other types of soil. As to be high speed railway foundation, it can fully promote the deep soil to commonly resist the external loads and effectively reduce the settlement.
4. The G/Gmax~log (γ) and D~log (γ) curves, the maximum shear modulus G_{max} and the maximum shear strength τ_y are given in the paper, which can provide basic datum and reference for the relavant engineering.

ACKNOWLEDGEMENT

Research presented in this paper was supported by the national natural science foudation of China (project numbers: 51027002).

REFERENCES

Chen Guoxing & Liu Xuezhu. 2004. Testing Study on ratio of dynamic shear moduli and ratio of damping for recently deposited soils in Nanjing and its neighboring areas. Chinese Journal of Rock Mechanics and Engineering 23(8):1403–1410.

CHEN Shang-xiong & Song Jian & Zhou Quan-neng etc. 2010. Theory and Practice for High-speed Railway Settlement and Deformation Observation and Evaluation. Beijing: China Railway Publishing House.

Duncan J.M. & Chang. 1970. Nonlinar analysis of stress and strain in soils. Journal of the Geotechnical Engineering Division ASCE 103(6): 517–533.

Estelle Delfosse-Ribay & Irini Djeran-Maigre & Richard Cabrillac etc. 2004. Shear modulus and damping ratio of grouted sand. Soil Dynamics and Earthquake Engineering 24: 461–471.

Hardin B.O. & Drnevich. 1972a. Shear modulus and damping in soils: measurement and parameter effets (Terzaghi Leture). Journal of the Soil Mechanics and Foundation Division ASCE 98(SM6): 603–624.

Hardin B.O. & Drnevich. 1972b. Shear modulus and damping in soils: design equations and curves. Journal of the Soil Mechanics and Foundation Division ASCE 98(SM7): 667–692.
Jia Pengfei. 2012. Behaviour and modeling of plastic strain accumulation of soil during low-amplitude high-cycle loading process. Wuhan: Institute of Rock and Soil Mechanics, Chinese Academy of Sciences.
Liu Xiaohong. 2011. Research on dynamic stability of red clay subgrade under ballastless track of high-speed railway. Changsha: Central South University.
Liu Xuezhu. 2003. The study and test on dynamic properties of recently deposited soils and vibratory liquefaction of sands in area of Nanjing and its nerghbor. Nanjing: Nanjing University of Technology.
Martin P.P. & Seed H.B. 1982. One-dimensional dynamic ground response analyses. Journal of Geotechnical Engineering Division ACSE 108(7): 935–952.
Pantazopoulos I.A. & Atmatzidis D.K. 2012. Dynamic properties of microfine cement grouted sans. Soil Dynamics and Earthquake Engineering 42: 17–31.
Senetakis Kostas & Anastasiadis Anastasios & Pitilakis Kyriazis. 2011. Dynamic response of sandy and gravelly soils: Effect of grain size characteristics on G-γ-D curves. In proceedings of the fifth international conference on earthquake geotechnical engineering. Santiago.
Senetakis Kostas & Anastasiadis Anastasios & Pitilakis Kyriazis. 2012. Dynamic properties of dry sand/rubber (SRM) and gravel/rubber (GRM) mixtures in a wide range of shearing strain amplitudes. Soil Dynamics And Earthquake Engineering 33(2): 38–53.
The Ministry of Railways of the people's Republic of China. 2010. TB10102-2010 Code for Soil of Railway Engineering. Beijing: China Railway Publishing House.
Vucetic. 1990. Normalized behaviour of clay under irregular cyclic loading. Canadian Geotechnical Jounal 27:29–46.
Wang Quan-min & Li Gang & Chen Zheng-han etc. 2005. Research on dynamic characteristics of sands in Xiamen city. Rock and Soil Mechanics 26(10): 1628–1632.
Wang Zhijie & Luo Yasheng & Wang Ruirui etc. 2010. Experimental study on dynamic shear modulus and damping ratio of undisturbed loess in different region. Chinese Journal of Geotechnical Engineering 32(9): 1464–1469.
Xie Ding-yi. 1988. Soil Dynamics. Xi'an: Xi'an Jiaotong University Press.
Xu Yimin & Jiang Pu. 1991. The application of resonant column in soil test. Journal of Hohai University 19(6): 36–41.
Yuan Xiaoming & Sun Rui & Meng Shangjiu etc. 2000. Laboratory experimental study on dynamic shear modulus ratio and damping ratio of soils. Earthquake engineering and engineering vibration 20(4): 133–139.
Zhang Yajun & Lan Hongliang & Cui Yonggao. 2010. Statistical studies on shear modulus ratios and damping ratios of soil in Shanghai area. World Earthquake Engineering 26(2): 171–175.

Modeling and Computation in Engineering II – Xie (ed)
© 2013 Taylor & Francis Group, London, ISBN 978-1-138-00058-2

Research on steel structures design method using energy dissipation brace for story lateral ductility ratio control

Zhong-Jun Li
Design Institute of Civil Engineering & Architecture of Dalian University of Technology Co., Ltd., China

ABSTRACT: In the high-rise steel structure buildings with bigger storey height and large-span, layer lateral ductility ratio does not meet standardize requirements limits, generally. In this paper, metal yield dampers in the beam and support node was proposed, and studied in an example of one training center for high-rise steel structure building, in order to control the structural the large interlayer lateral ductility ratio. And damper node design scheme was obtained. The calculation results show that the energy dissipation brace can effectively reduce the displacement response of the structure under seismic action and it can improve the seismic performance. It is contrast to ordinary steel support with good economy.

1 INTRODUCTION

Steel buildings are generally with smaller general stiffness and damping, sensitive to earthquake. Damage investigation results show that the lack of structural strength is not the main factor leading to structural damage. As long as the structural be maintained strength of the earthquake process, and the structure has a certain amount of bombs plastic deformation capacity, the structure will be able to survive in the earthquake (Shen 2006). At present, China Seismic Design of Steel mainly follow the principle that "no damage while minor earthquake, little damage while moderate earthquake, no collapse while severe earthquakes". It has been made in the design theory and methods more research results (Li 2006, Hen 2009, Xiong 2009, Peng 2007, Liu 2004, Xu2010). Technical specification for civil construction steel structures of tall building "(JGJ99-98 1998) (hereinafter referred as "high steel code") gives the two-stage design of the steel structure design method: In frequent earthquake, the first phase of a multi-elastic analysis is checking component of the carrying capacity and stability as well as the structure of the layer between the lateral displacement; second stage is case of the elasto-plastic deformation analysis under very rarely earthquake lateral displacement ductility ratio the side shift and the interlayer between the checking the structure of the layer.

Supporting steel structure is common lateral-force-resisting member, energy dissipation brace is a special support in the form, it belongs to the field of energy-dissipating technology, which is an important innovation in the building structure in the seismic. "Code for seismic Design of Buildings" in China (GB50011-2001) for the first time as an important content of the seismically and energy-dissipating technology isolated specifically set up a chapter. The current "Code for seismic Design of Buildings" (GB50011-2010) hereinafter referred as "seismic code", retained the original content about the technology of energy dissipation and the related provisions were partially modified. Visible energy dissipation technology will become the development trend of China's seismic field. In this paper, the energy dissipation brace applied for engineering practice to solve the problem of the steel structure design process middle lateral displacement ductility ratio does not meet the requirements, the improved ductility and energy dissipation capacity of the steel structure, steel structure design a new idea.

2 PROJECT OVERVIEW

Dalian, a training center for the 8-story steel frame structure, fortification intensity 7, structures as-signed to seismic Grade for 3, Site-classes II, for 2nd design earthquake group. The design characteristic period of ground motion shall be taken as 0.40 s, Basic wind pressure of 0.65 kN/m^2; Site the roughness class B, Basic snow load is 0.40 KN/m^2, live loads on Floor are as follows: Office is 2.0 KN/m^2, Corridor, hall, stairs is 3.5 kN/m^2, the Master roof is 2.0 KN/m^2, Toilet is 4.0 KN/m^2, Elevator, ventilation equipment room is 7.0 kN/m^2. There is no dynamic loads considered. Due to the use of functional requirements, the building is with the middle of the open span and high soft features. The structure plan view is shown in Figure 1. The ordinary steel supports is adopted to increase the lateral stiffness of the structure in the structural design of the original program setting and the two-stage design is done in accordance with the "high steel code". In the design process, rarely earthquake displacement between the layers of the structure along the Y-axis direction does not get better control, and does not meet the "high steel code" set forth in the layer lateral displacement ductility ratio requirements. Energy dissipation technology can effectively reduce the structural vibration response under seismic action, at a later design procedure; ordinary steel support of the Y-axis direction will be replaced with a metallic yield damper energy dissipation brace to solve the above mentioned problems.

3 ENERGY DISSIPATION BRACE DESIGN

Common energy dissipation braces include several kinds of scheme, such as constraint buckling support, dampers with ordinary steel support combinations, oblique corner support, etc. Energy dissipation and support joint use, as well as constraints buckling support is most common in the project. The current "seismic code" damper divided into two categories: Speed type and displacement type. Speed-type dampers usually include viscoelastic dampers and viscous dampers; Displacement-type dampers typically include friction dampers and metallic yield dampers. Currently, viscous dampers and metallic yield dampers are more applicable of projects in China. Especially metallic yield dampers based on economic, practical engineering

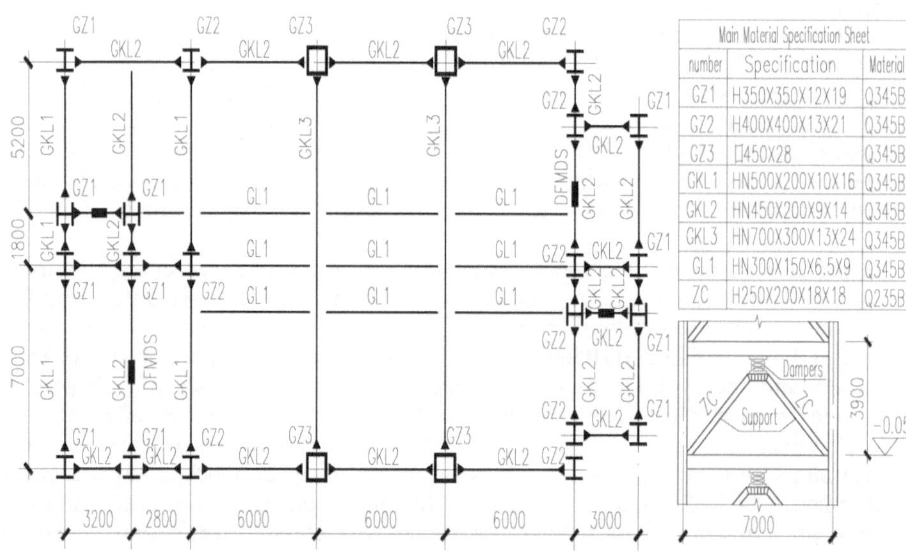

Main Material Specification Sheet		
number	Specification	Material
GZ1	H350X350X12X19	Q345B
GZ2	H400X400X13X21	Q345B
GZ3	□450X28	Q345B
GKL1	HN500X200X10X16	Q345B
GKL2	HN450X200X9X14	Q345B
GKL3	HN700X300X13X24	Q345B
GL1	HN300X150X6.5X9	Q345B
ZC	H250X200X18X18	Q235B

(a) The standard floor plan (b) Elevational view of supporting

Figure 1. Structure plane, the energy dissipation brace arrangement drawing.

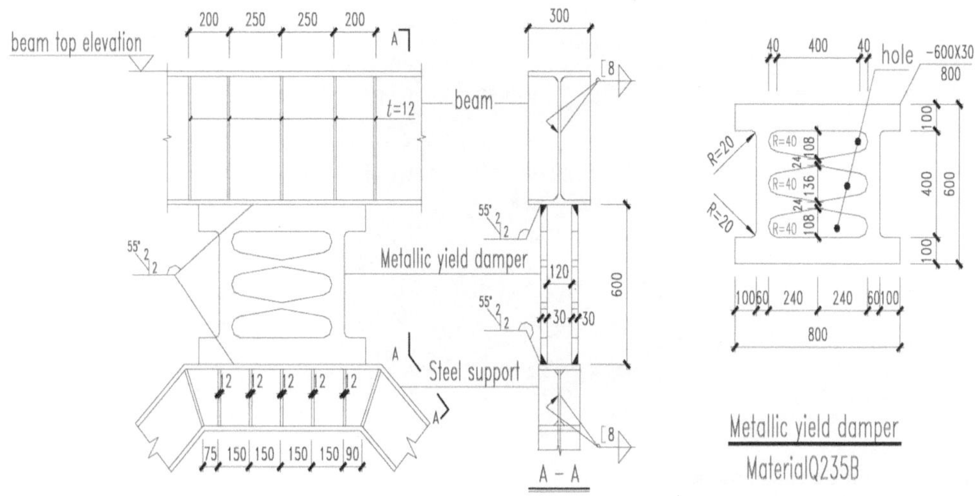

Figure 2. Node schematic diagram.

community widely accepted. A new type of metal yield dampers as energy dissipation devices developed (Li 2006, Li 2007) is used in this paper. Metallic yield dampers with steel support link constitute energy dissipation brace system. Energy dissipation support node as shown in Figure 2.

4 COMPUTATIONAL ANALYSIS

4.1 *Model*

In order to solve displacement reaction of steel structure equipped with energy dissipation brace under rarely earthquake displacement reaction, Finite element analysis model was built using ANSYS, also with APDL and embedded parameter language FORTRAN. Beams, columns are used Beam188 3D beam element, dead weight of the structure was simulated using Mass 21 element, and mass inertia force of the three directions of the structure of X, Y, Z was considered, the Metal damper was using Combin40 element. The frame beam column model considering steel elastoplastic, strengthen bilinear model simulation. The floor is concrete elastic material, to ignore the nonlinear deformation. Overall finite element model is shown in Figure 3; the damper beam connection node model is shown in Figure 4.

4.2 *Earthquake records*

Construction field assigned to Site-class II, fortification intensity 7. The calculation process selected two natural earthquake records on behalf of Site-class II venue and an artificial earthquake records (see Table 1). During the calculating process of the displacement response of the structure under rarely earthquake, peak acceleration is 220 gal, and vibrations were inputted along the X and Y directions.

4.3 *Modal analysis*

In order to verify the accuracy of the ANSYS finite element model, its modal analysis was compared with SATWE software analysis. When using ANSYS finite element software to calculate the modal, the material nonlinear behavior as an elastic material was not regarded. Table 2 gives the order of the top five for the fundamental period of the structure calculated using the ANSYS and SATWE software. The calculation results show that the closely

Figure 3. Finite element model.

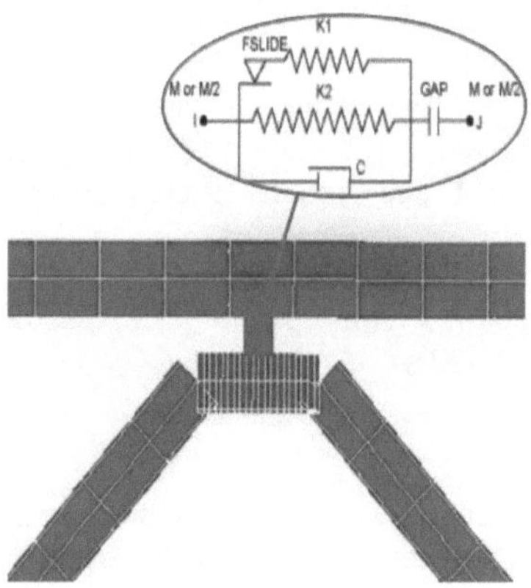

Figure 4. Finite element nodes figure.

Table 1. Earthquake records.

Seismic record number	Stations	Earthquake (time)	Components
1	0453 Taft Lincoln School	Kern county (1952/7/21)	Taf021(Y) TAF111(X)
2	LA–BALDWIN HILLS	Northridge (1994/1/17)	BLD090(Y)
3	–	Artificial wave	RGB090(Y) RGB360(X)

Table 2. Structural natural periods contrast.

Calculation model	1st	2nd	3rd	4th	5th
PKPM	1.52	1.11	1.06	0.55	0.35
ANSYS	1.47	1.09	0.99	0.58	0.39

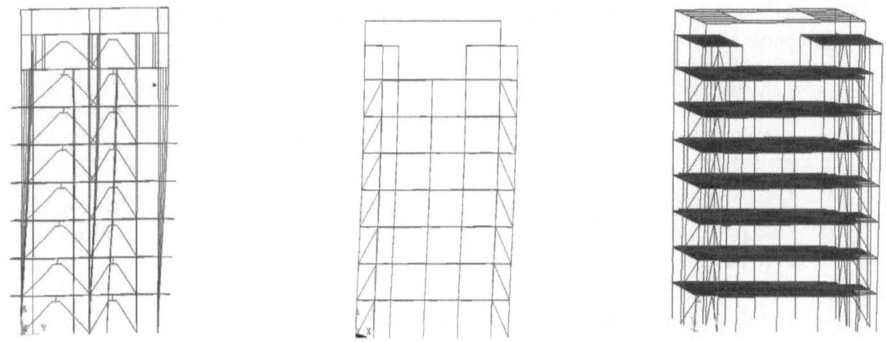

(a) 1st (b) 2st (c) 3st

Figure 5. The first 3 modal deformation figure under set energy dissipation structure.

Table 3. Under frequently earthquake the maximum story drift and story drift rotation of each layer.

Story	1	2	3	4	5	6	7	8
Story drift/mm	2.7	3.2	3.5	3.6	3.6	3.4	3.0	2.0
Story drift rotation	1/1945	1/1282	1/1139	1/1090	1/1084	1/1125	1/1222	1/1454

results of the two models of first five for the fundamental period of the structure, the visible, ANSYS the model with SATWE model, have a good consistency. ANSYS software analyzed first 3 modes deformation was as shown in Figure 5. As figure shows, first mode to Y to lateral displacement deformation, the second mode for the X direction lateral displacement deformation, the third modes torsional deformation.

4.4 *Earthquake response*

4.4.1 *Frequently earthquake*

The structural components and energy dissipation brace under frequently earthquake are in the elastic stage, ordinary steel structure design and analysis methods were utilized. Analysis result of SATWE software served as basis. Structure under frequently earthquake along the Y-direction maximum story drift and story drift rotation are shown in Table 3.

4.4.2 *Rarely earthquake action*

Structure equipped with energy dissipation brace steel frame under rarely earthquake show strong nonlinear. SATWE software can not consider the structure of the nonlinear, especially the non-linearity of the energy dissipation. Therefore, based on the results of finite element software ANSYS, Table 4 shows three kinds of seismic records under the layers of the structure along the Y direction maximum story drift and story drift rotation.

4.5 *Lateral-story drift ductility ratio*

"High steel code" mentioned the lateral-story drift ductility ratio concept, it is the ratio of elastic later-story drift to the maximum story drift, when the structural framework of eccentric support and its value shall not exceed 3.0. Table 5 shows the ordinary steel installed support and installation of energy dissipation brace steel structure along the Y-axis direction

Table 4. Under rarely earthquake the maximum story drift and story drift rotation along the Y direction.

Story	Story drift				Story drift rotation
	Seismic 1	Seismic 2	Seismic 3	Average value	
8	3.7	4	7.0	4.9	1/796
7	4.4	3.6	8.5	5.5	1/709
6	5.4	4.6	10.9	7.0	1/560
5	6.3	5.4	12.8	8.2	1/478
4	6.8	6	13.8	8.9	1/440
3	6.9	6.3	14.1	9.1	1/429
2	6.4	6	12.8	8.4	1/464
1	4.3	4.1	8.9	5.8	1/676

Table 5. Lateral-story drift ductility ratio of ordinary steel installed support and installation of energy dissipation brace for steel structure.

Story		1	2	3	4	5	6	7	8
Ordinary support	Story drift (frequently earthquake)	1/1797	1/1272	1/1187	1/1177	1/1194	1/1261	1/1399	1/1711
	Story drift rotation (rarely earthquake)	1/536	1/330	1/280	1/261	1/259	1/272	1/298	1/352
	Lateral-story drift ductility ratio	3.4	3.9	4.2	4.5	4.6	4.6	4.7	4.9
Energy dissipation brace	Story drift (frequently earthquake)	1/1945	1/1282	1/1139	1/1090	1/1084	1/1125	1/1222	1/1454
	Story drift rotation (rarely earthquake)	1/676	1/464	1/429	1/440	1/478	1/560	1/709	1/796
	Lateral-story drift ductility ratio	2.7	2.7	2.8	2.7	2.5	2.3	2.0	2.1

maximum story drift rotation, and lateral-story drift ductility ratio in the frequently and rarely earthquake. The data in the table show that, install the ordinary steel support structure lateral-story drift ductility ratio obviously does not meet the "high steel code" requirements, installation of energy dissipation brace can greatly reduce the story drift of the structure under rarely earthquake, structural lateral-story drift ductility ratio decreases too, its value to meet the limits specified in the "high steel code".

5 ECONOMIC ANALYSIS

This project uses the energy dissipation brace to solve problem of the Lateral-story drift ductility ratio does not meet the requirements, is a new attempt in the seismic design of the project to a better understanding of the effectiveness and the economy for this method, enlarging the cross-section dimensions of columns and supporting scheme for comparative analysis. Table 6 shows a comparison of the amount of steel, the data in the table can be seen, the application of energy dissipation support scheme is better than the traditional methods

Table 6. Comparison of the amount of steel structure of the first floor.

Member	Brace	Column
Increase section	H350 × 200 × 20	□500 × 30
Set energy dissipation brace	H250 × 200 × 18	□450 × 28
Save the amount of steel %	22	16

in solving the Lateral-story drift ductility ratio problems, which has significant economic on the amount of steel is significantly reduced.

6 CONCLUSIONS

Story drift of steel structures with ordinary steel support are often too large under rarely earthquake, which make the lateral-story drift ductility ratio does not meet the "high steel code" limits requirements in structural design process. In this paper, a training center for steel works was used to served as an study example, to solve Lateral-story drift ductility ratio of the energy dissipation brace does not meet the requirements of the tall steel structure design process, in the design process to solve the energy dissipation brace frame beams connection between the nodes, and earthquake analysis technology issues. The analysis results show that the energy dissipation brace can effectively reduce the displacement response of the structure under seismic action, and economic, practical reference for similar projects.

REFERENCES

Hen Xiaofeng, Deng kaiguo & Hao Jiping. Discussion on scismic design of steel structures, Building structure, J. Supp. Vol. 39, 2009, pp. 465–471.

Li Gang & Li Hong-nan. Research on the performance of a new type of mild steel damper, Journal of vibration and shock, Vol. 25, No. 3, 2006, pp. 66–73.

Li Gang, Li Hong-nan & Li Zhong-jun. Mild steel damper seismic design of reinforced concrete frame structure "dual function", Journal of Building Structures, J. Vol. 28, Apr. 2007, pp. 36–44.

Li Guoqiang & Sun Feifei. Some problems and suggestion on seismic design of high-rise steel structures. J. Earthquake engineering and engineering vibration, Vol. 26, Jun. 2006, pp. 108–114.

Liu Jie-ping, Li Xiao-dong & Zhang Ling-xin. Summary of seismic research of mega steel structure, World earthquake engineering, J. Vol. 20, Dec. 2004, pp. 42–46.

National Standard of the People's Republic of China, Technical specification for steel structure of tall building, JGJ. 99 - 98. China Architecture & Building Press, Beijing 1998.

National Standard of the People's Republic of China, Code for Seismic Design of Buildings, GB 50011 - 2001, China Architecture & Building Press, Beijing 2001.

National Standard of the People's Republic of China, Code for Seismic Design of Buildings, GB 50011 - 2001, China Architecture & Building Press, Beijing 2010.

Peng Guanshou & Gao Xuanneng. Performance-based seismic design theory and method for steel structures, Steel structure, J. Vol. 22, Jan. 2007, pp. 49–54.

Shen Zuyan & Sun Feifei. Discussion and recommendation on seismic design for steel structures, Building structure, J. Vol. 39, Nov. 2006, pp. 115–122.

Xu Hai-bo, Wang Guang-jian, Yi Fang-min,. etc. Comparison and analysis of seismic resign of steel structure between Chinese code and American code, J. Vol. 32, Apr. 2010, pp. 81–86.

Xiong Jun, Wang Yuan-qing, Shi Yong-jiu & Wang Bin-bin. Seismic Analysis of Eccentrically Braced High-rise Steel Frames, Journal of Zhengzhou University (Engineering Science), J. Vol. 30, Sep. 2009, pp. 1–4.

Modeling and Computation in Engineering II – Xie (ed)
© 2013 Taylor & Francis Group, London, ISBN 978-1-138-00058-2

Finite element inversion computation for surrounding rock mass parameters of large-span shallow-buried highway tunnel

Zhizheng Yin & Jiasheng Zhang
School of Civil Engineering, Central South University, Changsha, Hunan, China

ABSTRACT: There is no clear and practical code to guide the design of large-span shallow-buried highway tunnel, for the main reason of this it is difficult to determine the surrounding rock mass parameters. In this paper, based on the elastic-plastic theory and the displacement of the field loading tests, the calculation model of the displacement back analysis was set up. The theory of the finite element inversion computation was used and the four basic rock mass parameters were calculated by the finite element inversion theory. Then the four basic rock mass parameters were used in the finite element computation model to calculate the vault settlement of the each step of the excavation. The results of calculation agree with the measured values, which show that the identified surrounding rock mass parameters are credible.

1 INTRODUCTION

With the rapid development of the economy, the scale of highway construction is increasingly extended, and it has promoted the development of highway tunnel construction. Large-span shallow-buried highway tunnel in the form of distinctive characteristics caused a series of problems in engineering, such as support parameters, support patterns, construction methods, and stability of surrounding rocks. So there no practical, clear codes for guided to construct large-span shallow-buried highway tunnel.

The conventional design method of tunnel needs to know the magnitude of the load, and the load generally measured by the field loading tests. But field loading tests will consume a large amount of manpower, material resources and time, so the inversion computation method for load was presented and developed. How to obtain reasonable mechanical parameters are the key factors of load calculation. With the developing of the construction scale and complexity, surrounding rock properties, loading history and boundary conditions become more and more complex in tunnel engineering. As description by several authors (Baili 1990, Fakehimi 2004), relying solely on classical soil mechanics theory is insufficient to solve these problems, so the numerical method in tunnel engineering was promoted. From 1960s, finite element method in tunnel engineering field got rapid development, but determining calculation parameters in finite element method is very difficult. The parameters directly affect the calculation results of the reliability and accuracy, so it is necessary to use reasonable methods to obtain the parameters of tunnel surrounding rock mass.

A lot of research works (Hisatake 1995, Linde 1996, Fakehimi 2004, Hao 2007, Yunpeng 2008) was done on how to determine the parameters of surrounding rock mass. The common method is to use indoor and outdoor test or displacement back analysis calculation to get the required parameters of surrounding rock mass. For the displacement back analysis calculation, there are many optimization iteration process such as simplex method, simulated annealing method, complex method, conjugate gradient method, genetic algorithm, and penalty function method used by Hehua 2003, Zhaoli 2003, Chuanhua 2005, and Peilin 2008.

Based on the elasto-plastic finite element theory, the displacement back analysis method to determine parameters large-span shallow-buried highway tunnel surrounding rock mass was put forward. In this method, firstly the each test point displacement on the surface of the bedrock was measured by the bedrock surface static load test, then application of elastic-plastic finite element back analysis theory, selection of simplex method, genetic algorithm and simulated annealing algorithm for optimization iteration process is integrated to derive parameters of surrounding rock mass.

2 ROCK MASS FIELD LOADING TESTS

2.1 *Test design*

Yonglong tunnel is a bidirectional six lane large-span highway separated tunnel connecting the Yonghe Development Zone and Zhenlong town. There is about 300 m v class shallow-buried surrounding rock mass in the left line of the Yonglong tunnel. The complex geological conditions, developed fissures, poor rock integrity and heavy unload weathered rock have brought many engineering geological troubles, such as the surrounding rock stability of underground works, stability of ground works, foundation bearing conditions and under unsymmetrical pressure. The fissures and joint were filled with wet clay, and the wet clay was turn to mud under the action of the underground fissure water. Due to the poor rock mass integrity, using conventional method to determine the mechanical properties of rock mass parameters are extremely difficult, so the field loading tests were used to determine the mechanical parameters of rock.

Considering the actual field condition, three test points were selected in v class of the left line. All these test points are located at the bottom of the undisturbed bedrock. Devices of load test were consisting of a counterforce frame, a jack, a diameter of 300 mm circular loading plate and some dial indicators (Figure 1a), and the test points were installed about 15 dial indicators for measure the displacement, wherein 4 for horizontal displacement (Figure 1b).

2.2 *Test results*

Through the counterforce frame and jack, multi-stage loading was realized on the circular loading plate. The multi-stage loading show as pressure is 707 kPa, 1175 kPa, 1651 kPa, 2123 kPa, 2592 kPa and 3063 kPa, the load continue increase till the bedrock turn into the state of plastic yield. When the loading to 2592 kPa, the bedrock at the 1# test point turn into the state of plastic yield, the 2# test point is 3063 kpa, and the 3# test point is 3063 kpa. The P-S curve of these states was shown in Figure 2. The plastic deformation was generated when the surrounding rock turn into the state of plastic yield. The load corresponding test point data for the surrounding rock parameters will be used in inversion computation.

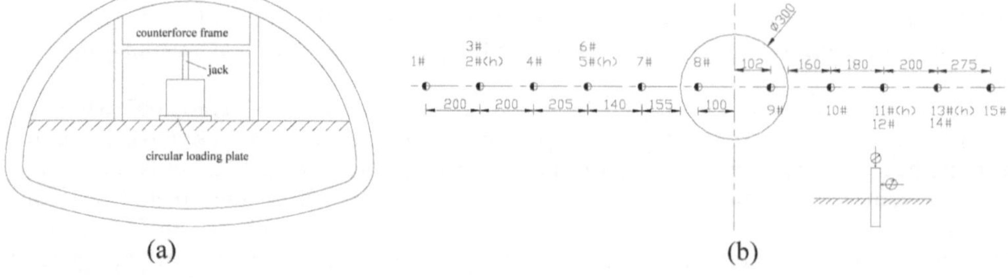

Figure 1. Device of load test on the tunnel bedrock
(a) Device of load test, (b) Test points arrangement. unit: mm.

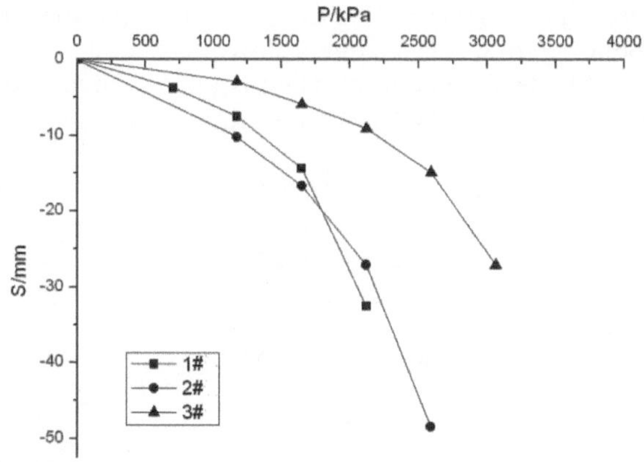

Figure 2. P-S curve of the load plate.

3 INVERSE CALCULATION MODEL

The rock mass was consider as isotropic elastic-plastic material, in which Drucker-Prager criterion is taken as the yield function, hardening and softening model is adopted with a plastic potential function to Drucker-Prager criterion write as Q, which can be written as

$$Q = \frac{2\sin\phi}{\sqrt{3}(3-\sin\phi)} I_1 + \sqrt{J_2} - \frac{6c\cos\phi}{\sqrt{3}(3-\sin\phi)} = 0 \tag{1}$$

where I_1 and J_2 are the stress invariants; and ϕ is friction angle.

Because the stress strain relationship of plastic deformation, no relationship based on the total deformation theory of plasticity can be established by the Hooke's law. In order to facilitate the finite element interative calculation, associated flow plastic rule was adopted, that is $Q = f$, then:

$$\{d\varepsilon\} = \{d\varepsilon^e\} + \{d\varepsilon^p\} \tag{2}$$

$$\{d\varepsilon\} = D^{-1}\{d\sigma\} + \lambda\frac{\partial Q}{\partial\sigma} \tag{3}$$

$$\{d\varepsilon\} = D^{-1}\{d\sigma\} + \lambda\frac{\partial f}{\partial\sigma} \tag{4}$$

So the stress strain relation using increment theory elastic plastic can be expressed as

$$\{d\sigma\} = \left[D - (1-\lambda)\frac{D\left\{\frac{\partial Q}{\partial\sigma}\right\}\left\{\frac{\partial f}{\partial\sigma}\right\}^T D}{\left\{\frac{\partial f}{\partial\sigma}\right\}^T D\left\{\frac{\partial Q}{\partial\sigma}\right\}} \right]\{d\varepsilon\} \tag{5}$$

where D is a constant elasticity matrix; ε^e is elastic strain; ε^p is plastic strain; and λ is a positive value can be show as

$$\lambda = -\frac{f_0}{f_1 - f_0} \tag{6}$$

where f_0 is yield function value corresponding to the elastic initial stress state; and f_1 is yield function value corresponding to the plastic initial stress state.

Objective function of the displacement inversion is the minimum modules of the difference about the calculated values and the test values.

$$f = \sum_i \|\bar{u}_i - u_i\| \tag{7}$$

$$\min_{x \in D} f = \min \sum_i \|\bar{u}_i - u_i\| \tag{8}$$

The constraint conditions of the objective function are:

$$g(x) \leq 0 \tag{9}$$

$$x = (E, \mu, \ldots)^T \tag{10}$$

where $\bar{u}_i$ is the displacement of i point in FEM; and u_i is the test displacement of i point, the unknown parameters of rock mass by elastic-plastic finite element model as the variable, using the interative calculate method to search to get the convergent solution.

The initial values of parameters selection on the displacement back analysis is very important, in this paper, base on the geologic survey report and the previous engineering experience, the initial values of the rock mass parameters are given in table 1.

According to the characteristics of the tunnel, the finite element analysis model is simplified as a plane strain model. For the convenience of computation, the displacements of two stages loading before failure in compression test were used in the finite element inverse analysis. Based on the elastic-plastic finite element inversion theory, the plane strain back analysis model under the point load was established. The rock mass was assumed as isotropic elastic-plastic material, and was divided into 3286 triangular plane strain calculation unit, at the same time, the tectonic stress was not considered in calculation, only the self-weight stress was considered.

4 RESULTS OF THE INVERSE CALCULATION

The initial value of calculate parameter was substituted into the finite element back analysis. The optimal solutions of rock mass parameters were calculated by using the simplex method, simulated annealing, and genetic algorithm, respectively. The inverse calculation results of the 1#, 2#, 3# test point displacement as shown in Figure 3, Figure 4, and Figure 5, and the inverse computation rock mass parameters were shown in table 2.

Table 2 show the coefficient of variation of the three test point's inverse calculation is small. The V class surrounding rock mass parameters were get by the mean values, namely $E = 1853.6$ MPa, $\mu = 0.302$, $c = 0.164$ MPa, $\varphi = 30.7°$.

Table 1. Initial values of the rock mass parameters for FEM inverse computation.

Calculation parameters	Symbol	Unit	Value range	Initial value of calculation
Elastic modulus	E	MPa	1200~2200	1500
Poisson's ratio	μ	/	0.2~0.35	0.25
Adhesion	C	MPa	0.01~0.5	0.25
Internal friction angle	ϕ	°	25~45	30
Rock density	γ	kN/m³	22	22

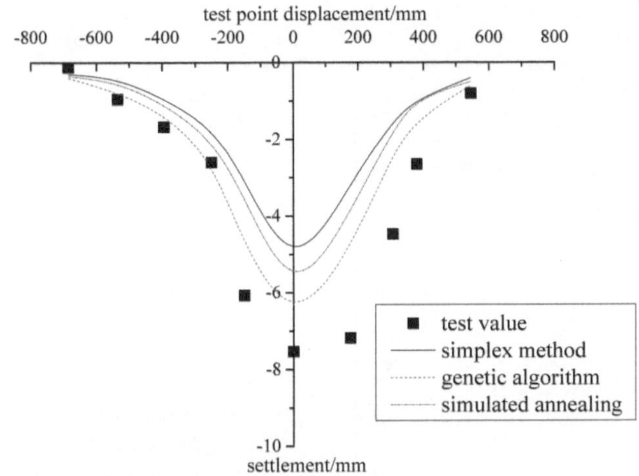

Figure 3. FEM inverse computation results of No. 1 test point.

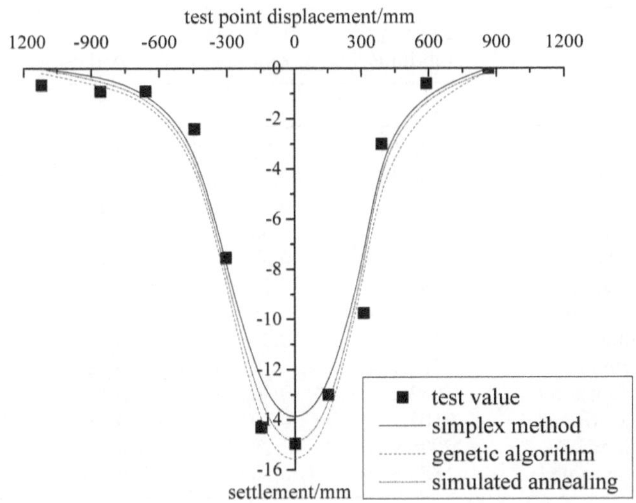

Figure 4. FEM inverse computation results of No. 2 test point.

5 PARAMETERS VERIFICATION

In order to verify the accuracy of the parameters by back analysis, substitute these parameters to the finite element computation model to calculation the vault crown settlement. The scheme of the tunnel's FEM computation steps as shown in figure 6. The computed values and measured values of the vault crown settlement as shown in table 3.

Computed values and measured values of the vault crown settlement are close to each other for each step, and have the same trend, it show the parameters of inverse analysis are credible.

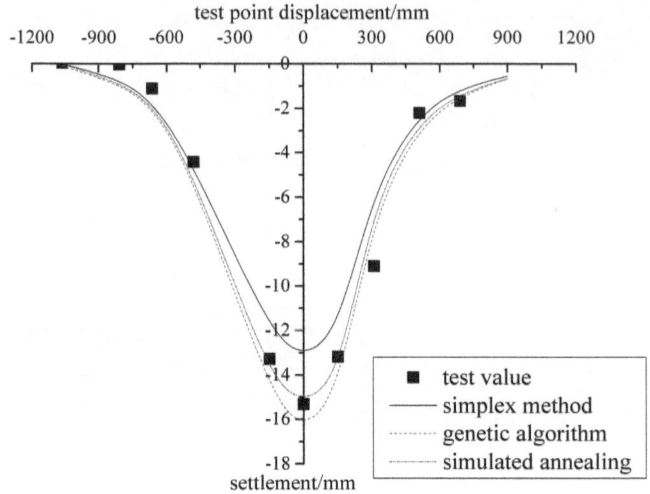

Figure 5. FEM inverse computation results of No. 3 test point.

Table 2. Value of FEM inverse computation rock mass parameters.

Test point and methods		Elastic modulus	Poisson's ratio	Adhesion	Internal friction angle
		MPa	/	MPa	°
1#test point	Simplex method	1825.2	0.29	0.15	28.2
	Simulated annealing	1830.0	0.30	0.16	31.1
	genetic algorithm	1795.0	0.31	0.15	29.7
	Average value	1816.7	0.30	0.153	29.6
2#test point	Simplex method	1628.5	0.32	0.16	25.6
	Simulated annealing	1644.3	0.30	0.17	23.3
	Genetic algorithm	1670.0	0.31	0.15	27.9
	Average value	1647.6	0.31	0.16	25.6
3#test point	Simplex method	2152.0	0.29	0.18	36.4
	Simulated annealing	2033.2	0.30	0.17	39.3
	Genetic algorithm	2105.7	0.30	0.185	35.5
	Average value	2096.9	0.296	0.178	37.1
	Total average value	1853.6	0.302	0.164	30.7

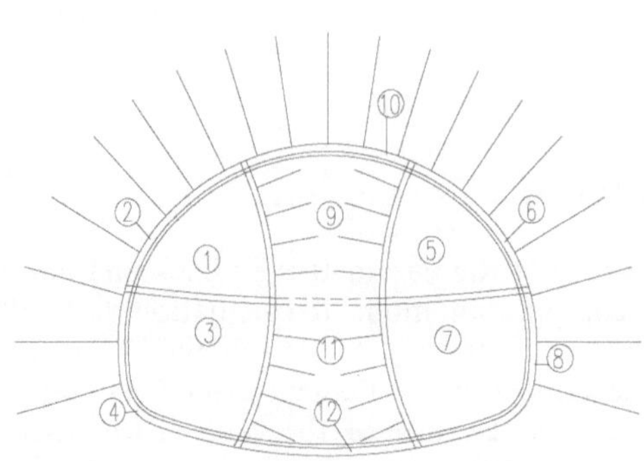

Figure 6. Scheme of the tunnel's FEM computation steps.

Table 3. Computed values and measured values of the vault crown settlement.

Step of calculations	Calculated values mm	Measured values mm
1	7.32	6.79
3	16.74	15.81
9	24.76	23.15
13	26.03*	24.59*

*Demolition the temporary steel arch.

6 CONCLUSIONS

Based on the elastic-plastic foundation theory and the displacement of the field loading tests, the calculation model of the displacement back analysis was set up. Then the optimal solutions of rock mass parameters were calculated by using the simplex method, simulated annealing, and genetic algorithm, respectively.

Substitute the parameters from the back analysis to the finite element computation model to calculation the vault crown settlement. The measured vault crown settlements appear to be reasonably close to those determined from the computed values. It shows the parameters of inverse analysis are credible.

These calculations have some application value and theoretical significance for the design of the large-span shallow-buried highway tunnel.

ACKNOWLEDGEMENTS

The paper supported by the National Natural Science Foundation of China (41272324).

REFERENCES

Baili, ZHU, & Zhujiang, SHEN. 1990. *Calculation of soil mechanics*. Shanghai: Shanghai Science and Technology Press.

Chuanhua, XU, & Qingwen, REN. 2005. Displacement back analysis based on support vector machine and simulated annealing. *Chinese Journal of Underground Space and Engineering* 24(22): 4134–4138.

Fakehimi, Salehi & Mojtabai. 2004. Numerical back analysis for estimation of soil parameters in the Resalat Tunnel project. *Tunneling and Underground Space Technology* 19(1): 57–67.

Hao, WU, & Zixin, ZHANG. 2007. Application of the theory of optimization back analysis in back analysis of tunnel surrounding rock parameters. *Chinese Journal of Underground Space and Engineering* 3(6): 1162–1167.

Hehua, ZHU, & Xuezeng, LIU. 2003. Comparison study of mixed optimal methods based on genetic algorithm in back analysis. *Chinese Journal of Underground Space and Engineering* 22(2): 197–202.

Hisatake. 1995. Back analysis in tunneling. *International Journal of Rock Mechanics and Mine Science & Geomechanics* 32(3): 117A.

Linde, YANG. 1996. *Back analysis theory and engineering practice of geotechnical problems*. Beijing: Science Press.

Peilin, NING. 2008. *Bearing capacity analysis and design optimization of the multi-arch tunnel lining*. Guangzhou: Guangdong University of Technology.

Yunpeng, LI, & Changling, HAN. 2008. Back analysis on mechanical parameters of surrounding rock of tunnel with small spacing using orthogonal design. *Journal of Highway and Transportation Research and Development* 25(9): 107–111.

Zhaoli, WANG. 2003. *Study on rock mechanics parameter back analysis and construction mechanics problem in underground chamber*. Nanjing: Hehai University.

Modeling and Computation in Engineering II – Xie (ed)
© 2013 Taylor & Francis Group, London, ISBN 978-1-138-00058-2

Research on numerical simulations of structure reasonableness for GINA water stops in immersed tunnel

Zhi-nan Hu, Yong-li Xie, Xia-bing Yue, Hong-guang Zhang & Sheng-lin Bin
College of Highway, Chang'an University, Xi'an Shanxi, China

Guo-ping Xu
CCCC High way Consultants Company Limited, Beijing, China

ABSTRACT: Through comparing the structures of different type GINA water stops, three main structure differences between the original GINA water stop and improved GINA water stops were found. Building finite element models and comparing the contract stress and deformation laws between the original one and the improved ones under different compressions, the mechanical characters and water tightness caused by structure changes could be analyzed,which could provide supports for GINA optimum design and engineering selection.

1 INTRODUCTION

In 1966, GINA water stop was first used in the construction of Rotterdam north-south subway line in Netherlands. It is possible for the flexible waterproof of joints and hydraulic compression method by using GINA water stops. To some extent, GINA water stop promoted the development of immersed tunnel. What's more, with the development of the immersed tunnel, GINA water stops have also been improved. Primary GINA water stops composed by top tip rib, body, bottom rib and flange (Wang Meng-shu 2010). The application field of this GINA water stop was very wide, which has been used in the zhujiang immersed tunnel and the construction of immersed tunnel of Hong Kong-Zhuhai-Macau bridge engineering in China. The improved GINA water stops are shown in Fig. 1: TRELLEBORG type was researched by Dutch company; PHOENIX type was developed by German company; The others were exploited by Japanese companies (Xue Yong 2005, Xu Jun 2005). The stress and deformation laws of TRELLEBORG type GINA water stop had been analyzed by finite element software, under the working conditions of axial compression, shearing compression, long-term creep and stress relaxation in the work of Huang Fan (Huang Fan 2010). It had showed that this kind of water stop fulfill the long-term water waterproof requirements. Liu Zheng-gen (Liu Zheng-gen 2009, 2011) carried on the residual amount of compression study and three dimensional numerical research for the improved type GINA water stop in Yongjiang immersed tunnel, the contact stress was got and the safety grade of this type GINA water stop was established.The sealing effect of the traditional type and the TRELLEBOGR type were analyzed and some inferences were made in the work of Guan Min-xin (Guan Min-xin 2004). Even though several types of GINA water stops have some applications, there are not enough proofs to prove which kind of GINA water stop has more reasonable structure. This paper compared the deformation characteristics of various types GINA water stops under different compressions, in order to find the basis of structural optimization to GINA water stops.

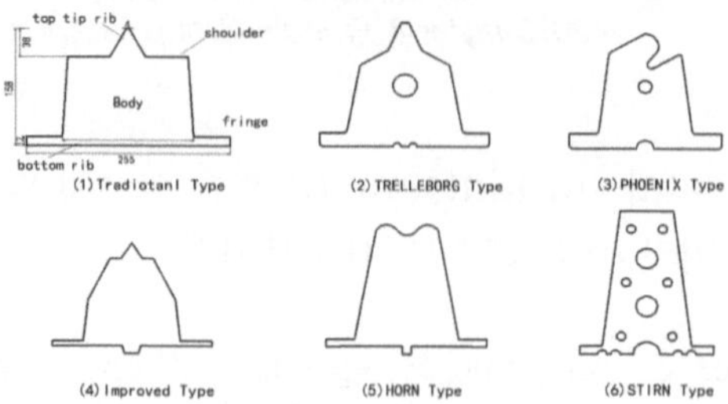

Figure 1. GINA water stops in foreign (mm).

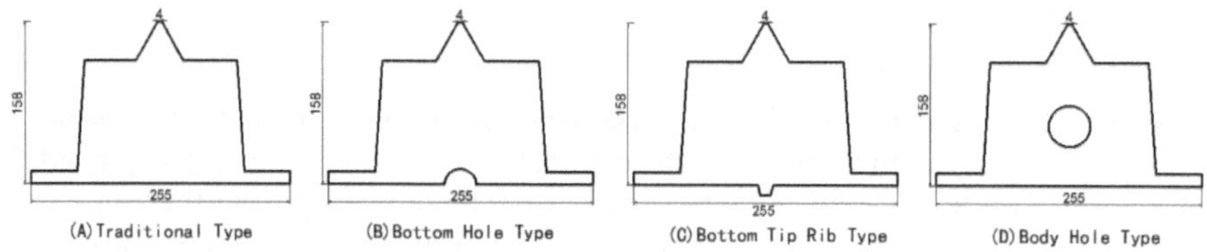

Figure 2. GINA water stops used in research (mm).

2 RESEARCHING TYPES

It can be easily found from the structures of these GINA water stops, sharp ribs play a role in preliminary stage of tube contacting and prepare for the hydraulic compression; the body rubbers undertake main lateral water pressure so that they are keys to long-term sealing capacity of GINA. Bottom ribs generally use soft rubber in order to contact with the bottom floor completely; Flanges mainly play a role of auxiliary construction. Therefore, comprehensive studying of several GINA water stops, the GINA water stops optimization mainly exist in following three aspects: ①Setting tip ribs bellow the bottom ribs. ②Making holes among the body. ③Disposing semi-circular holes under bottom ribs. Therefore, the traditional GINA water stops and the other three type GINA the water stops (fig. 2) are taken as the main object of study.

3 PARAMETER SELECTION AND MODEL BUILDING

3.1 *Rubber material parameter selection method*

There are two methods to obtain rubber parameters. One is the analytical method which combines constitutive relationship with the strain energy density function, inputting the uniaxial tension and compression or shear test data (Zheng Ming-jun 2001, Wang Wei 2004, Huang Jian-long 2008), then the material constants $a_{10,}$ a_{01} could be got, the details can be seen in the references.

GINA water stops are mainly undertaking compression stress in the course of working. Taking the compression test of rubber pad which Shore hardness is 55° and the uniaxial compression stress vs. strain curve is shown in Fig 3. Super-elastic material constitutive relationship:

$$t_{ij} = \frac{\partial W}{\partial I_1}\frac{\partial I_1}{\partial \gamma_{ij}} + \frac{\partial W}{\partial I_2}\frac{\partial I_2}{\partial \gamma_{ij}} + \frac{\partial W}{\partial I_3}\frac{\partial I_3}{\partial \gamma_{ij}} \tag{1}$$

where *tij* = Kirchhoff stress tensor; *rij* = Green strain tensor.

Two Parameter Mooney-Rivlin strain energy density function:

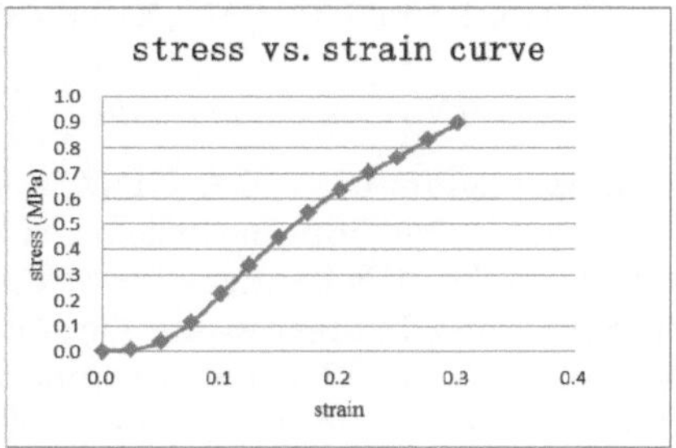

Figure 3. Stress vs. strain curve of rubber.

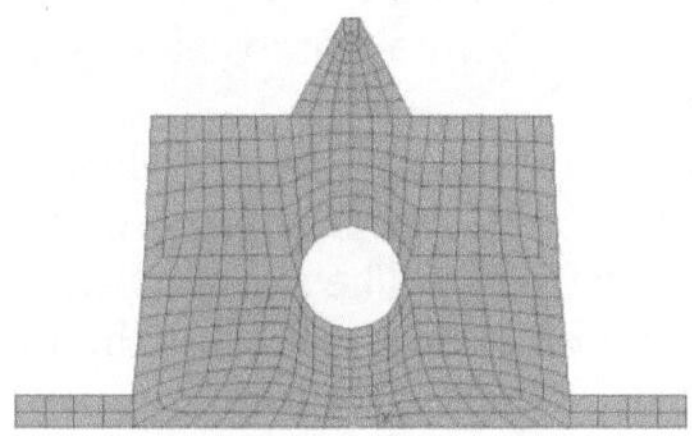

Figure 4. Mesh model of body hole type GINA water stop.

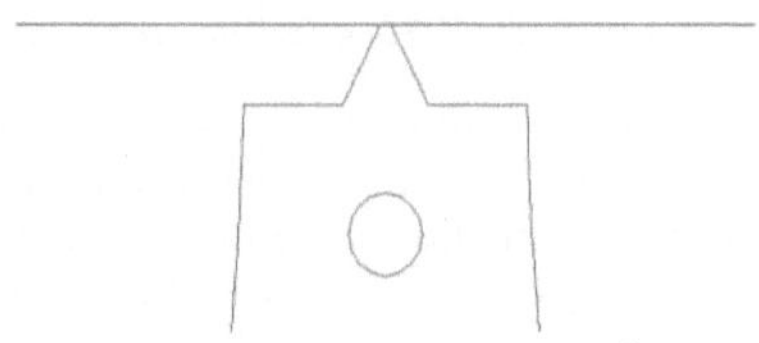

Figure 5. Contact pairs of body type GINA water stop.

$$
\begin{aligned}
W &= a_{10}(I_1 - 3) + a_{01}(I_2 - 3) \\
I_1 &= J^{-2/3}(\lambda_1^2 + \lambda_2^2 + \lambda_3^2) \\
I_2 &= J^{-2/3}(\lambda_1^2\lambda_2^2 + \lambda_2^2\lambda_3^2 + \lambda_3^2\lambda_1^2) \\
I_3 &= J^{-2/3}(\lambda_1^2\lambda_2^2\lambda_3^2) = 1
\end{aligned}
\tag{2}
$$

J is the volume ratio of initial position vs. final position; *I1, I2, I3* is the first, second and third strain invariants. λ_i is the i-direction stretching ratio; a_{10}, a_{01} is the material constants[8–10].

Another method is inputting test data into the finite element software ANSYS which can calculate the constant. Rubber is the almost incompressible material, so Poisson ratio is taken as 0.499.

3.2 *Construction of model*

The model was meshed by hyper56 super elastic element which was defined by rubber material parameters. Adopting target169 and contat171 elements to simulate the contact between models, the "hard-soft" contact pairs were set among roof, bottom floor and GINA water stops and the "soft-soft" contact pairs were set at the hole of body rubber. As the main role of the flange rubber was securing, the X and Y constrict was applied at the flange of the water stop. The mesh model and contact pairs of the body hole type GINA water stop is showed in fig. 4 and fig. 5.

4 ANALYSIS OF CALCULATION RESULTS ON AXIAL COMPRESSION STATE

4.1 *Analysis of traditional (type A) GINA water stop results*

Analyzing the equivalent strain nephogram of type A GINA water stop can get results as follows: When compression amount is small, bottom rib deformation is little so that the construction factors have large effects on the GINA water stop. Stress transfer of type A GINA water stop is basically like a reverse arrow. In other words, the stress firstly transfers from tip rib to the bottom floor center and two edges. Then with compression increasing, the stress gradually transfers to other parts of the bottom floor and the stress on the edge of the bottom rib is greater than that on the center. There is a relatively small deformation area between the position of the edge and the center. With the increasing of the compression, there will be a deformational concentration area on the body due to the large compression of tip rib. However, as the together contact deformation of the body shoulders and the tip rib, which makes the contact of GINA with roof more homogeneous. So this large deformation area does not affect the sealing of the roof.

Analyzing the contact stress diagram of traditional GINA water stop under different compressions can get following results: with the increasing of the compression, contact stress of tip rib growth in a straight line basically. The shoulder of body begin to contact after the compression reaches 40 mm, its contact stress is far less than that of tip rib. When compression is less than 60 mm, contact stress on the bottom floor center is close to that on the edge of bottom rib. When compression is up to 80 mm, the contact stress of the bottom edge is about two times than that of bottom rib center.

4.2 *Analysis of bottom hole type (type B) GINA water stop result*

Analyzing the equivalent strain nephogram in the compression of 70 mm and contact stress diagram of type B GINA water stop under different compression, when compression is 70 mm, it can be seen that, when the compression is less than 50 mm, the contact stress in the center of bottom rib is close to that on the edge; When the compression is more than 50 mm, deviation value between the contact stress on the center of bottom rib and that on the edge is gradually increasing; But the deviation value is smaller than that of type A, this shows that the contact effect of bottom floor is better than that of traditional type; The contact of water stop with roof is homogeneous, which is similar to that of traditional one.

4.3 *Analysis of bottom tip rib type (type C) GINA water stop result*

Analyzing the equivalent strain nephogram in compression of 70 mm and the contact stress diagram of type C GINA water stop under different compressions, it can be seen that, when the compression is small, the contact stress on the tip rib of bottom rib is bigger than that on the edge of bottom rib. This shows that, when the compression is small, top tip rib and bottom tip rib is the main sealing area. But the contact area of GINA with roof and bottom is small, which is unfavorable to the initial sealing of immersed tunnel. When the compression is up to 70 mm, the contact stress on the edge of bottom rib is close to that on the tip. But the contact stress on the sides of bottom tip rib is small, from the view of bottom contact, its sealing effect is worse than that of traditional one.

4.4 *Analysis of body hole type (type D) GINA water stop result*

Analyzing the equivalent strain nephogram in compression of 70 mm and the contact stress diagram of type D GINA water stop under different compressions, it can be seen that, the contact of water stop with roof is poorer than the other type GINA water stops and the distribution of stress is uneven. The sealing of bottom rib mainly depends on the edge and

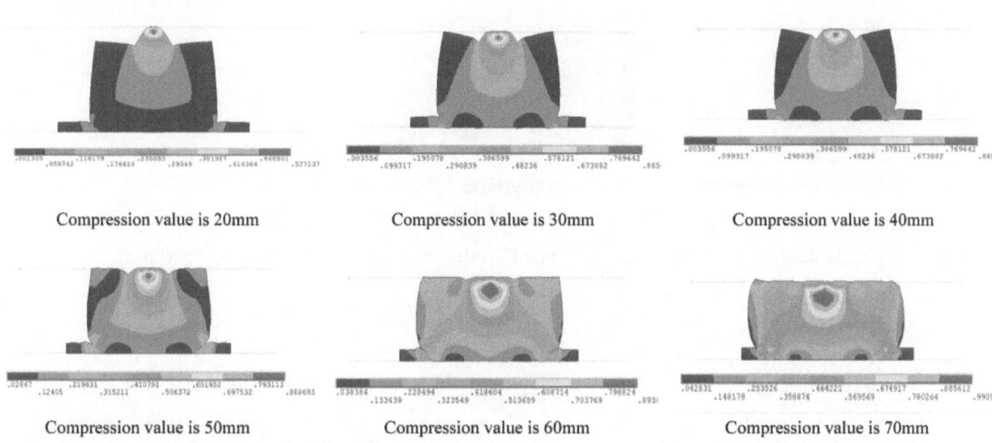

Figure 6. Equivalent strain nephogram of type A GINA water stop under different compression.

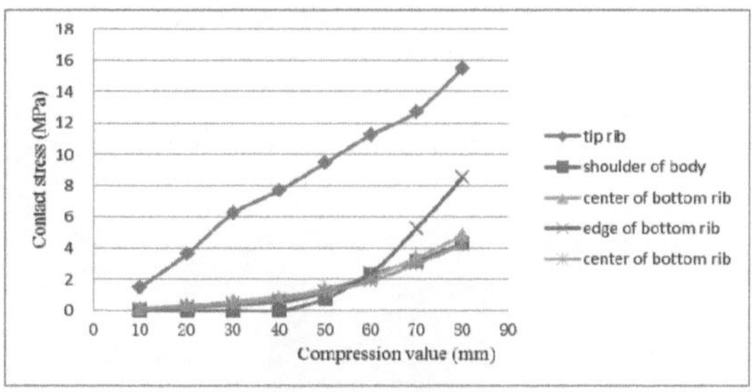

Figure 7. Contact stress of type A water stop under different compression.

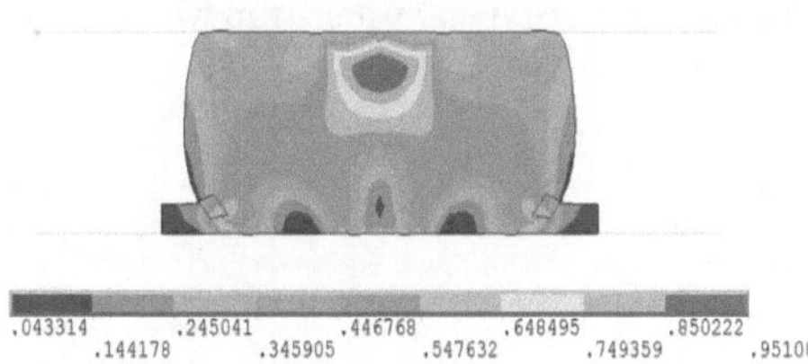

Figure 8. Equivalent strain nephogram of type B water stop when compression is 70 mm.

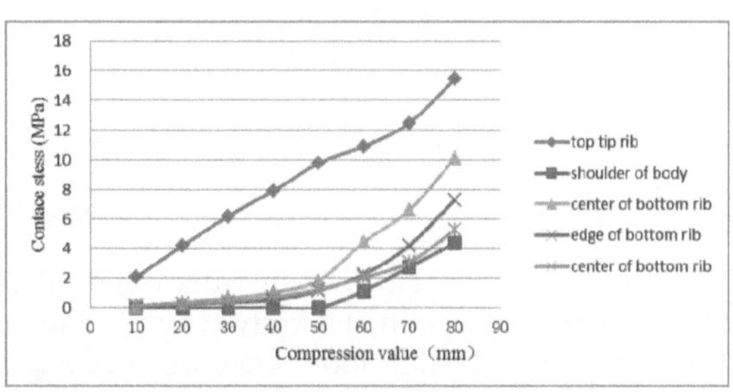

Figure 9. Contact stress of type B water stop under different compression.

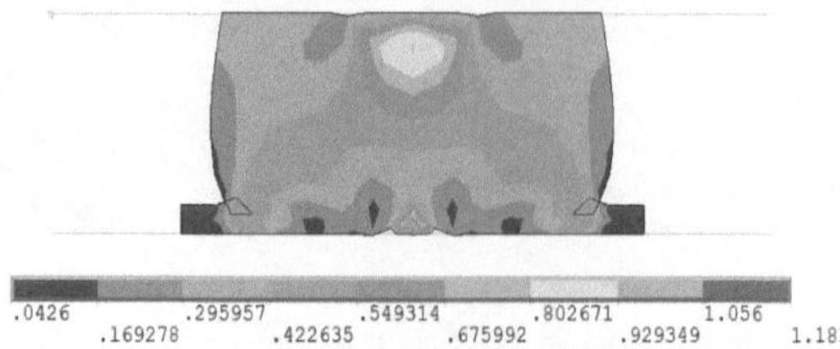

Figure 10. Equivalent strain nephogram of type C water stop when compression is 70 mm.

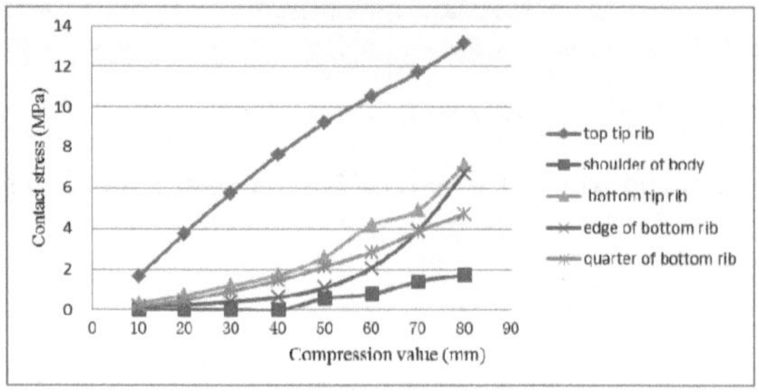

Figure 11. Contact stress of type C water stop under different compressions.

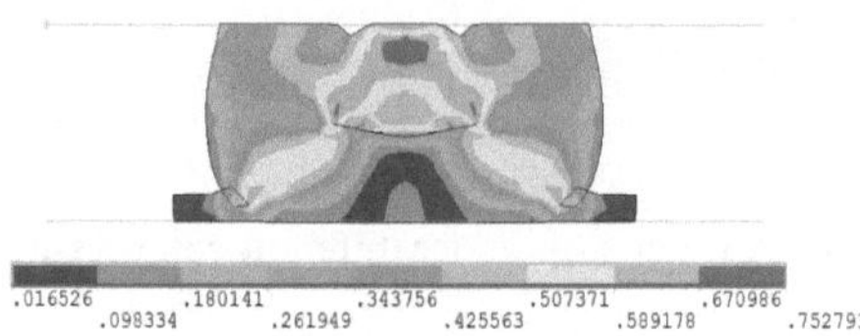

Figure 12. Equivalent strain nephogram of type D water stop when compression is 70 mm.

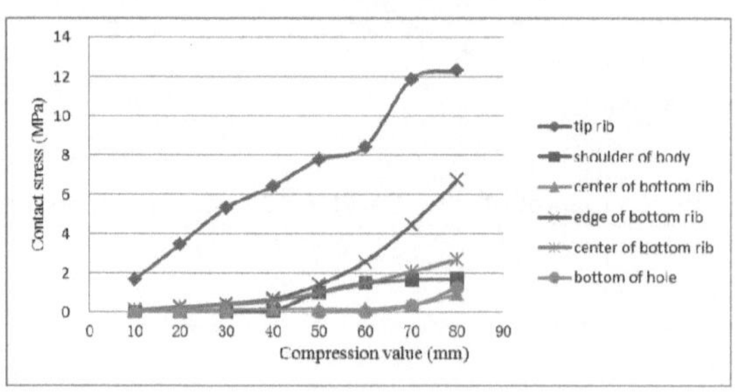

Figure 13. Contact stress of type D water stop under different compression.

the contact stress on the center of bottom rib is smaller than that of the other types. As the existence of body hole, deformation of tip rib and body is reduced under the same compression. Therefore, the body rubber can sustain more axial compression. With the increasing of compression, the contact of GINA water stop with roof and bottom floor is tending to uniformity.

5 CONCLUSION

1. Contact stress of bottom ribs on 4 type GINA water stops are small, indicating that the bottom ribs is the weak area of GINA water stops. The contact of type B with bottom floor is the best in the above mentioned types, it is proved that opening holes under bottom ribs can improve the waterproof performance of bottom ribs.
2. In the compression initial stage of water stops, type C GINA water stop mainly relies on the sealing of top and bottom tip rib, which contact area is small so that overturning and slipping are easy to happen. Tip ribs are essential to the sealing of roof and excessive contact stress may affect the durability of the rubber. It can be seen that type A have the maximum tip rib contact stress. However, the type C has lower tip rib contact stress as the together work of top and bottom tip rib, which is favorable for the long-term sealing of water stops.
3. Type D GINA water stops can improve the deformation properties of body and reduce rib tip stress concentration, which is advantageous to the long-term effects of GINA water stops, but its contact with roof is worse than the other types and its sealing of bottom rib is mainly depend on the edges.
4. From contact of the 4 type GINA water stops with the roofs, it can be seen that the body shoulder contact stresses of type C and type D are smaller than those on the other two types, so their roof sealing effect is worse in this view.

REFERENCES

GUAN Min-xin. 2004. Waterproofing of the tube sections and joints of river-crossing submerged tube tunnels. Modern Tunneling Technology 41 (06): 57–59.

HUANG Fan. 2010. Numerical Simulation Analysis of GINA Rubber Water Stop. Structural Engineers 26 (01): 96–102.

HUANG Jian-long, XIE Guang-juan, et al. 2008. FEA of Hyper Elastic Rubber Material Based on Mooney-Rivlin Model and Yeoh Model. Rubber Industry 55 (08): 467–471.

LIU Zheng-gen & HUANG Hong-wei. 2009. Performance Evaluation and Safety Pre-warning of GINA in Immersed tunnel. Chinese Journal of Underground Space and Engineering 5 (02): 347–353.

LIU Zheng-gen, HUANG Hong-wei & ZHANG Dong-mei. 2011. 3D Nonlinear Numerical Simulation on Immersed Tunnel Joint, Chinese Journal of Underground Space and Engineering 7 (4): 691–694.

WANG Wei & DENG Tao. 2004. Determination for Material Constants of Rubber Mooney-Rivlin Model. Special Purpose Rubber Products 25 (04). 8–10.

WANG Meng-shu. 2010. Tunneling and Underground Engineering Technology in China. Beijing: China Communications Press: 721–722.

XUE Yong. 2005. Technical Progresss of the Tube Tunnel. Special Structures 22 (01): 70–72.

XU Jun, LAN Jin, et al. 2009. Water Proofing Technology of Immersed Tunnel. Communications Standardization. 206 (19): 156–161.

ZHENG Ming-jun & ZHENG Ji-long. 2001. Finite Element Analysis of Large Deformation of Compressed Rubber Component. Journal of Northern Jiaotong University 25 (01): 76–79.

Modeling and Computation in Engineering II – Xie (ed)
© 2013 Taylor & Francis Group, London, ISBN 978-1-138-00058-2

Study on strength characteristics of critical state for expansive soil

Wei Fu & Jinqing Lv
China Communications the Second Highway Consultants Co., Ltd., Wuhan, China

Bin Zhao
Yunnan Highway Development & Investment Co., Ltd., Kunming, China

ABSTRACT: The stress-strain characteristics of saturated soil are researched systematically under different confining pressure, initial dry density, initial water content, shearing rate and drainage condition. The inherent unity of diversity of shearing strength for the same samples measured by different experimental methods is indicated according to the normalization of critical state test results. And the failure lines in space of remoulded saturated expansive soil under consolidated drained and undrained condition are attained.

1 INTRODUCTION

Undrained strength indices are often used in analysis of strength and stability for the difficult in the determination of excess pore water pressure in practical engineering. The critical strength should be adopted for design basis (Wood, 1990). The K0 compression curves of normally consolidated soil and critical state lines can be simplified to be parallel lines in coordinate system. The stress ratio of critical state met certain relationships in triaxial compression and tension condition (Scofield, 2006). There was a unique relationship, which did not change with stress path, among of normally consolidated clay and weak consolidated remoulded clay (Li, 2012).

2 TEST METHOD

Test apparatus is an automatic triaxial shearing apparatus which is produced by KTG company. The apparatus equips with programmable stepless variable speed control and stepping motor loading. Samples are selected at Mengzi County in Yunnan province. The content of montmorillonite, which is higher, is about 22% to 48%. The free swelling ratio of samples is about 30% to 60%. The highest is up to 130%.

3 ANALYSIS OF TEST RESULTS

3.1 *Effect of confining pressure*

The relationship between stress and strain is presented in Fig. 1. The bigger the confining pressure is, the bigger the shearing strength is. The sample which has bigger shearing strength is easy to exhibit a kind of strain softening characteristic. The phase transition behavior from

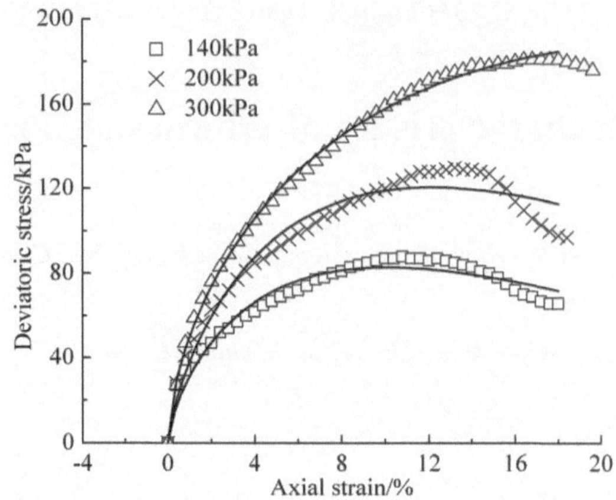

Figure 1. Fitting stress-strain curves for samples.

strain hardening to strain softening appears gradually with the confining pressure increasing. The phenomenon of shear dilatation occurs easier when the density of consolidated sample is larger and larger with consolidated pressure increasing. The topsoil is easy to collapse for the lower confining pressure and shear strength in slope of expansive soil (Li, 2013).

3.2 *Effect of initial dry density*

It has great effect for initial dry density to stress-strain characteristics of samples under the same initial water content and shearing rate in CU tests. The steady state line of sample with the bigger initial dry density is underneath the line of sample with the smaller initial dry density in the relationship between average effective normal stress and specific volume in Fig. 2. The two steady state lines of samples with different dry density show the parallel relationship. The bigger the initial dry density is, the bigger the shearing strength is. The stress-strain behaviors of samples change from strain hardening to strain softening with dry density increasing.

3.3 *Effect of initial water content*

It has great effect for initial water content to stress-strain characteristics of samples under the same initial dry density and shearing rate in CU tests. The steady state line of sample with the bigger initial water content is above the line of sample with the smaller initial water content in Fig. 2. The two steady state lines of samples with different water content show the parallel relationship. The lower the initial water content is, the bigger the shearing strength is.

3.4 *Effect of shearing rate*

It has great effect for shearing rate to stress-strain characteristics of samples under the same initial dry density and water content in CU and CD tests. The steady state line of sample with the bigger shearing rate is above the line of sample with the smaller shearing rate. The two steady state lines of samples with different shearing rate show the parallel relationship. The larger the shearing rate is, the larger the shear strength of sample is in CU test. The larger the shearing rate is, the smaller the shear strength of sample is in the CD test. The rule is just opposite for shearing drainage condition. The higher shearing rate causes strain softening phenomenon and the lower shearing rate causes strain hardening of sample. The phenomenon indicates that the bifurcation appears in shearing process.

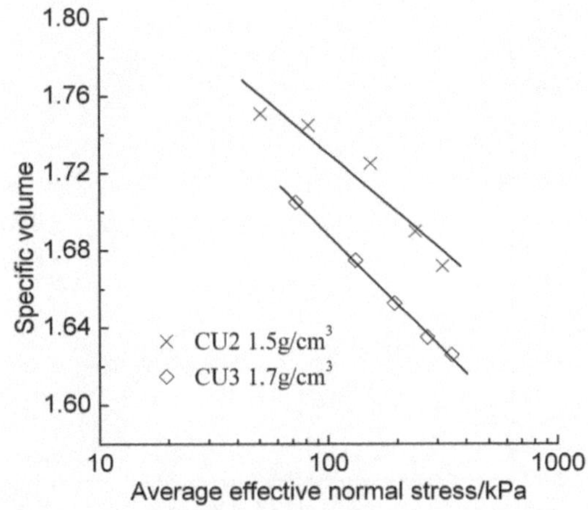

Figure 2. Relationship between average effective normal stress and specific volume under different parameters.

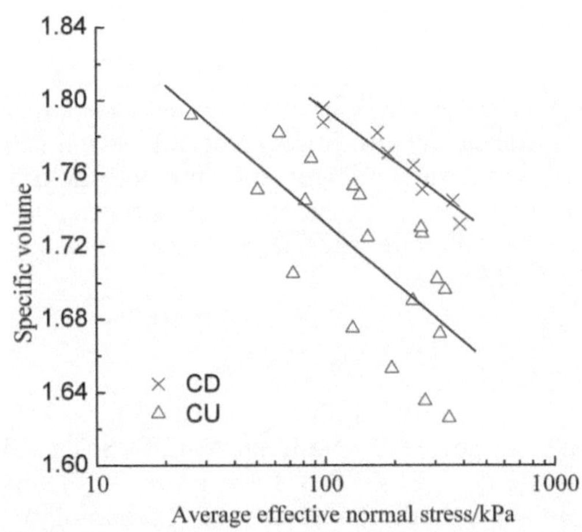

Figure 3. Comparison between p' and v under different drainage conditions.

3.5 *Effect of drainage condition*

The steady state line under drainage condition was still higher the line under undrainge condition below 200 kPa pressure. The steady state line of sample under drainage condition which exhibits a higher steady state is above the line under undrainage condition in Fig. 3. The effect of drainage condition to triaxial shearing test exhibits mainly the influence of excess pore water pressure during shearing test.

4 TREATMENT OF NORMALIZATION

The normalization method can be used to achieve the unique relationship between average effective normal stress and specific volume. The initial dry density and water content are used to be normalization parameters in the same shearing rate condition. The linear relationship after normalization can be expressed by the formula as showed in Fig. 4.

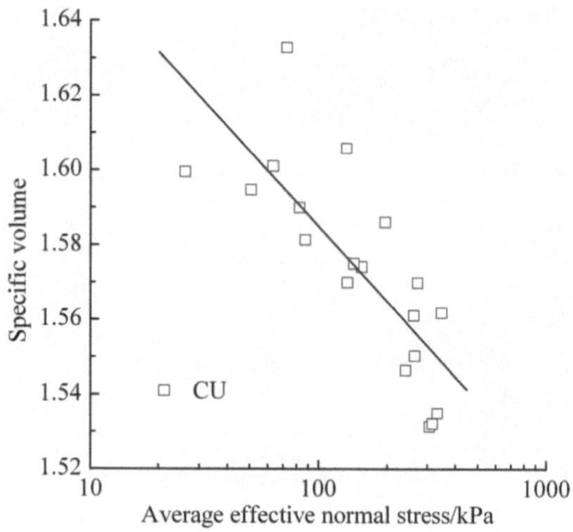

Figure 4. Relationship between p' and v after normalization.

5 CONCLUSIONS

According to Mengzi expansive soil, the saturated strength tests of consolidated drainage (CD) and undrainage (CU) are carried on. It has great influence for the factors, ei confining pressure, initial dry density, initial water content, shearing rate and drainage condition to the characteristics of stress-strain of soil. The inherent unity of diversity of shearing strength for the same samples measured by different experimental methods is indicated according to the normalization of critical state test results.

ACKNOWLEDGEMENTS

This research is financially supported by National Natural Science Foundation of China (NO. 41102186); Opening fund of State Key Laboratory of Geohazard Prevention and Geoenvironment Protection (Chengdu University of Technology) (SKLGP2011 K001); Opening fund of State Key Laboratory for Geomechanics and Deep Underground Engineering (China University of Mining & Technology) (SKLGDUEK 1006); the Open Research Fund of State Key Laboratory of Simulation and Regulation of Water Cycle in River Basin (China Institute of Water Resources and Hydropower Research), Grant NO: IWHR-SKL-201217.

REFERENCES

LI Zhiqing, TANG Chuan & HU Ruilin. 2012. Research on Deformation and Strength Characteristics of Unsaturated Soil, Advanced Science Letters, 15(8): 83–85.

LI Zhiqing, TANG Chuan, HU Ruilin & ZHOU Yingxin. 2013. Research on model fitting and strength characteristics of critical state for expansive soil. Journal of Civil Engineering and Management, 19(1): 9–15.

Schofifld, A.G. 2006. Interlocking and peak and design strengths, Geotechnique, 56(5): 357–358.

Wood, D.M. 1990. Soil behavior and critical state soil mechanics. London: Cambridge University Press.

Modeling and Computation in Engineering II – Xie (ed)
© 2013 Taylor & Francis Group, London, ISBN 978-1-138-00058-2

The analysis of real estate industry contribution to Shenyang economic

X.J. Qi, Y.X. Zhou & P. Xiao
School of Resources and Civil Engineering, Northeastern University, Shenyang, Liaoning, China

L.J. Deng
Shenyang Institute of Engineering, Shenyang, Liaoning, China

ABSTRACT: This paper firstly reviews the macro-control policies in our country for the real estate industry and the fundamental situation of Shenyang real estate industry in recent years. It elaborates the relations between the real estate industry and the national economy, which can be analysis from the backward linkage, the forward linkage and the circumferential association from the perspective of qualitative analysis. It elaborates the contribution and stimulating impact of real estate industry to Shenyang economy as a mainstay industry from the perspective quantitative analysis. At last, it proposes an idea of coordinated development of the real estate industry and other industries.

1 THE REVISION OF THE MACRO-CONTROL POLICIES AND THE REAL ESTATE INDUSTRY OF SHENYANG

1.1 *Reviews of recent macro-economic control policies*

Ever since 2005, governmental authorities began to implement macro-economic controls on real estate fields. Generally, this paper divided regulation rules since 2007 into three stages, that is, "tighten, easing, tighten". It is explained in details as follows:

1.1.1 *Tightened real estate policy—central government imposes restrictions*

During the periods between early 2007 and June 2008, China economy experienced historically high CPI, excess liquidity and rapidly rising house purchasing prices. Just based on these negative backgrounds, authorities chose to regulate national economy by implementing multiple policies, such as monetary policy, fiscal policy, land-using adjustment, etc. Mainly, central bank of China raised interest rates by five times consecutively and rose up monetary reserve rates by five times simultaneously. What's more, some important regulation documents also came into effects. They are as follows: Notice on strengthening loan management of commerce real estate institutions, "9.27 rules about mortgaging management, Notice on making using of limited land resources effectively.

1.1.2 *Loose real estate policy—local government rescues markets*

From early 2008, due to the international financial crisis, the rapid decline in GDP, the risk of deflation and central government tightened policy, China's real estate market (especially the first-tier cities) experienced a significant decline, causing extensive concerns. Central government and local governments "rushed" to change policy from "tight" to "loose". The purpose is to release liquidity to support the economy and to deal with the financial crisis, to maintain economic growth, expand domestic demands. The main policies include that the central bank announced to lower lending rates and deposit rates (2008.9.15), lowered the loan reserve ratio

Table 1. Basic data table for Shenyang real estate industry from 2004s to 2011s.

	2004s	2005s	2006s	2007s	2008s	2009s	2010s	2011s
GDP (¥ Billion)	177.293	208.413	251.963	322.115	378.087	426.851	501.754	591.571
GDP growth rate (%)		17.55%	20.90%	27.84%	17.38%	12.90%	17.55%	17.90%
Value of housing investing (¥ Billion)	34.262	41.357	53.829	73.036	101.091	118.870	145.008	186.472
Growth rate of housing investing (%)		20.71%	30.16%	35.68%	38.41%	17.59%	21.99%	16.18%
Amount of houses sold (¥ Billion)	15.263	31.745	41.992	53.932	60.469	68.436	94.505	127.402
Growth rate of houses sold (%)		107.98%	32.28%	28.43%	12.12%	13.17%	38.09%	34.81%
Housing prices (Yuan/m^2)	2911.12	3186.93	3376.17	3688.81	4127.46	4464.38	5411.05	5884.18
Growth rate of housing prices		9.47%	5.94%	9.26%	11.89%	8.16%	21.21%	8.74%

and the deposit reserve rate (2008.10.9), issued by the State Council on < Start to stimulate real estate development policies in the country (2008.10. 17).

1.1.3 *Second tightened real estate policy—the stability of prices, investment inhabitation*
From the end of 2009 to early 2010, due to the economic recovery, housing prices rebounded in some areas (especially some first-tier cities like Peking, Shenzhen and Guangzhou) and the real estate market became overheated. From the end of 2009, central governments has launched some related regulatory policies including the repeal of the Second Suite preferential and repeal of the taxation preferential for the house owned less than 5 years and the other 11 new policies issued by the state council. To stabilize the housing price, inhibit investment, increase securities are main strategies. Reviews of recent real estate industry development in Shenyang.

1.2 *Reviews of recent real estate industry development in Shenyang*

From charts above, it is concluded that it has experienced a stable increase in investing real estate industry, value of houses sold and housing prices during the recent five years. The growth rates kept at 20% around. Due to economic crisis in 2008, growth rates of GDP saw some decreases in 2008 and 2009. Meanwhile, growth rates of housing investing and sold had some fluctuation to some extent. However, generally speaking, real estate industry in Shenyang has been in a stale raise during the last six years even if having some ups and downs.

2 THE QUALITATIVE ANALYSIS OF THE RELATIONSHIP BETWEEN THE REAL ESTATE INDUSTRY AND THE ECONOMY DEVELOPMENT IN SHENYANG

The real estate industry as an independent industry which is formed with the development process of the industrialization, urbanization and modernization, and it also promotes the development process of the industrialization, urbanization and modernization, and has become an important part of modern socio-economic system. The real estate industry is not only a long industrial chain, high associate degree, can directly or indirectly affect the development of related industries, but also in the operation cycle of the national economy, the real estate development is the precondition of most reproduction of various industries, relative to the emergence of the economic cycle each stages, the real estate industry plays a good demonstration and guiding role. This paper argues that the basic relationship of the real estate industry and other industrial and socio-economic environment is shown in Figure 1.

2.1 *The backward linkage of real of the real estate industry to the national economy*

The backward linkage, commonly known as "investment-pull", refers to that has promoted impact to the industries which provide factors of production for it because of the requirement of its own development, such as building materials industry, metallurgical industry, finance and insurance industries. In terms of Shenyang, the real estate industry has a huge demand for products of related industries, the requirement of real estate funds is large, the real estate industry is an important material basis of urban economic construction.

2.2 *The forward linkage of the real estate industry to the national economy*

The forward linkage, commonly known as "sale-pull", refers to that the real estate industry pulls industries which rely on it to develop because of their own development needs, such as the textile industry, household appliances, decoration, kitchen equipment, finance and insurance industries. In terms of Shenyang City, the real estate industry can promote the related industries and the financial industry, the real estate industry is the precondition of social labor producing and improving quality.

2.3 *The circumferential association of the real estate industry to the national economy*

Flanking association, also known as the circumferential association, because of the need of the real estate industry development, the real estate industry promotes those industries which are indirectly stimulated. For example, transportation, automobile manufacturing and sales, city utilities, cultural education. In terms of Shenyang City, the real estate industry provides material conditions for the development of the national economy. The real estate industry is an important source of accumulated funds. The development of the real estate industry can optimize the structure of urban consumption.

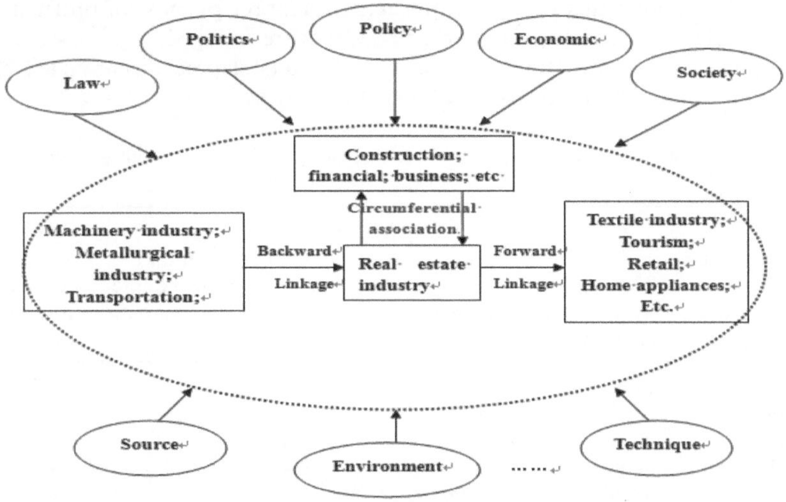

Figure 1. The basic relationship of the real estate industry and other industrial and socio-economic environment.

3 THE QUANTITATIVE ANALYSIS OF THE RELATIONSHIP BETWEEN THE REAL ESTATE INDUSTRY AND THE ECONOMY DEVELOPMENT IN SHENYANG

3.1 *The basic algorithm set by National Bureau of Statistics*

According to the related definition of National Bureau of Statistics, the contribution rate of an annual economic growth of the real estate industry in Shenyang City, the percent of contribution rate to national economic growth, which can be calculated as follows:

$$C_r = \frac{Gr}{G} \tag{1}$$

where Cr means a year real estate industry's contribution to economic growth; Gr is this year real estate industry's growth; and G is this year GDP growth.

$$P_r = R \times P_1 \tag{2}$$

where P_r means a year's percentage point of the real estate industry's contribution to economic growth; R is this year's contribution rate of the national economic growth; and P_1 is this percentage point of the real estate industry to economic growth.

According to Shenyang Statistical Yearbook (2005~2011), the estimated process and results of the contribution of the real estate industry to Shenyang economic growth in recent years, are listed in Table 2.

Therefore, from 2007 to 2011, the contribution rate of Shenyang real estate industry to economic growth are 3.70%, 2.98%, 3.61%, 8.62% and 5.36%, and the average contribution rate is 4.85%; above five years, Shenyang City real estate industry's contribution to economic growth are 1.03, 0.52, 0.47, 1.51 and 0.96 percentage, the average is 0.90 percentage.

3.2 *The shortcoming of the National Bureau of Statistics' caliber tend*

According to the existing provisions of the National Bureau of Statistics, there are the following problems or defects in the accounting of added value of the real estate industry, resulting in the added value of the real estate industry accounted for according to the current statistics caliber tend to be less than its actual value:

1. Due to data limitations, there are not rental activities for the purpose of profit and income which non-real estate businesses and urban residents are engaged;
2. The urban and rural residents' own house current annual depreciation rate are 2% to 4%, in the case of urban demolition, renovation faster, may be low;

Table 2. Estimates of contribution rate of recent real estate industry on economic growth in Shenyang.

Years	The real estate industry's growth value (¥ Billion)	The growth value of GDP (¥ Billion)	Year-on-year growth of GDP (%)	The contribution rate of the real estate industry on economic growth (%)	Contribution percentage points of the real estate industry for economic growth
2005	1.883	31.120	17.55	6.05	1.06
2006	2.049	43.550	20.90	4.70	0.98
2007	2.598	70.152	27.84	3.70	1.03
2008	1.671	55.972	17.38	2.98	0.52
2009	1.758	48.764	12.90	3.61	0.47
2010	6.455	74.903	17.55	8.62	1.51
2011	4.813	89.817	17.90	5.36	0.96

3. AS the historical cost (original value) as calculated housing depreciable basis, ignoring the actual market value (transaction assessment price) after housing value adding, which cause serious low depreciation charges;
4. As actual income as total output, which underestimates value of the welfare housing (affordable housing, low-rent housing) from government subsiding. This affects the total output of the real estate industry.

Moreover, the most important problem is that the growth of house services of the urban residents is underestimated. Therefore, it need adjust the growth of residential housing services to amend the underestimated growth of real estate industry from the current accounting methods of the National Bureau of Statistics.

3.3 *The revision of the National Bureau of Statistics' caliber tend*

For the fact growth of urban housing services is underestimated, the revised growth of the real estate industry can be calculated as follows:

$$G_0 = G_1 - G_2 + G_3 \tag{3}$$

where G_0 means real estate industry's growth; G_1 means now statistical coverage growth of real estate industry; G_2 means now statistical coverage growth of residential housing services; and G_3 is the actual growth of residential housing services.

$$G_2 = V_0 \times P_V = S \times U \times P_V \tag{4}$$

where G_2 means now statistical coverage growth of residential housing services; V_0 means now statistical coverage total output value of residential housing services; P_V is the proportion of growth in total output value of residential housing services; S is now statistical coverage annual per capita spending on housing services; and U is Urban population.

$$G_3 = G_T \times P_V \tag{5}$$

where G_3 means the actual growth of residential housing services; and G_T is the proportion of growth in total output value of residential housing services.

$$T = P_c \times S_r \times K \times U \times A \tag{6}$$

where P_c means commercial housing prices; S_r is sale rent ratio; K means 12 Month; A is per capita housing area.

The proportion of the growth of the total output in the residential housing services is 50%; sale-rent ratio refers that the ratio of the monthly rental price (rent) and the price of commercial house, generally considered reasonable interval is 1:90 to 1:150, because the current price of commercial house is high, this paper uses 1:150.

In order to use the above formula to estimate the growth of the real estate industry, now the basal data of Shenyang in recent years that are sorted out from Shenyang Statistical Yearbook (2005–2011) are listed in Table 3.

Thus, according to the data in Table 3, using the formula (4) can calculate the total output of residents housing services based on the existing statistical caliber, according to formula (6) can calculated the total output of the actual residential housing services, then using the formula (4) and (5) can the calculate current statistical growth of residential housing services and the actual growth of residential housing services, and then according to equation (3) get the revised growth of Shenyang real estate industry in recent years, which are listed in Table 4.

Using the above calculation results to adjust GDP, adjusted GDP = GDP—the growth of the real estate industry based on existing statistical caliber + actual growth of the real estate

Table 3. The basic data of estimating the real estate industry's growth.

Years	Commercial housing prices (Yuan/m²)	Urban population (million)	Per capita housing area (m²)	Now statistical caliber annual per capita spending on housing (Yuan)
2005	3026.64	4.959	23.36	797.50
2006	3184.00	4.999	25.24	974.21
2007	3524.80	5.049	27.01	1129.28
2008	3856.19	5.090	29.02	1544.79
2009	4197.11	5.122	30.91	1523.83
2010	5108.95	5.154	32.84	1364.94
2011	5612.96	5.191	33.12	1502.46

Table 4. The revised growth of Shenyang real estate industry in recent years (¥ Billion).

Years	2005	2006	2007	2008	2009	2010	2011
Now statistical caliber growth of the real estate industry	10.672	12.722	15.320	16.991	18.749	25.204	30.061
The revised growth of real estate industry	24.752	27.619	32.289	35.674	39.955	51.367	66.667

Table 5. The contribution and pull of Shenyang real estate industry on economic growth.

Years	2006	2007	2008	2009	2010	2011
The contribution rate of contribution rate of economic growth on economic growth (%)	10.33	9.74	9.48	9.61	10.86	10.61
The pull of the real estate industry on economic growth (percentage points)	2.12	2.67	1.64	1.27	2.00	1.86

industry, finally, the revised contribution of Shenyang City real estate industry to economic growth in recent years and the pull of economic growth (GDP) percentage points can be calculated according to the formula (1), formula (2), and the results are summarized in table 5.

This paper argues that the conclusions are calculated by the revised the method of the National Bureau of Statistics are more scientific, and 2007 to 2011 five-year average data, Shenyang real estate industry's contribution to economic growth is 10.06%, for economic growth pull is 1.89 percent which is more realistic.

As can be seen from the figure 2, the real estate industry's contribution to national economic is closely related to the macro-control policies. When policy is easing, the real estate industry's contribution ratio to the socio-economic and pulling will rise; when policy is tightening, the contribution rate and pulling will go down. However, this leading role is not completely synchronized, after the introduction of the policy, it takes about half year for the real estate industry the real estate industry's contribution to socio-economic to response after the policy are introduced.

3.4 *Real estate is the major industry of the social economics in Shenyang*

Major industries are those which develop rather fast and play a role in leading and promoting the whole national economy. Those industries have a strong effect which induces the rise of new industries, and have a profound and extensive influence on economic structure,

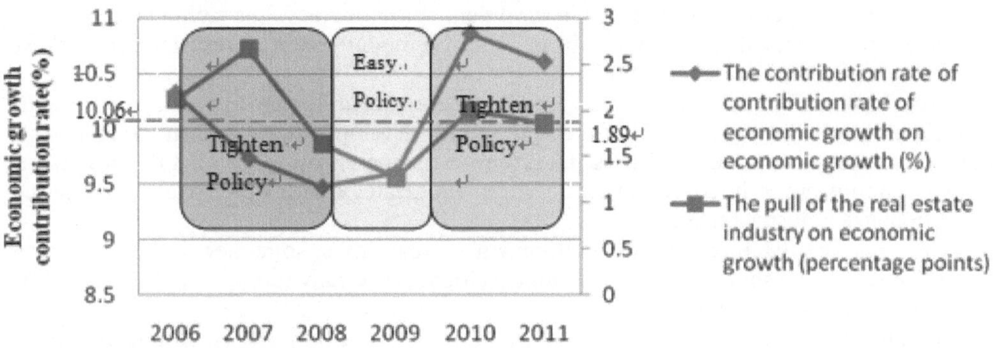

Figure 2. Shenyang real estate industry contribution to economic growth the relationship of Shenyang real estate industry and macro-control policy.

change and development in the departments and regions which provide them the means of production. China applies nine reference standards, such as added value, exportation, number of employees, correlation of industry, industry concentration, technological maturity, the income elasticity of demand, economic efficiency, and sustained growth rate, and also puts forward that added value should account for more than 5% of GDP.

As income levels and living conditions improve with the development of social economics, people will maintain strong demand for improvement of working standards and living conditions. Moreover, the acceleration of urbanization, the revitalization of the old industrial bases, and the consolidation of advantages in central regions of Shenyang will each play a huge role in the promotion of said improvements. Therefore, the real estate industry will continue to contribute to economic development and the improvement of social conditions.

At different steps in social economic development, because of the change in main driving factors, the major industries will change. The major industries of the developed countries of the West have gone through land, resources, agriculture, traditional processing industries, iron and steel, chemicals, automobiles, and machinery and other capital-intensive industries, as well as funds for the support of financial, high-tech and information technology and bioengineering. In China, knowledge-intensive industries, funding of knowledge and technology to support the financial sector, software and computer services, information transmission services, advertising, and real estate development are replacing traditional industries and have become competitive industries in many cities.

Based on the above analysis, the real estate industry has been and will continue to be the major industry of the city's economy. Utilized properly, and avoiding a negative impact on the real estate industry, the community will see further integration of funding, technology, human resources and other advantages, improvement of industrial comparative advantage and urban competitiveness, at the same time extension of the industrial chain, strengthening industrial relations, and improvement of supporting industries.

4 SUMMARY

Recognizing that real estate is a major industry, and can contribute to the development of the social economy, the following measures should be taken to improve the harmony between real estate and other industries:

First, create a good operating environment for the development of the real estate industry, such as the development of the real estate property and trading relations, to convert the role of government, improve laws and regulations to regulate market behavior.

Secondly, the optimization of the real estate industry promotes the industry's upgrade. Mainly from real estate development and investment, distribution, service and consumption of the four major areas to expand the development space; increase the technical content of the real estate industry through the establishment of a modern enterprise system, to improve the quality of the real estate industry.

Third, the optimization of the structure of the real estate and related industries ensure the smoothness of the industrial chain. In the interactive development process in the real estate industry and other related industries, to promote their own optimization, the real estate industry should properly coordinate a development strategy with other related industries, and rationalize industrial relations.

Finally, taking advantage of the development of real estate, some new major industries relating to real estate should be built up and developed, especially those industries exhibiting the national competitive spirit.

In summary, the real estate industry is the major industry of the socio-economic Shenyang, but should promote its sustainable development, actively cultivate other industries, and promote the coordinated development of economy and society.

REFERENCES

Hua Wei, & Shen Ningyan. 2012. The real estate industry's regional innovation under the macro-control policies, *Exploration and Free views*, 10: 57–59.

Liu Junmin. 2006. The Relationship of Chinese Real Estate Industry Development and Macro-economic Stability, *China Real Estate Industry*, 2: 6–9.

Liang Yongsheng. 2010. The relationship of Macro economic development and the real estate industry development relationship, *ChengShi JianShe LiLun Yan Jiu*, 26: 1–13.

Niu Xuehua. 2010. The research of Chinese real estate development, *China Economist*, 10: 291.

Ren Murong, & Su Guoqiang. 2010. The goal system research to real estate industry macro-control policies of our country, *Economic Review*, 4: 52–55.

Shenyang Municipal Bureau of Statistics. 2005–2011. Shenyang Statistical Yearbook. Beijing: China Statistics Press.

Ye Jianping, & Xie Jingrong. 2005. Coordinated Development Research for the Real Estate Industry and the Socio-economic. Beijing: China Renmin University Press.

Ya Feng. 2006. The real estate has become the pillar industry, China economy is showing polis trend, *China Industrial Economic Information Network*, 6: 2.

Modeling and Computation in Engineering II – Xie (ed)
© 2013 Taylor & Francis Group, London, ISBN 978-1-138-00058-2

Calculation methods of wave forces on vertical wall

Dechun Li & Jijian Lian
State Key Laboratory of Hydraulic Engineering Simulation and Safety, School of Civil Engineering, Tianjin University, Tianjin, China

ABSTRACT: The unified formula of wave forces on vertical wall presented by Gada is suitable for every wave pattem and every wave condition, also is widely used in Japan and European countries. The formula of wave forces on vertical wall given by Chinese Code of Hydrology for Sea Harbour is different from Goda's formula in representing forms and useful conditions. According to the comparison of wave forces on vertical wall calculated by two methods for every wave pattem and every wave condition, the differences between Goda's formula and Chinese code's formula are discussed. It is useful for coastal project design and standard revision.

1 INTRODUCTION

Wave force is important load needing consideration during the design of waterfront structures and substantive researches have been made for this (Xiulun Wang 1983, Ning Shu 2003). The unified formula (Communications Press 1998), which is set up by Japanese scholar Goda Yoshizane for the calculation of wave force on a vertical wall under different states and conditions, is simple in expression and use, and is widely used in Japan and Europe. China's *Code for Sea Port Hydrology* divides wave into three states, namely vertical wave, broken wave and breaking wave. It gives the computing formulas of wave force on vertical wall according to the different wave forms and different wave parameters. The computing formulas are complex and the use conditions are complicated; besides, there are no appropriate computing formulas under certain wave conditions. This paper aims to find out how is the computing result of the wave force computing formula specified in China's Code for Sea Port Hydrology different from that of Goda Formula based on study.

There have been some researches on the differences and similarities between the computing result of the wave force computing formula specified in China's Code for Sea Port Hydrology and that of Goda Formula. Li Yucheng et al. (Yucheng Li 2002) employs a series of aggregate analysis methods including theoretical study, dimensional analysis and case study based on the physical model test of irregular waves to test the applicability of Goda formula, and arrives at the calculation method of breaking wave force. Shu Ning et al. (Ning Shu 2003) introduces the calculation methods of wave force specified in Goda formula adopted by British Standard BS6349, and the calculation method specified in Japan standard, Korea standard, and China code, and makes comparative analysis based on specific projects. Zhang Zongliang et al. (Zhang Zongliang 2006) employs Fourier numerical approximation solution and makes calculations about the wave force of vertical waves under different conditions, and compares it with the calculation method specified in China's *Code for Sea Port Hydrology* and Goda formula.

The above said researches on the formula specified in China's Code for Sea Port Hydrology and Goda formula neglect the requirements of Goda formula that the wave height used shall be the maximum wave height ($H1/250$); instead, the standard limit of wave height ($H1\%$) specified in China's *Code for Sea Port Hydrology* is used. As a result, the wave force worked out through Goda formula is obviously smaller, and the results acquired are not so precise. Besides, the above said comparative studies are only targeted at a single wave form like breaking wave or vertical wave, and fail to make systematic comparative analysis to the two calculation methods used for vertical wave, broken wave and breaking wave.

This paper analyzes the similarities and differences between the wave force calculated through the formula specified in China's *Code for Sea Port Hydrology* and Goda formula in three wave forms including vertical wave, broken wave, and breaking wave under different wave conditions according to the wave height limit specified in different methods ($H1/250$ in Goda formula and $H1\%$ in China code), which is of referential meaning to further improve the calculation method of wave force on vertical wall in China code.

2 CALCULATION METHOD OF WAVE FORCE SPECIFIED IN CHINA'S CODE FOR SEA PORT HYDROLOGY

According to the wave parameters of incident wave, water depth in front of structures, bottom slope and water depth of foundation bed surface in front of the wall, China's *Code for Sea Port Hydrology* (JTJ213-98)[4] divides the wave in front of a vertical wall into three forms, namely vertical wave, broken wave and breaking wave, and gives the computing formula of wave force on vertical wall according to the different wave forms and different wave parameters.

In the calculation of wave force, first of all, it decides the wave form according to the standard in the code; then it calculates the wave force using the corresponding formula according to the wave form decided as well as the wave parameters of incident wave and water depth in front of the structure.

For vertical wave, wave parameters and water depth conditions in front of the structure are divided into five types: namely $d \geq 1.8\ H$, $d/L = 0.05\sim0.12$; $d \geq 1.8\ H$, $d/L = 0.12\sim0.139$ and $8 < T^* < 9$; $H/L \geq 1/30$, $d/L = 0.139\sim0.2$; $H/L \geq 1/30$, $d/L = 0.2\sim0.5$; $d/L \geq 0.5$. The formula of wave force is then given under each condition.

See Reference [4] for the details of calculation of wave force in the Code for Sea Port Hydrology (JTJ213-98).

3 GODA FORMULA FOR WAVE FORCE

Goda Yoshizane puts forward a general formula which can be used to calculate both the vertical wave force and the wave force of breaking wave and broken wave according to substantive model tests and tests of in-situ breakwater, and considers the impacts of oblique wave. Goda formula is convenient to use and has been adopted by Japan Design Standard for Sea Port Structures and is widely accepted by many European countries.

See figure 1 for Goda wave pressure computation model. Suppose the wave pressure imposed on the wall surface is in straight line distribution and wave pressure at z above the still water level is zero; wave pressure p_1 on the still water level is the maximum.

Work out z value according to the following formula:

$$z = 0.75(1+\cos\beta)\,\lambda_1 H_{\max} \tag{1}$$

Wave pressure on the still water level is:

$$p_1 = \frac{1}{2}(1+\cos\beta)\left(\lambda_1\alpha_1 + \lambda_2\alpha^*\cos^2\beta\right)\gamma H_{\max} \tag{2}$$

Sea bottom wave pressure is:

$$p_2 = \frac{p_1}{\cosh(2\pi d/L)} \tag{3}$$

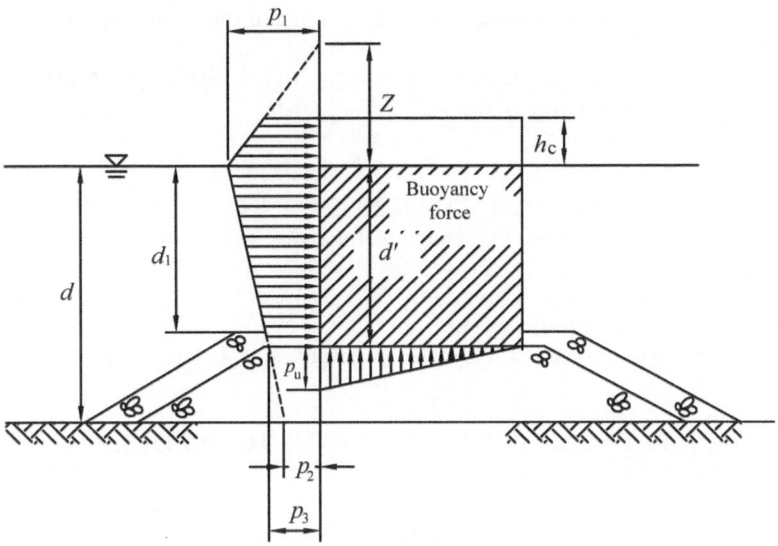

Figure 1. Wave pressure computation model of goda formula.

Wall bottom wave pressure is:

$$p_3 = \alpha_3 p_1 \tag{4}$$

Wall crown wave pressure is:

$$p_4 = \alpha_4 p_1 \tag{5}$$

Uplift pressure at levee toe is:

$$p_u = \frac{1}{2}(1+\cos\beta)\,\lambda_3\alpha_1\alpha_3\gamma H_{\max} \tag{6}$$

Where in,

$$\alpha_1 = 0.6 + \frac{1}{2}\left[\frac{4\pi d/L}{\sinh(4\pi d/L)}\right]^2 \tag{7}$$

$$\alpha_2 = \min\left\{\frac{h_b - d_1}{3h_b}\left(\frac{H_{\max}}{d_1}\right)^2,\ \frac{2d_1}{H_{\max}}\right\} \tag{8}$$

$$\alpha_3 = 1 - \frac{d'}{d_1}\left[1 - \frac{1}{\cosh(2\pi d/L)}\right] \tag{9}$$

$$\alpha_4 = 1 - \frac{h_c^*}{z} \tag{10}$$

$$\alpha^* = \max\{\alpha_2, \alpha_1\} \tag{11}$$

$$h_c^* = \min\{z, h_c\} \tag{12}$$

wherein: β—Line-to-line angle between the wall in normal direction and the predominant wave direction;

H_{max}—Maximum wave height; in deep water area $H_{max} = H1/125 = 1.8\ H1/3$; in broken water area, H_{max} is 5 $H1/3$ in front of the wall;

L—Wave length;
γ—Water density;
h_b—Water depth at 5 $H1/3$ in front of the wall;
h_c—Height of breakwater above the design water level;
λ_1—Correction factor of creeping pressure;
λ_2—Correction factor of impact pressure;
λ_3—Correction factor of uplift pressure.
λ_1, λ_2, and λ_3 are related to structural style. Generally, $\lambda_1 = \lambda_2 = \lambda_3 = 1.0$.

4 COMPARISON BETWEEN THE FORMULA IN CHINA CODE AND GODA FORMULA

The wave force of vertical wave, broken wave and breaking wave is respectively calculated in the method specified in the Code for Sea Port Hydrology (JTJ213-98) and Goda formula under different wave conditions. When use the method in the Code for Sea Port Hydrology (JTJ213-98) to calculate the wave force, the wave height takes $H1\%$; when calculating the wave force using Goda formula, the wave height takes $H1\%$ and $H1/250$ respectively to make a comparison.

4.1 *Comparison of the wave force of vertical wave*

Vertical wave force can be expressed in the following dimensionless form:

$$\frac{P}{\gamma H d_1} = f\left(\frac{d_1}{d}, \frac{d}{L}, \frac{H}{L}\right) \tag{13}$$

wherein, P is the resultant force of wave force; meanings of other variables are the same as those stated above; see figure 1.

a. when $d \geq 1.8\ H$, $d/L = 0.05\sim0.12$

Figure 2 shows the curve of dimensionless wave force $P/\gamma Hd_1$ changing with H/L when $d \geq 1.8\ H$, $d/L = 0.05\sim0.12$, and $d_1/d = 1.0$ and $d/L = 0.11$.

From figure 2 it is observed that, the dimensionless wave force $P/\gamma Hd_1$ worked out with Goda formula increases as H/L increases; the dimensionless wave force $P/\gamma Hd_1$

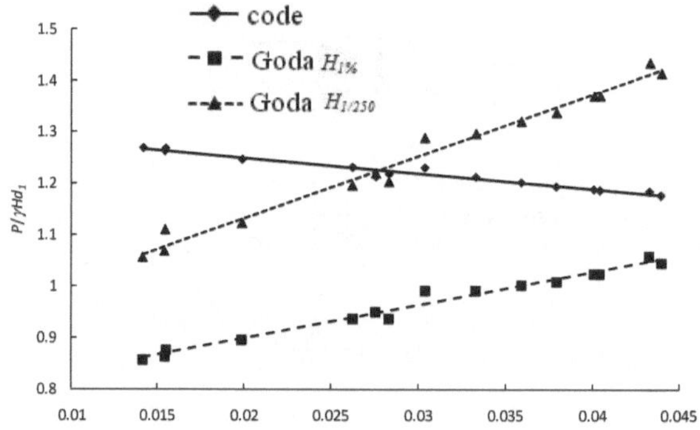

Figure 2. Changes of the dimensionless vertical wave force along with the changes of H/L when $d \geq 1.8\ H$ and $d/L = 0.05\sim0.12$.

worked out in the method specified in the Code for Sea Port Hydrology decreases as H/L increases. When the wave height is $H1\%$, the wave force worked out with Goda formula is about −10%~ −33% smaller than that worked out with the method in the Code for Sea Port Hydrology. When the wave height is $H1/250$, the wave force worked out with Goda formula is about 0~ −17% smaller than that worked out with the method in the Code for Sea Port Hydrology when H/L is smaller, and is about 0~22% bigger than that worked out with the method in the Code for Sea Port Hydrology when H/L is bigger.

b. when $d \geq 1.8\,H$, $d/L = 0.12$~0.139 and $8 < T^* < 9$

Figure 3 shows the change curve of dimensionless wave force $P/\gamma Hd_1$ along with the change of H/L when $d \geq 1.8\,H$, $d/L = 0.12$~0.139 and $8 < T^* < 9$ and when $d_1/d = 1.0$ and $d/L = 0.13$.

From figure 3 it is observed that, the dimensionless wave force $P/\gamma Hd_1$ worked out in the method specified in the Code for Sea Port Hydrology and Goda formula all increases as H/L increases. When the wave height is $H1\%$, the wave force worked out with Goda formula is about −15%~ −26% smaller than that worked out with the method in the Code for Sea Port Hydrology. When the wave height is $H1/250$, the wave force worked out with Goda formula is about 0~ −8% smaller than that worked out with the method in the Code for Sea Port Hydrology when H/L is smaller, and is about 0~18% bigger than that worked out with the method in the Code for Sea Port Hydrology when H/L is bigger.

c. when $H/L \geq 1/30$ and $d/L = 0.139$~0.2

Figure 4 shows the change curve of dimensionless wave force $P/\gamma Hd_1$ along with the change of H/L when $H/L \geq 1/30$, $d/L = 0.139$~0.2 and when $d_1/d = 1.0$ and $d/L = 0.15$.

From figure 4 it is observed that, the dimensionless wave force $P/\gamma Hd_1$ worked out in the method specified in the Code for Sea Port Hydrology and Goda formula all increases as H/L increases. When the wave height is $H1\%$, the wave force worked out with Goda formula is about −22%~−24% smaller than that worked out with the method in the Code for Sea Port Hydrology. When the wave height is $H1/250$, the wave force worked out with Goda formula is about 0~−1% smaller than that worked out with the method in the Code for Sea Port Hydrology when H/L is smaller, and is about 0~9% bigger than that worked out with the method in the Code for Sea Port Hydrology when H/L is bigger.

d. when $H/L \geq 1.8\,H$ and d/L=0.2~0.5

Figure 5 shows the change curve of dimensionless wave force $P/\gamma Hd_1$ along with the change of H/L when $H/L \geq 1.8\,H$, $d/L = 0.2$~0.5 and when $d_1/d = 1.0$ and $d/L = 0.32$.

From figure 5 it is observed that, the dimensionless wave force $P/\gamma Hd_1$ worked out in the method specified in the Code for Sea Port Hydrology and Goda formula all increases as H/L increases. When the wave height is $H1\%$, the wave force worked out with Goda

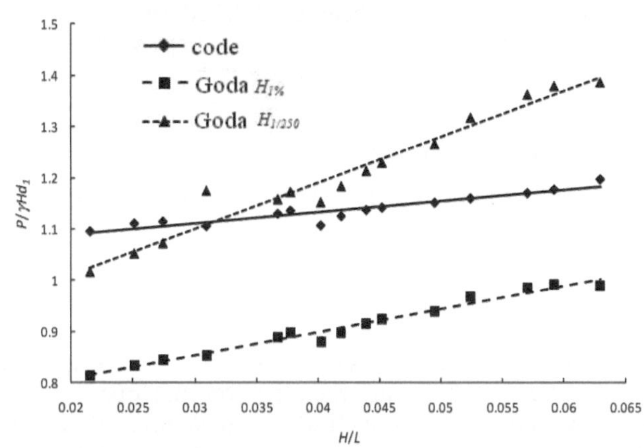

Figure 3. Changes of the dimensionless vertical wave force along with the changes of H/L when $d \geq 1.8\,H$, $d/L = 0.12$~0.139 and $8 < T^* < 9$.

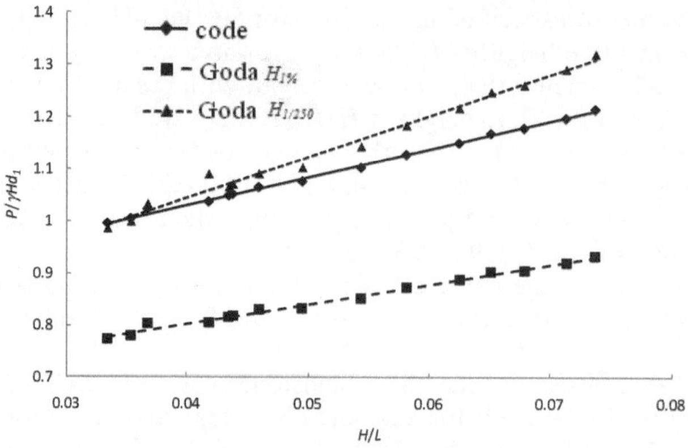

Figure 4. Changes of the dimensionless vertical wave force along with the changes of H/L *when* $H/L \geq 1/30$ and $d/L = 0.139{\sim}0.2$.

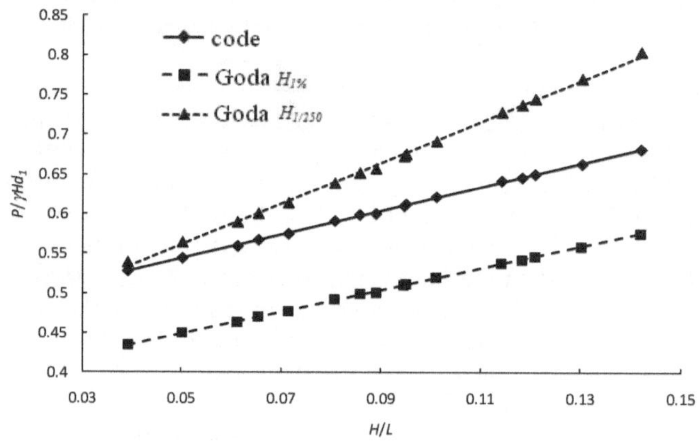

Figure 5. Changes of the dimensionless vertical wave force along with the changes of H/L *when* $H/L \geq 1.8\ H$ and $d/L = 0.2{\sim}0.5$.

formula is about −15%~ −18% smaller than that worked out with the method in the Code for Sea Port Hydrology. When the wave height is $H1/250$, the wave force worked out with Goda formula is about 2~18% bigger than that worked out with the method in the Code for Sea Port Hydrology.

4.2 *Comparison of the wave force of broken wave*

Broken wave force can be expressed in the following dimensionless form:

$$\frac{P}{\gamma H d_1} = f\left(\frac{d_1}{d}, \frac{H}{d_1}, \frac{d}{L}, \frac{H}{L}, i\right) \tag{14}$$

Figure 6 shows the change curve of dimensionless wave force $P/\gamma Hd_1$ along with the change of H/L when $d_1/d = 1.0$ and $H/d_1 = 0.11$ and when the bottom slope $i = 0$.

From figure 6 it is observed that, the dimensionless wave force $P/\gamma Hd_1$ worked out in the method specified in the Code for Sea Port Hydrology and Goda formula all increases as H/L decreases. When the wave height is $H1\%$, the wave force worked out with Goda formula is about −10%~ −24% smaller than that worked out with the method in the Code for Sea Port

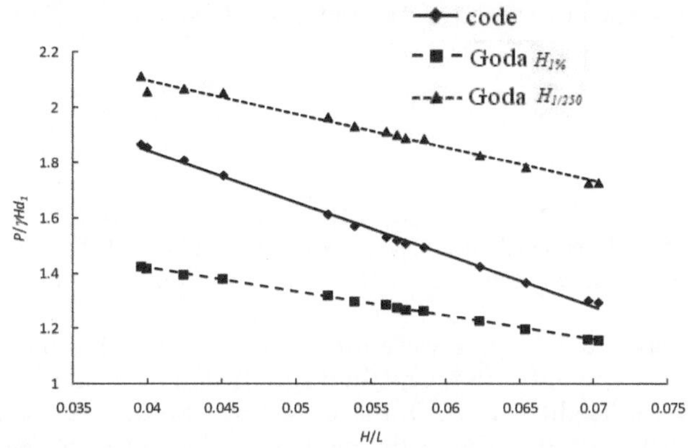

Figure 6. Changes of the dimensionless wave force of broken wave along with the changes of H/L.

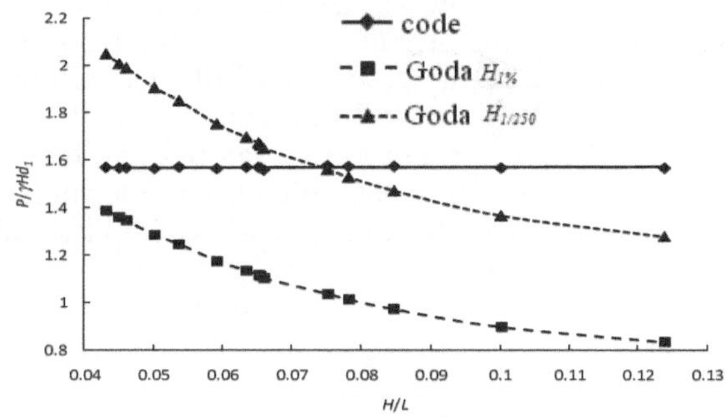

Figure 7. Changes of the dimensionless wave force of breaking wave along with the changes of H/L.

Hydrology. When the wave height is $H1/250$, the wave force worked out with Goda formula is about 10~34% bigger than that worked out with the method in the Code for Sea Port Hydrology.

4.3 *Comparison of the wave force of breaking wave*

Breaking wave force can be expressed in the following dimensionless form:

$$\frac{P}{\gamma Hd_1} = f\left(\frac{d_1}{d}, \frac{H}{d_1}, \frac{B_m}{L}, \frac{H}{L}, i\right) \tag{15}$$

Figure 7 shows the change curve of dimensionless wave force $P/\gamma Hd_1$ along with the changes of H/L when $d_1/d = 0.5$ and $H/d_1 = 0.75$ and when the relative base shoulder width $B_m/L = 0$, and when the bottom slope $i = 0$.

From figure 7 it is observed that, the dimensionless wave force $P/\gamma Hd_1$ worked out with Goda formula decreases as H/L increases; the dimensionless wave force $P/\gamma Hd_1$ worked out in the method specified in the Code for Sea Port Hydrology does not change obviously as H/L changes and is basically a constant. When the wave height is $H1\%$, the wave force worked out with Goda formula is about −11%~ −47% smaller than that worked out with the method in the Code for Sea Port Hydrology. When the wave height is $H1/250$, the wave force worked out with Goda formula is about 0~31% bigger than that worked out with the method in the Code for Sea Port Hydrology when H/L is smaller, is about 0~ −19% smaller than the work force worked out with the method in the Code for Sea Port Hydrology when H/L is bigger.

5 CONCLUSION

1. The standard for wave height selection in the wave force calculation method specified in the Code for Sea Port Hydrology and Goda formula is different; in the Code for Sea Port Hydrology, the wave height is $H1\%$ and in Goda formula, the wave height is $H1/250$.
2. When the wave height is $H1\%$, the wave force worked out with Goda formula is much smaller than that worked out with the method specified in the Code for Sea Port Hydrology; When the wave height is $H1/250$, the work force worked out with Goda formula is smaller than that worked out with the method in the Code for Sea Port Hydrology under some wave conditions, but is bigger than that worked out with the method specified in the Code for Sea Port Hydrology under other wave conditions, so it needs further studying.
3. The Goda formula unifies the calculation method of work force of vertical wave, broken wave and breaking wave in one formula, and is not limited to the wave conditions, so it's convenient to use. The Code for Sea Port Hydrology of China gives the formulas to calculate the work force of vertical wave, broken wave and breaking wave respectively; for vertical wave, different formulas are given under different wave conditions, so it's not convenient to be used in projects. Whether the formulas for calculation of wave force in the Code for Sea Port Hydrology can be simplified also needs studying further.

REFERENCES

Xiulun Wang, Xuewen Hu & Manrong Yang. 1983. Japan's transport ministry Harbor Bureau harbor Institute, Port facilities technical standard and explanation, *Communications Press* 23–30, Beijing: China.

Ning Shu & Manying Wang. Application of Goda Formula for Wave Pressure in British Standard [J], *Harbour Engineering*. 2003, 2(1):18–22, China.

Seaport hydrology criterion [S]. *Communications Press* 1998. Beijing, China.

Yuxiu Yu. Random wave and its application in Engineering [M], *Dalian University of Technology press* 2000. Dalian.

Yucheng Li & Bin Teng. Wave action on Maritime Structures [M], Ocean Press 2002, Beijing.

Yucheng Li, Dazhong Liu & Xiaojun Su. The irregular Breaking Wave Forces on Vertical Walls [J], *Journal of Hydrodynamics* 1997, 12(4):457–469.

Zongliang Zhang & Qinghe Zhang. Discussion on Computation of Standing Wave Forces on Vertical Walls [J]. *Port Engineering Technology*, 2006, 3(1):5–7.

Modeling and Computation in Engineering II – Xie (ed)
© 2013 Taylor & Francis Group, London, ISBN 978-1-138-00058-2

Effects of corrosion on the mechanical behavior of corroded steel plate

Bin Qiu & Shanhua Xu
Department of Civil Engineering, Xi'an University of Architecture & Technology, Xi'an, China

ABSTRACT: The objective of this article is to provide data to engineers on the mechanical behavior of corroded steel plates. All the specimens were corroded by the accelerated corrosion process, three different kinds of accelerated corrosion process, the method of acidic soil accelerated corrosion process, the method of acidic atmospheric accelerated corrosion process and the method of constant temperature and humidity. The results indicate that the yield strength and the ultimate strength are not affected by the presence of corrosion, but for the plasticity, it does affected by the presence of corrosion.

1 INTRODUCTION

Existing steel structures exposed to the environment are subject to corrosion. Even galvanized steel can experience corrosion after galvanic protection is consumed, and their material performance is changed accordingly. In recent years, systematic study on corrosion of steel provided by Larrabee et al. (1962) and Shastry (1988) in U.S.A., Kucera (1987) in Sweden and Caifeng Liang (1999) in china, etc.

From 1983 (Cao Chu-nan 2003, Huang Guiqiao 2001, Xia Lanting 2002, Xiao Yi-de 2005), a series of research on steel corrosion under air, sea water and soil have been carried out in china by some scientists on metal material, and got valuable achievements and datum, which is sponsored by Natural Science Foundation of China (NSFC).

Micheal Bruneau et al. (1997) made a study of three samples, and get the conclusions that the yield strength and ultimate strength of steel plate members can be doomed not significantly affected by the presence of corrosion, and the difference between corroded and uncorroded remains within statistical expectations.

The phenomenon of corrosion on steel bars is well studied, but very little research has been done on how the Material behavior of steel plate members is affected once corrosion has developed. There is a lack of experimental data on the relation between weight loss due to corrosion and the yield strength and the ultimate strength of steel plate. The present study was undertaken as a first step toward developing an integrated experimental to quantitatively account for the effects of corrosion on the Material performance of structural steel plate.

In this study, the qualitative investigation on the effect of corrosion, produced by three different kinds of accelerated corrosion process that are the method of acidic soil accelerated corrosion process, the method of acidic atmospheric accelerated corrosion process and the method of constant temperature and humidity, on the yield strength and the ultimate strength of steel plate was investigated by using Electric-Fluid—Servo Universal Testing Machine.

2 EXPERIMENTAL PROCEDURE

2.1 *Material and specimen*

The material used in this study was Q235 steel. In order to get the specimens for mechanical behavior studies, the specimens for corrosion must be longer and wider. According to GB/T 3075-2008 (2008), the specimens for corrosion were 280 mm in length, 50 mm in width and 8 mm in thickness, All the corroded specimens weighted and numbered before corrosion.

2.2 *Accelerated corrosion procedure*

The steel members were corroded by the accelerated corrosion process, three different kinds of accelerated corrosion process. The method of acidic soil accelerated corrosion process, the method of acidic atmospheric accelerated corrosion process and the method of constant temperature and humidity. All the specimens for different kinds of accelerated corrosion process were divided into 6 groups for corrosion, each group concluded 3 specimens.

2.2.1 *The method of acidic soil*

We put and compacted the depth of 250 mm soil in the bottom of the plastic box, which size is 1000 mm × 500 mm × 500 mm. Then laid all the steel members on the top of soil and covered by the depth of 50 mm soil and the soil on the top should be loose for air supply.

The plastic box was exposed to the natural environment and regularly sprayed with HCL solution (18% by volume). To guarantee a similar degree of corrosion damage on every sides for each specimens, the soil on the top should be loosened and all the specimens should be turn over every week. All the specimens were submitted to the process for durations about 6 months, and each group of specimens are removed every one month.

2.2.2 *The method of acidic atmospheric*

All the steel members were exposed to the natural environment and regularly sprayed with acid mist (HCL solution 18% by volume) every week. To guarantee a similar degree of corrosion damage on every sides for each specimens, each specimens should be turn over every week. All the specimens were submitted to the process for durations about 6 months, and each group of specimens are removed every one month.

2.2.3 *The method of constant temperature and humidity*

Because of hygroscopicity of salt, it can greatly shorten the wetting time of metal surface (Fei-feng Deng 2008), and the salinity can enhance the conductivity of the surface, which is the main factors influencing corrosion ratio (Xue-qing Liu 2004). The method of constant temperature and humidity with the temperature is 55°C can accelerate corrosion process (Xing-cai Liang 2001). Therefore, We chose a combinatorial method that immerse the specimens in NaCL solution for about two minutes every five days, then place them into the box of constant temperature and humidity with the temperature is 55°C, and the relative humidity (RH) is 95 ± 3%. All the specimens were submitted to the process for durations about 6 months, and each group of specimens are removed every one month.

2.3 *Mechanical behavior testing*

For mechanical behavior testing, the groups with larger corrosion damage level were chosen, and the corrosion ratio is almost larger then 5%.

All specimens for mechanical behavior studies were 240 mm in length, 30 mm in width and 8 mm in thickness. The details of the plate dog-bone specimens can be found in (GB/T15248-2008& GB/T3075-2008 2008, GB/T3075-82 1982).

Mechanical behavior testing was carried out in Electric-Fluid Servo Universal Fatigue Testing Machine (NO. CSS—WAW300DL), which was to determine the yield strength, the tensile strength and other mechanical behavior.

Before testing, we determined the minimal cross section of the specimens with vernier caliper (measured at three different locations and chose the minimal cross section) and recorded.

3 RESULTS AND DISCUSSION

3.1 *Corrosion ratio measuring*

The corrosion level was measured in terms of weight loss, can be calculated by Eqs. (1). All the specimens for corrosion level measuring was weighted before and after corrosion, and the specimens after corrosion was cleaned from corrosion.

$$\eta_S = \frac{m_0 - m_c}{m_0} \times 100\% \tag{1}$$

here: η_s = Corrosion ratio;
M_0 = the mass of specimens before corrosion (g);
m_c = the mass of specimens after corrosion (g).

The mechanical properties and corrosion ratio for each specimen are shown in Table 1.

Making a study of Table 1, the following comments can be made:

The yield strength and ultimate strength of un-corroded specimens are almost equal. The yield strength and ultimate strength of corroded steel plates, under different corrosion environment, different corrosion damage level, are kind of different, but deviation is little.

3.2 *The stress-strain curve*

Figure 1 shows the stress-strain curve of the specimens selected after and before corrosion. The corrosion damage level is about 6% for A51 and E22, for the rest of the specimens in Figure 1a is greater than 10%. The corrosion damage level for the specimens in Figure 1b, which corroded under acidic atmospheric, is between 5.77% and 10.34%.

Table 1. The mechanical properties and corrosion ratio.

No	σ_y (MPa)	σ_u (MPa)	E (GPa)	η_s (%)	No	σ_y (MPa)	σ_u (MPa)	E (GPa)	η_s (%)
1	272.7	447.5	107	0					
2	273.0	442.6	151	0	3	266.4	442.5	129	0
A41	276.1	425.9	288	5.62	B11	322.9	444.8	218	5.77
A42	251.0	423.6	203	5.33	B12	266.4	421.6	161	5.69
A43	288.0	425.6	325	5.31	B13	300.8	438.0	168	6.34
A51	265.4	425.3	177	6.29	B21	308.7	435.9	200	10.34
A52	273.1	432.5	184	2.98	B22	295.3	459.0	221	10.53
A53	323.1	445.9	247	3.81	B23	300.5	437.8	283	9.99
E11	269.0	429.8	244	5.43	B31	263.5	420.2	147	13.15
E12	289.6	423.0	193	5.98	B32	259.3	423.9	187	12.85
E13	265.7	420.7	256	4.43	B33	280.9	422.1	158	14.84
E21	269.5	429.3	196	7.11	B41	317.7	438.5	208	8.17
E22	298.3	437.1	187	7.10	B42	267.2	420.2	182	11.62
E23	267.0	423.0	152	7.79	B43	261.8	422.0	226	2.60

(No = the number of the specimens; σ_y = The yield strength; σ_u = The ultimate strength)

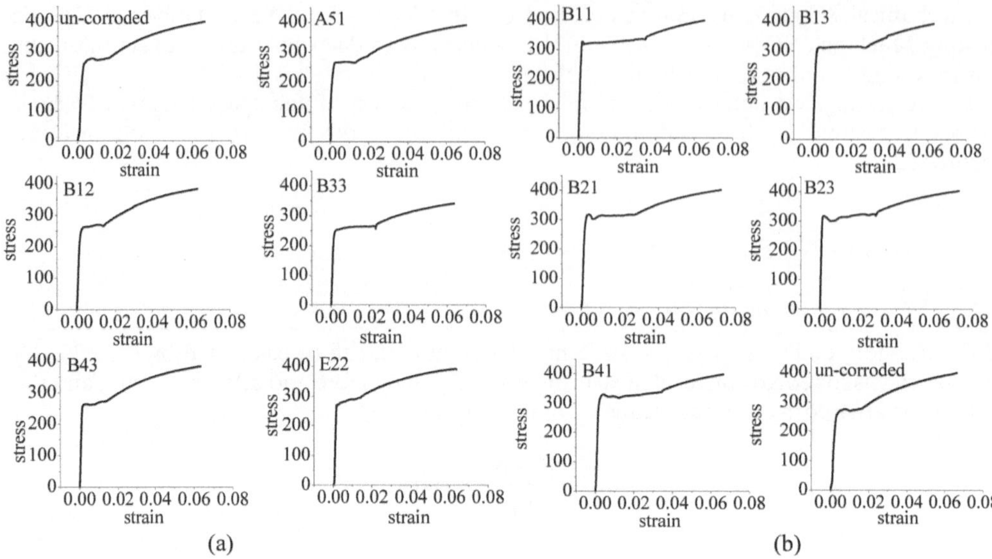

Figure 1. Stress-strain curves for specimens after and before corrosion.

Notes:

1. The descending branch of the stress-strain curve is not tested for we focused on the degeneration of the yield strength and ultimate strength.
2. Because of the limitation of length, here only the stress-strain curve of 11 representative specimens are listed, one un-corroded and ten corroded ones.
3. All the conclusions below is a comprehensive analysis of all the stress-strain curves for all specimens.

Stresses in Figure 1 were obtained by dividing the applied load by the cross-sectional at the minimum area. When making a study of Figure 1, the following comments can be made:

1. In elastic stage, the stress-strain curve before and after is the same; In plastic stage, the stress-strain curve before and after is different; elastic behavior is not affected by the presence of corrosion damage but the plastic behavior is significantly affected by the presence of corrosion damage.
2. The yield strength and ultimate strength of specimens, before and after corrosion, are kind of differ. But the difference remains within statistical expectations, so the conclusion that the yield strength and the ultimate strength can not be affected by the presence of corrosion significantly can be got.
3. A well-defined yield plateau becomes shorter or does not exit for the corroded specimens. Like specimens numbered B12 and B43, its yield plateau become shorter obviously, and the specimen numbered E22, its yield plateau does not exit any more.
4. The ultimate strain for corroded specimens is almost the same as the ultimate strain for the un-corroded ones. Although, the ultimate strain before and after corrosion is kind of differ, the difference remains within statistical expectations.
5. In elastic stage, the stress-strain curve before and after is the same; In plastic stage, the stress-strain curve before and after is different. It indicates that the elasticity modulus is not affected by the presence of corrosion, but the secant modulus is affected by corrosion.

It is of interest to note that for the corroded specimens in Figure 1b, the yield strength is above that of un-corroded ones, the yield plateau is longer than that of un-corroded ones; The ultimate strain before and after corrosion is almost the same, but the gap between yield strength and ultimate strength for corroded specimen significantly narrowed.

It should be noted that the stress-strain curves in Figure 1b were different from the others and therefore needs do exist to further study the effects of corrosion on the stress-strain curves under different corrosion environment and different corrosion damage level.

3.3 *Discussion*

1. A well-defined yield plateau becomes shorter or does not exit for the corroded specimens. This is logical because the cross-sectional area varies continuously along the length of the specimen due to randomness in the corrosion attack (Micheal Bruneau 1997). Such a discontinuity of cross-sectional area causes a high stress concentration at specimen and the ductility of specimen is significantly affected by stress concentration, so the yield plateau becomes shorter or disappear for the corroded specimens.
2. The ultimate strain and ultimate strength of specimens, before and after corrosion, are kind of differ. This difference mostly can be attributed to a slightly premature necking initiation and a shorter descending branch of the stress-strain curve past the point (Micheal Bruneau 1997), But it remains within statistical expectations. Thus, we can say that the ultimate strain and ultimate strength can not be affected by corrosion.
3. The ultimate strain, before and after corrosion, is kind of differ. First, it can be attributed to the systematic residuals which is from experimental instrument, experimental method, experimenter and computer data processing system; second, it also can be attributed to the high stress concentration produced by corrosion pit, the stress concentration can cause the decrease of the strength of steel surface; Third, it must be recognized that the true local minimum thickness could have been missed or simply been beyond reach because of the finite size of the micrometer's head (Micheal Bruneau 1997), then, it will cause calculation residuals by dividing the applied load by the cross-sectional at the minimum area.
4. The reason why different effects of corrosion on different stage for stress-strain curve is as follows: In elastic stage, the whole cross section of the specimen is in elastic and the stress on cross section is very small, thus, the material behavior in this stage is not affected by the stress concentration; In yield stage, the stress on some specific location that its stress is the maximum reaches the yield strength of steel, the stress on the other location of the cross section does not reach the yield, thus, due to the redistribution of stress, the material behavior in this stage is affected little by the stress concentration; In strain hardening stage, the stress on the whole cross section reaches the yield stress, with high stress concentration around corrosion pit, the stress around corrosion pit will exceed the ultimate strength and it will fracture firstly, while the stress on the other location of the cross section is far away from the ultimate strength and the load applied on the specimen is also far away from the ultimate load, thus, the material behavior in this stage is affected by the presence of corrosion damage significantly.

4 CONCLUSIONS

This article presents the results of material performance test of 9 groups, 2 of them being corroded at the method of acidic soil, 4 of them being at acidic atmospheric and 2 of them being at constant temperature and humidity. Besides that, there was another un-corroded group was used as a control. Each group contains 3 specimens. With a comprehensive analysis of all the stress-strain curves and all the datum for all specimens, the following conclusions can be made:

1. The yield strength, ultimate strength and ultimate strain of steel plate members can not significantly affected by the presence of corrosion.
2. The ductility of steel plate members can be significantly affected by the presence of corrosion for a well-defined yield plateau become shorter or does not exist for corroded specimens.

3. It has different effects of corrosion on different stage for stress-strain curve. In elastic stage, the stress-strain curve affected little by corrosion, but in non-elastic stage, it affected significantly by corrosion.
4. The elasticity modulus of steel plate members can be doomed not significantly affected by the presence of corrosion, but the secant modulus can be significantly affected by the presence of corrosion.

REFERENCES

Cao Chu-nan etc. An introduction of the achievements of "the data accumulation and laws of materials corrosion in the natural environment" of china [J]. Bulletin of National Natural Science Foundation of China, 2003, 6: 323–325 (In China).

Fei-feng Deng, Feng Pei & Yong-jun Liu. Effect of Air Relative Humidity on Metal Corrosion of Outdoor Electrical Cabinets and the Measure to Control Humidity [J]. Central China Electric Power, 2008, 21(2): 57–61 (In China).

Huang Guiqiao. Corrosion Behaviour of Carbon Steels Immersed in Sea Areas of China [J]. Corrosion Science and Protection Technology, 2001, 13(2): 81–84 (In China).

Hou, W.T. & Liang, C.F. Eight-year atmospheric exposure of steels in China [J]. Corrosion, 1999:1.

Kucera Knotkova, D., Gullman, J. & Holler, P. Cormsion of structural metals in atmospheres with different corrosivity at 8 years, exposure in Sweden and Czechoslovakia A. Proc.10th Int. Cong. Met. Corros. C. Oxford and IDH [M], Madras, India, 1987: 167.

Larrabee, C.P. & Coburn, S.K. The atmospheric corrosion of steels as influenced by changes in chemical composition A. Proc. lst Int. Cong. Met. Corros. C. Butterworth [M], London, UK, 1962: 276.

Micheal Bruneau & Seyed Mehdi Zahrai. Effect of severe corrosion on cyclic ductility of steel [J]. Journal of structural engineering, 1997, 9: 1478–1486.

National Standard of the People's Republic of China. Metallic materials-Fatigue testing—Axial—force—controlled method [S]. (GB/T3075-2008 (in China))

National Standard of the People's Republic of China. Method of axial force controlled fatigue testing of metals [S]. (GB/T3075-82 (in China))

National Standard of the People's Republic of China. The test method for axial loading constant—amplitude low-cycle fatigue of metallic materials [S]. (GB/T15248-2008 (in China))

Shastry, Friel, J.J. & Townsend, H.E. Sixteen-year corrosion Performance of weathering steels in marine, rural and industrial environments A. Degradation of Metals in the Atmosphere C. ASTM STP965 [M], West Conshohocken, PA, ASTM. 1988: 5.

Xia Lanting etc. Present Status of Research on Sea-Water Corrosion of Metal in China [J]. China Foundry Machinery & Technology, 2002, 6: 1–4 (In China).

Xiao Yi-de etc. Recent development in atmospheric corrosion study of materials in China [J]. Equipment environmental engineering, 2005, 2(5): 3–9 (In China).

Xing-cai Liang. Research on correlation of the artificially accelerated test of material and products with the weathering exposure tests [J]. Environmental Technology, 2001(4): 4–7 (In China).

Xue-qing Liu. Investigation on the corrosion behavior and corrosion prediction model of engineering steel used in marine environment [D]. PhD thesis. Graduate University of Chinese Academy Sciences, Qingdao, China, 2004 (In China).

Modeling and Computation in Engineering II – Xie (ed)
© 2013 Taylor & Francis Group, London, ISBN 978-1-138-00058-2

Mechanical performances of multi-cell girders with corrugated steel webs

Lei Ma, Shangmin Zheng & Shui Wan
College of communication, Southeast University, Nanjing Jiangsu, China

ABSTRACT: With the finite element method, several geometrical parameters influencing the mechanical performances of multi-cell girders with corrugated steel webs under eccentric loading were analyzed. The ratio of height and span, the ratio of width and span, corrugation configuration of steel web and the thickness of steel web were analyzed. The results show that decreasing the ratio of height and span, the ratio of width and span and increasing the thickness of steel webs can improve its mechanical performances under eccentric loading. The effect of the rotation of corrugation configuration of steel web is complex. In order to choose the proper corrugation configuration of steel web, FEM analysis must be used.

1 INTRODUCTION

Compared with the concrete box girder, the torsional rigidity of composite box-girder bridge with corrugated steel webs is reduced to 40%. Lateral bending stiffness and longitudinal bending stiffness decreased respectively by 10% and 25% (Li 2003 & Sayed-Ahmed 2005). The spatial effect of multi-cell girders with corrugated steel webs is prominent because of the large width of bridge deck (Xu 2005). The longitudinal warping normal stress which caused by constrained torsion and distortion can reach to considerable level under eccentric loading.

With the finite element method and the test, several geometrical parameters influencing its mechanical performances were analyzed for providing reference for the design of multi-cell girders with corrugated steel webs.

2 ENGINEERING SITUATION

Taking Weihe bridge (Wu 2010) for an example, it is a three spans of continuous composite box-girder bridge with corrugated steel webs and constant cross-section. The spans are 47 m, 52 m and 47 m. As shown in fig. 1, the top flange width of the box-girder is 16.85 m, the thickness is 25 cm; the width of the floor is 11.85 m, the thickness is 22 cm; the height is 3.2 m; The webs are corrugated steel webs whose thickness is 12 mm. The concrete is C50 which proportion is 2650 kg/m^3, the elastic modulus is 3.5×10^4 MPa and Poisson's ratio is 0.1667; the webs are steel Q345D, which proportion is 7850 kg/m^3, the elastic modulus is 2.06×10^5 MPa and Poisson's ratio is 0.3.

3 MODEL DESCRIPTION

3.1 *Element types*

The finite element analysis program ANSYS version 12.0 is used to model mechanical performances of multi-cell girders with corrugated steel webs under eccentric loading. For the

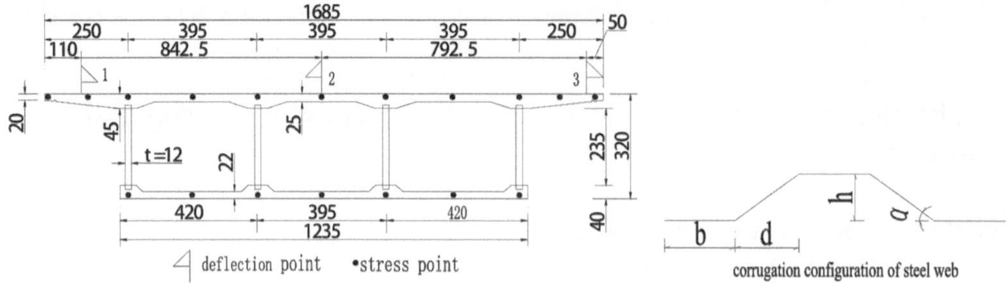

Figure 1. Geometry of cross-section of the beam and arrangement of test (unit: cm).

Figure 2. Finite element model.

main regions of concrete top and bottom flanges (the middle span as shown in Fig. 2), an eight-node 3-D Reinforced Concrete Solid (SOLID45) element is used. Each node has three degrees of freedom, namely three translations in the nodal *x, y* and *z* directions, respectively. The corrugated steel webs and steel flanges are simulated by a four-node quadrilateral layered shell element SHELL163. This element is selected for its displacements compatibility with the concrete element SOLID45. Each node has six degrees of freedom, namely three translations in the nodal *x, y* and *z* directions, respectively, and three rotations about the nodal *x, y* and *z* axes, respectively. This element also has bending stiffness in addition to the membrane stiffness, which is particularly important in simulating the behavior of thin-walled structures and local buckling. The prestressed tendons are modeled by 3D two-node spar element LINK8. This element is a uniaxial tension-compression element with three degrees of freedom at each node: translations in the nodal *x, y,* and *z* directions. An initial strain is given in LINK8 element to simulate prestressing through Real Constant option. The slip between reinforcing bars and concrete is ignored through sharing the nodes of SOLID45 element with those of LINK8 element.

3.2 *Mesh sensitivity*

The sensitivity of the multi-cell girders with corrugated steel webs response to various mesh configurations of corrugated webs is also investigated. Two different meshes were generated in ANSYS. The sub-panel is meshed by 1 and 4 elements while the vertical edges are meshed by 6 and 12 elements in Mesh 1 and Mesh 2, respectively. The results of the mesh sensitivity analyses show that mesh refinement in the corrugated webs has negligible effect on bending stresses. In order to reduce the computing time, Mesh 1 is used in remaining studies. The finite element mesh of concrete top and bottom flanges is assigned in such a way that all elements maintain reasonable aspect ratios to avoid shear lock. The specimen is meshed by 81084 SOLID45 elements; 7232 LINK8 elements; 10808 SHELL163 elements and 131567 NODEs in all. The global *x, y,* and *z* axes are defined in the transverse, vertical and longitudinal direction as shown in Fig. 2.

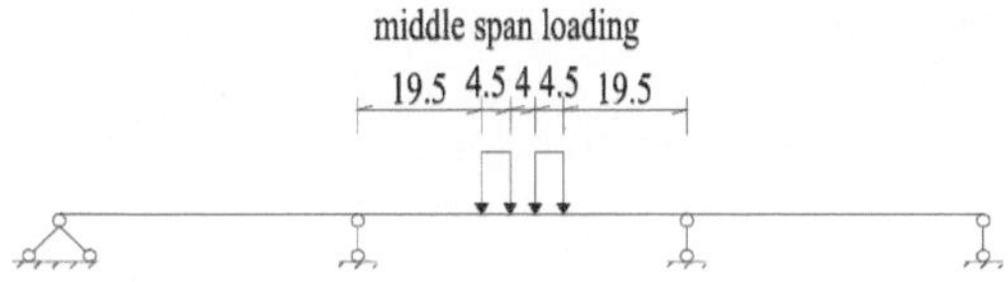

Figure 3. Axial direction loading layout (unit: m).

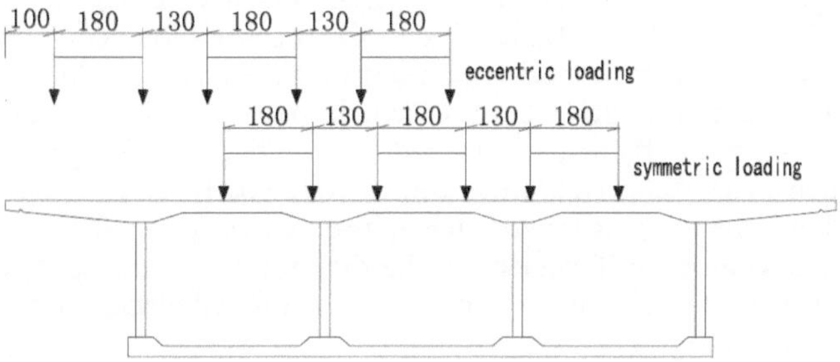

Figure 4. Transverse loading layout (unit: cm).

Table 1. Deflection and stress comparison between test and FEM results.

	Deflection/mm		Stress/MPa	
Conditions	Test	FEM	Test	FEM
Symmetrical load	–4.62	–4.89	1.1	1.31
Eccentric load	–6.27	–6.72	1.31	1.87

3.3 *Loading and boundary condition*

The load test is conducted for the above project by using six trucks with payloads of 30 *t*. Loading layout along the bridge in axial direction is shown as fig. 3 and transverse loading layout shown as fig. 4. The connection between steel flanges and concrete top and bottom flanges is regarded as rigid by coupling translations in the nodal *x, y,* and *z* directions. *Ux, Uy* and *Uz* displacement at the fixed hinge support are restricted; *Ux* and *Uy* displacement at the sliding hinge support are restricted.

3.4 *Model verification*

The FEM described above is used to predict the mechanical performances of multi-cell girders with corrugated steel webs under eccentric loading. Table 1 shows the comparison of deflection and stress between test and FEM results. It should be noted that, in Table 1, the deflection result is the value of point 2 (fig. 1) at the symmetrical load condition and point 3 (fig. 1) at the eccentric load condition; The stress results are the value at the right side corner.

It can be seen from table 1 that the test (MA 2012) values and the FEM values are agreement. This indicates that the FEM established in this paper can be used to predict the mechanical performances of multi-cell girders with corrugated steel webs under eccentric loading efficiently.

4 PARAMETRIC STUDY

A series of FEM are used in the parametric study to examine the mechanical performances of multi-cell girders with corrugated steel webs under eccentric loading. These parameters are the ratio of height and span, the ratio of width and span, corrugation configuration of steel web and the thickness of steel web.

According to the above FEM, the diaphragms in the span and the prestressed tendons are neglected. The load conditions are shown as fig. 5. (P_1 = 100 kN, P = 200 kN) The law of the change of eccentric load coefficients and rotation according to the parametric change are discussed as the following.

4.1 *Ratio of height and span*

In order to investigate the effect of the ratio of height and span(H/L), eight types of height, are considered as shown in Table 2. Table 2 shows the eccentric load coefficients.

It can be seen that the eccentric load coefficient increases as the value of parameter H/L increases. When H/L is greater than 1/10, the eccentric load coefficient increases to more than 2.0, the effect of eccentric load can't be ignored in the structural design. And when H/L is relatively small, such as less than 1/20, the eccentric load coefficient is less than 1.75. If considering the dead load, the warping stress causing by eccentric load will be a small proportion in the normal stress. Then even if neglecting the diaphragms in the span, the effect of the eccentric load is little. It is a sensitive parameters of multi-cell girders with corrugated steel webs under eccentric loading.

4.2 *Ratio of width and span*

Four types of width are considered as shown in Table 3 to investigate the effect of the ratio of width and span (B/L).

It can be seen that the eccentric load coefficient increases as the value of parameter B/L increases. When B/L is greater than 1/4.4, the eccentric load coefficient increases to more than 1.8, the effect of eccentric load can't be ignored in the structural design. And when B/L is relatively small, such as less than 1/8, the eccentric load coefficient is less than 1.4. It is a sensitive parameters of multi-cell girders with corrugated steel webs under eccentric loading.

4.3 *Corrugation configuration of steel web*

When the length of the horizontal section b = 330 mm, $\alpha \leq 20°$, the rotation of the girder decreases as folding angle α increases; $20° \leq \alpha \leq 40°$, the rotation of the girder change little.

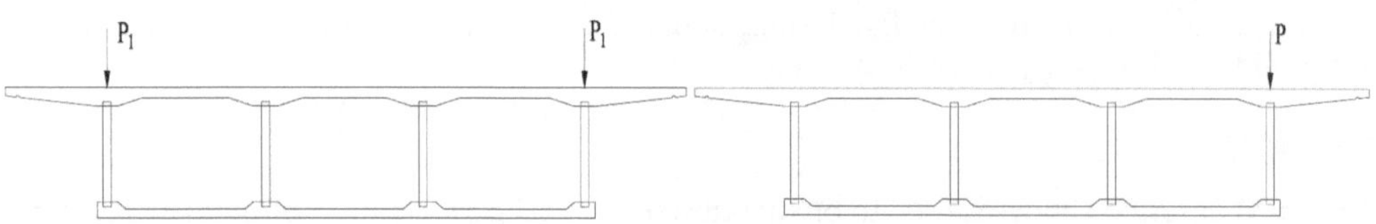

Figure 5. Loading layout.

Table 2. Eccentric load coefficients for the ratio of height and span.

FEM	H/(m)	H/L	Normal stress under eccentric loads/(MPa)	Normal stress under eccentric loads/(MPa)	Eccentric load coefficient
1	5.2	1/10	0.1290	0.2610	2.0233
2	4.16	1/12.5	0.1716	0.3318	1.9336
3	3.47	1/15	0.2136	0.3956	1.8521
4	3.2	1/16.25	0.2349	0.4270	1.8181
5	2.81	1/18.5	0.2728	0.4826	1.7690
6	2.6	1/20	0.2982	0.5197	1.7430
7	2.31	1/22.5	0.3412	0.5828	1.7082
8	2.08	1/25	0.3846	0.6466	1.6811

Table 3. Eccentric load coefficients for the ratio of width and span.

FEM	B/(m)	B/L	Normal stress under symmetrical load/(MPa)	Normal stress under eccentric loads/(MPa)	Eccentric load coefficient
1	11.85	1/4.4	0.2349	0.427	1.8181
2	10.4	1/5	0.2537	0.4453	1.7554
3	8.67	1/6	0.2709	0.443	1.6355
4	6.5	1/8	0.3296	0.4669	1.4165

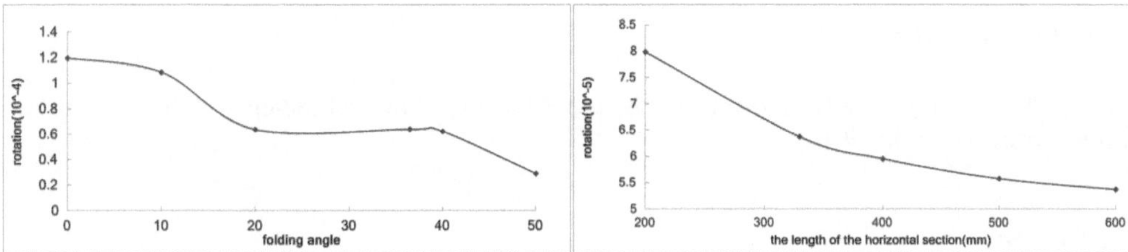

Figure 6. Rotation curve of corrugation configuration of steel web.

Table 4. Eccentric load coefficients for the thickness of web.

Thickness of web/(mm)	Normal stress under symmetrical load/(MPa)	Normal stress under eccentric loads/(MPa)	Eccentric load coefficient
20	0.2283	0.4066	1.7805
18	0.2304	0.4117	1.787
16	0.2329	0.4179	1.7944
14	0.236	0.4255	1.8031
12	0.2401	0.4354	1.8135
10	0.2456	0.4485	1.8262
8	0.2536	0.467	1.8418

$\alpha > 40°$, the rotation of the girder continues to decrease. When the folding angle $\alpha > 365°$, the rotation of the girder decreases as the horizontal section b increases; 200 mm $\leq b \leq$ 400 mm, the rotation of the girder decreases quickly as b increases; 400 mm $\leq b \leq$ 600 mm, the rotation of the girder decreases slowly. In order to choose the corrugation configuration of steel web, FEM analysis must be used detailed and accurate.

4.4 *Thickness of web*

The thickness of web is shown as figure 6. Table 4 shows the eccentric load coefficients.

It can be seen that the eccentric load coefficient increases as the thickness of web decreases. But the change is not obvious. It is a proof of the conclusion that the flanges provide the flexural strength of the beam with no contribution from the corrugated web (Ezzeldin et al., 2001). It is not a sensitive parameters of multi-cell girders with corrugated steel webs under eccentric loading.

5 SUMMARY AND CONCLUSIONS

A three dimensional finite element model of multi-cell girders with corrugated steel webs is established to investigate the behavior under eccentric loading. The deflection and stress predicted by FEM agree well with that obtained from filed test. Subsequently, a parametric study is carried out and the following conclusions are drawn:

The ratio of height and span (H/L), the ratio of width and span (B/L) are sensitive parameters of multi-cell girders with corrugated steel webs under eccentric loading. The eccentric load coefficient is increased as either H/L or B/L increases. The effect of the rotation of corrugation configuration of steel web is complex. In order to choose the proper corrugation configuration of steel web, FEM analysis must be used. Specimens with thick web have lower eccentric load coefficient. It's not a sensitive parameter of multi-cell girders with corrugated steel webs under eccentric loading.

ACKNOWLEDGEMENTS

This work was financially supported by the National Natural Science Foundation of China (Grant No. 50078014).

REFERENCES

Ezzeldin, Yazeed & Sayed-Ahmed. 2001. Behavior of steel and (or) composite girders with corrugated steel webs. *Canadian Journal of Civil Engineering*, 28(4):656–672.

Li Hongjiang. 2003. Experimental Study and Analysis on Torsion and Distortion of Box-girder with Corrugated Steel Web. College of communication, Southeast University:17–29.

Ma Lei, Jin Jiugui & Wan Shui. 2012. Experimental investigation of one box with three rooms PC composite box-girder bridge with corrugated steel webs. Journal of Highway and Transportation Research and Development, 29(3):74–79.

Ma Lei & Wan Shui. 2012. One box with three rooms PC composite box-girder bridge with corrugated steel webs Monitoring. Construction Technology, 41(378):11–14.

Sayed-Ahmed E.Y. 2005. Plate girders with corrugated steel webs. Engineering Journal, 42(1):1–13.

Wu Jifeng, Tang Yi & Wu Ping. 2010. The design of the girders with corrugated steel webs-Weihe. HIGHWAY, 1(1):57–62.

Xu Keying. 2005. Design method of the multi-cell box girder with arc-shape bottom. Shanghai: Tongji University.

Modeling and Computation in Engineering II – Xie (ed)
© 2013 Taylor & Francis Group, London, ISBN 978-1-138-00058-2

Design of adding floors reconstruction of brick & concrete structure building

Zhihui Bian
School of Civil Engineering, Tianjin University, Tianjin, China
Hebei Academy of Building Research, Shijiazhuang, China

Sujuan Fu
Hebei Academy of Building Research, Shijiazhuang, China

Jia Li
Hebei Building Research Technology Co., Ltd., Shijiazhuang, China

ABSTRACT: Through the adding floors reconstruction of an office building, this paper explores the feasibility of adding floors, and elaborates reinforcement construction points. In order to reduce the load increment, part of the wall mortar on the one storey is replaced by higher strength mortar. Meanwhile the original building needs to add ring beams and columns. According to scientific and rational seismic strengthening, results show that the integrity and service life of multistory masonry structure increases effectively. Through design analysis of above mentioned process, some effective methods are introduced, which can be used as reference for the reinforcement and reconstruction of similar projects.

1 INTRODUCTION

With the rapid development of China's urbanization, construction of the increasing size, low-rise, low construction area and other reasons made original building unable to meet the demand now. The demolition and reconstruction is not only a waste of money and resources, but also to interrupt the use of the original building, which is not economical. Therefore, adding floors for the old building has aroused more social attention. Retrofitting of old buildings, you should take full account of the components in the original structure, and select a suitable reinforcement way, considering economic rationality and to ensure the safety of retrofitting structures. This paper introduces the design of a brick and concrete office building adding floors transformation.

2 GENERAL SITUATION

An office building was built in the seventies of the last century. The structural form is three floors above ground, with four floors in part. The plane is a "line" type.

The building is 13.2 meters from north to south and 38.4 meters from east to west, with a area of 1600 square meters. The floor is precast hollow plate. The outer wall is 370 mm, and interior wall is 240 mm. Detection and identification results show that, the original structural are lack of construction measure including ring beams and columns, mortar strength of each floor from field-based test is shown in Table 1.

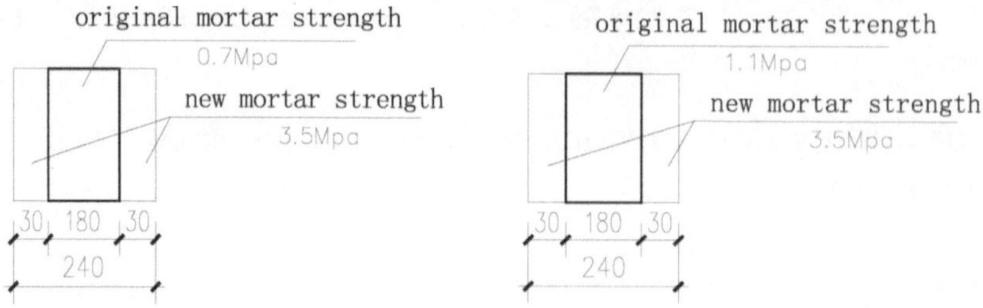

Figure 1. Weighted strength wall.

Table 1. Mortar strength of each floor after mortar strength weighted.

Floor	1	2	3	4
Mortar strength (MPa)	1.4	1.7	1.2	1.2

3 DESIGN OF ADDING STOREY PROJECT

According to the requirement of owner, this office building needs to increase one layer. Due to the increase of load and insufficient strength, modifying earthquake resistance measures of new code for seismic design of buildings, the original building needs to increase ring beams and columns, the wall needs mortar surface layer of steel mesh reinforcement.

3.1 *Code for seismic measures and structural measure*

According to the code for seismic design of buildings, the project is located in the area where fortification intensity is 7. Since the building is built in 1970s, and the seismic code is different from now, the original building is lack of seismic measure in ring beams and columns. According to the current seismic code, it is necessary to add columns in exterior wall corners, junction of cross wall and longitudinal wall, and big openings. Ring beams should be set in every floor.

3.2 *Accounting of structural bearing capacity headings*

In order to ensure the safety of the infrastructure, it is essential to account structural bearing capacity. Reinforcement should be carried for those inadequate bearing capacity components.

According to test report which includes the measured drawings, field detection building dimensions, and Mortar strength, the Wall bearing capacity is calculated. Since the original mortar strength is on the low side, the wall needs to Mortar surface layer of steel mesh reinforcement. Part of the wall mortar on the one storey is replaced by higher strength mortar. The result is as follows:

Wall strength of the first storey

$$f_1 = \frac{3.5 \times 30 \times 2 + 0.7 \times 180}{240} = 1.4\ \text{Mpa} \tag{1}$$

Wall strength of the second storey

$$f_2 = \frac{3.5 \times 30 \times 2 + 1.1 \times 180}{240} = 1.7\ \text{Mpa} \tag{2}$$

Layers of mortar strength are listed in the Table 1.

Taking the first storey as an example, some walls are lack of seismic strength after adding floor. The result shows that both the seismic capacity and the compression bearing capacity are insufficient, but also lack of local pressure, as shown in Figure 2 to Figure 4.

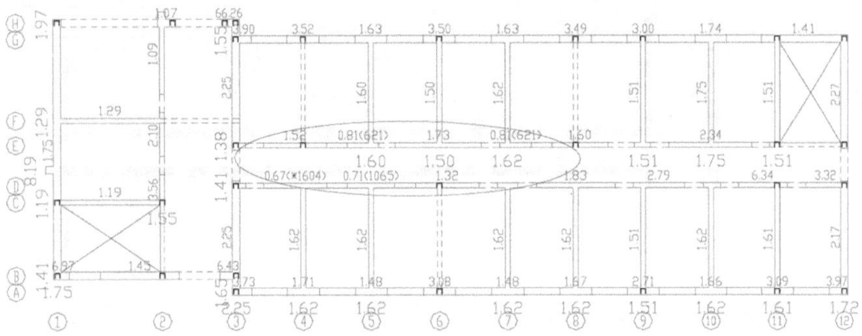

Figure 2. Seismic checking computation of first storey.

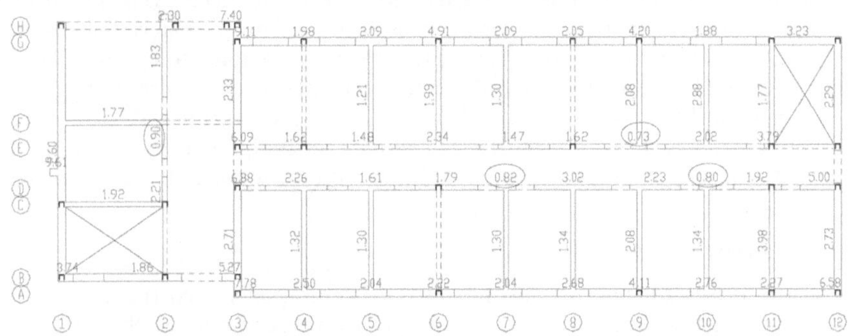

Figure 3. Compressive capacity computation of first storey.

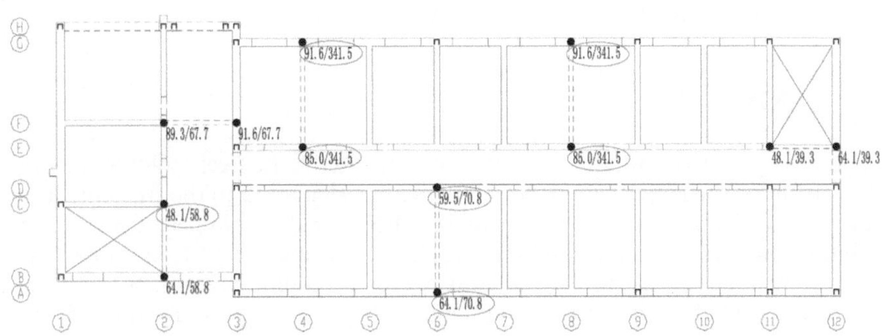

Figure 4. Compression bearing capacity of first storey.

4 STRENGTHENING MEASURES

4.1 *Constructional column addition*

Since the lack of building earthquake resistance measures, constructional columns are necessary to add in suitable place like the corners of the building and the stairwell. In order to strengthen the constructional column with the original structure of the connection, concrete pin which is 180 × 180 mm and concrete grade of C20 is set along the height of 1/3 storey, and lacing bar with more than 1500 mm length is connected with wall.

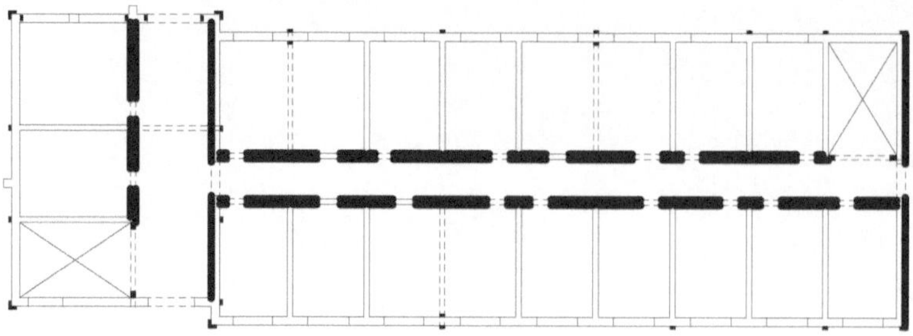

Figure 5. Reinforcement parts in the first storey.

4.2 *Ring beam addition*

In the case, considering the shortness of construction period and decrease of the influence as far as possible on the original building, ring beams additional to the external walls are made up of concrete from first to forth storey. While ring beams additional to the interior walls are reinforced encryption where are mesh reinforcement cement mortar reinforced wall, the other parts are adopted steel rod as ring beams. In order to ensure the integrity of the structure, a 180 × 180mm pin key is set every 1 to 2 meters along the length of the new cast in situ concrete beams, and the end of steel rod should be further into new concrete cast-in-place beam.

4.3 *Wall reinforcement practice*

Because the wall bearing capacity is insufficient, it is commonly used the method of reinforcement steel mesh reinforced mortar surface layer to strengthen the wall.

Considering the structure holistic concept design, although a few tablet wall are insufficient in the bearing capacity, as showed in Figure 2 and Figure 3, in order to avoid the structure torsion failure caused by strengthening stiffness with uneven seismic force concentration, both the vertical walls and cross walls are overall strengthening practice. Taking the first storey as example, reinforcement parts are showed in Figure 5.

5 CONCLUSIONS

To reduce the weight of the adding floor, it should be used light-steel building material. For the reinforcement of the main body portion, modeling calculation to the structure after adding floor is carried out. Following the principle of short construction period and low cost, to ensure reinforcement plan reliable, different reinforcement measures are employed in different parts. Adding floor reconstruction, creating a good economic benefits, not only increases the area of the original building, but also eliminates the security risks of old buildings.

REFERENCES

Lin Li, 2004. Design of adding floors reconstruction of brick & concrete structure building. *Shanxi Architecture* 30(24): 34–35.

Tu Zhichuan, 2004. The changing desing for added-level of multi-floor and mixed structure of building. *Fujian Architecture & Construction* 90(5): 69–70.

Zhou Qi & Zhao Cheng, 2010. Reconstruction design for an office building storey adding and floor reinforcement. *Engineering construction* 42(3): 24–28.

Wang Zhuang & Li Yuchun, 2010. A study of seismic strengthening of masonry structure of existing school building. *China Civil Engineering Journal* 43(supplement): 436–441.

Yin Tao. 2010. On the reinforcement and earthquake-resistant evaluation of a multi-storey masonry structure classroom building. *Shanxi Architeture* 36(13): 53–55.

Modeling and Computation in Engineering II – Xie (ed)
 ISBN 978-1-138-00058-2

Kinematic analysis of footwork for return of serve on world's elite tennis players

Yirong Li & Jihe Zhou
Chengdu Sport University, Chengdu, Sichuan, P.R. China

ABSTRACT: Step movement for return of service in tennis is a general term for various actions adopted by player to control body, and change position, direction and speed. The movements of footwork for return of serve of three players in the finals and semifinals of 2011 Chengdu Open-ATP Champions Tour were filmed and analysis with three-dimensional video analysis. The actions are divided into three stages, preparation phase, skip step phase and stroke phase, so as to analyze the main kinematic parameters. The kinematics characters of footwork were presented to providing the basis for tennis training and match. Research shows us: In the preparation phase, the average horizontal angles between the two feet of Sampras, El Aynaoui and Moya are 12.1 deg., 2.2 deg. and 45.6 deg. respectively, and the distances between the two feet are 0.465 m, 0.874 m and 0.583 m respectively. In the skip step phase, the maximum jumping heights of Sampras, El Aynaoui and Moya are 0.101 m, 0.175 m and 0.091 m respectively, the horizontal moving distances of their left feet are 0.501 m, 0.400 m and 0.437 m respectively, and the average moving speeds of their left feet are 1.139 m/s, 1.333 m/s and 0.491 m/s respectively. In the stroke phase, the moving distances of left feet of Sampras and Moya are greater (0.892 m and 0.667 m respectively) and the moving speeds of Sampras and Moya are faster, being 2.23 m/s and 1.96 m/s respectively.

1 INTRODUCTION

In the modern tennis, the serve receive technical occupies an important position. In the tennis matches, serving side is usually in the active status. Therefore, in a tennis tournament, serving side is generally in initiative position, and therefore if receiving side wants to occupy a leading position in the whole tournament, they shall "break" the service game of opponent so as to lay a foundation for the final win. Then, if one wants to acquire initiative in the service game of opponent, the only way is to improve the quality of link of receiving. Since a rapid, flexible and accurate footwork is the key factor for the success of receiving, the research on footwork to receive seems to be extremely important.

Elite players technology model is a biological mechanics rules and principles of the model of rational technology, to outstanding players action technology structure of thorough research, looking for technical structure characteristics and laws, the technology training practice has important reference value. Through the access relevant information found, at home and abroad at present in the tennis training receiving technology, training, and more research in the game to receive moving research is less. Therefore, this article through 3d camera analytic method, to 2011 chengdu open—ATP tournament championship final and semi-final three players receiving moving on the kinematics analysis, reveals the tennis project elite players receiving moving the kinematics characteristics, for tennis training project to provide the reference.

2 METHODS

This research mainly uses three dimensional camera analytic methods. (The research object shown in table 1).

Two JVC9800 cameras (50 fps) was used to shoot the serve receive technical movements of players in Chengdu ATP tour (camera position is shown in figure 1), and selecting video footage of three players champion, 2nd place and 3rd place, three return of services for each player. The selection standard as follows: Firstly, receiving position in the same area; Secondly, return the first service; Thirdly, receiving action with backhand chops, Fourthly, receiving success.

Video resolution using Signal TEC V2.0C three-dimensional video analytical software, coordinate system setting shown in figure 2, and choose Japanese Matsui's human body model (16 links, 21 articulation point).

Because of the frequency of the camera was only 50 Hz, we have not catch the picture of the moment that the ball get in touch with the racket in shooting process. Cubic Spline Interpolation Method was used during the data processing to interpolated the original data to make the data output frequency increased to 100 Hz. Smoothing the original data with optimization line low pass filter and the truncation frequency is 8 Hz.

Table 1. The research object basic.

Players	Nationality	Age	Height	Weight	Took hold	ATP highest ranking
Pete Sampras	USA	41	185 cm	80 kg	right	1
Younes El Aynaoui	Morocco	31	193 cm	84 kg	right	13
Carlos Moya	Spain	36	190 cm	84 kg	right	1

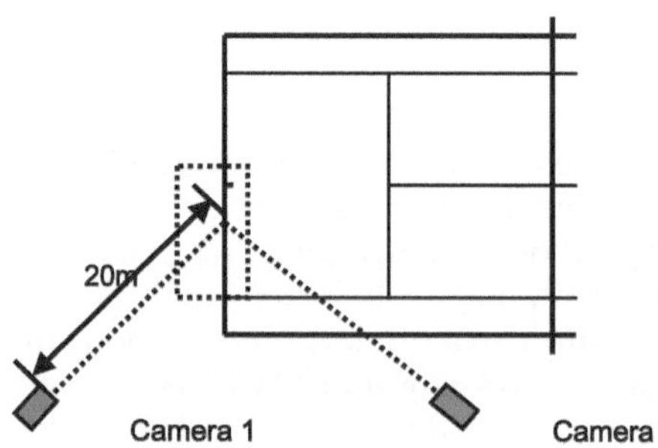

Figure 1. Camera position.

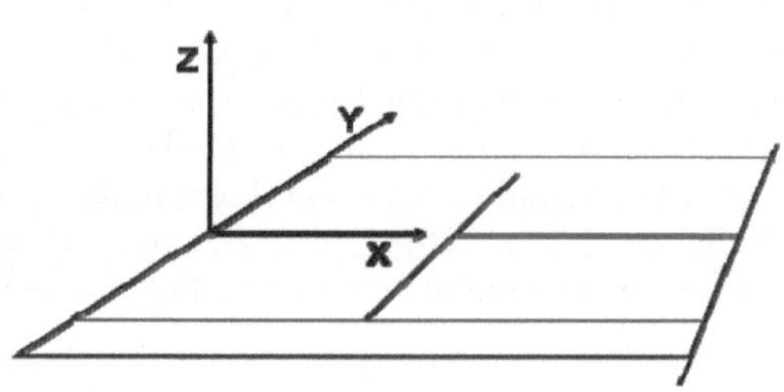

Figure 2. Coordinate system.

3 RESULTS AND DISCUSSION

Tennis Return footwork is a general term for a variety of action methods that the players control body, change position, direction and speed. In order to meet the needs of the fierce competition, footwork perfecting is more and more recognised by coaches and players. No flexible footwork, cannot take favorable position to hit the ball in a timely and effective response to the ball.

3.1 *Preparation phase*

The preparatory phase refers to the period from the time that player chooses suitable stance and prepares well to receive a service to the time that the player takes off one foot from ground when beginning to move the step. At this phase, stance is the important aspect of starting, it is on the bottom line near the single sideline in general, player lower the center of gravity, natural and relaxed use a variety of small hop in situ or weight transfer to ensure fast starting when bit the ball. There is not any particular provision for body posture and then quickly start.

The research *Comparative Analysis tennis open and closed the application of the baseline drives* shows that: "Use two feet palm connect with the bottom line angle size to division two stroke stance posture has gradually reached an agreement, it means when the angle is less than 30 degrees position for open stance, more than 30 degrees for enclosed." (Sun Yanming & Chang Chunfang. 2010.) It can be concluded from Table 2 that, in this stage, the average horizontal angles between both feet of Sampras, El Aynaoui and Moya are 12.1 deg., 2.2 deg. and 45.6 deg., respectively, and the distances between both feet are 0.465 m, 0.874 m and 0.583 m, respectively. It shows that Sampras and El Aynaoui have the approximately same stance, with both feet paralleling nearly, while Moya has the slightly larger angle between both feet. Distance between both feet is affected by height of player. The distance for Sampras is smaller relatively. The stance for return of service takes making body start quickly as the best, which has small influence on return of service.

3.2 *Skip step phase*

The skip step stage is the period from the time that player takes off one foot from ground when beginning to move step after preparation to the time that the player lands with both feet after skip step before stroking.

Skip step is the player make a small pad step before starting up, then starting to left or right. Its main function from physiological point view can be explained as the stretch reflex principle. It is pointed out in the *Analysis of Ttennis Serve Players Starting Action Structure*: "When muscle stretching passively, muscle proprioceptors, muscle spindle, was pulled, and back to the cerebral cortex dominated muscles work. It enhanced the excitability of the cerebral cortex, increased myodynamia. Within the scope of the muscles can withstand, pulling speed is proportional to muscle contraction force after pulling, and pulling pause is inversely proportional to the muscle contraction force." (Wang Qining. 1997.) Therefore, in order to get larger muscle force, muscles should be pulled as fast as possible.

Table 2. Comparison of player's step characteristics in the preparation stage.

Players	Horizontal angle between both feet	Distance between both feet	Period for preparation stage
Pete Sampras	12.1 degrees	0.465 m	0.62 s
Younes El Aynaoui	2.2 degrees	0.874 m	0.54 s
Carlos Moya	45.6 degrees	0.583 m	0.60 s

The starting of tennis receiving player include reaction time and actual movement time. A right skip step should be determined the direction of the ball before step landing. One leg active stretching in the land instead of legs, the other side leg not touch ground but have to move directly to the direction of movement. Grasp the step opportunity is measured as the key to complete quality of ready for action. But this opportunity is affected by jumping height, dropping height and serve player's throwing height and hitting speed. Determine whether the opportunity was grasped is based on determined the direction before ball landing. This can be done with a single leg form, maximum use of pedal leg strength, take full advantage of the available time, accelerates starting speed.

There is information showing that the height of skip step shall not be more than 0.15 m, and the distance between anterior soles of both feet shall be slightly wider than shoulder width when landing on the ground. It can be concluded from Table 2 that, in this stage, the maximum jumping heights of Sampras, El Aynaoui and Moya are 0.101 m, 0.175 m and 0.091 m, respectively, and the distances between both feet are 0.832 m, 1.001 m and 0.746 m, respectively. The jumping heights of Sampras and Moya are reasonable, but El Aynaoui's jumping height is slightly higher, which influences the quality of service.

After the skip step, the horizontal moving distances of left foot for Sampras, El Aynaoui and Moya are 0.501 m, 0.400 m and 0.437 m, respectively. All the distances for the three persons are longer than the horizontal moving distances of right foot, especially Sampras. The average moving speeds of left foot are 1.139 m/s, 1.333 m/s and 0.491 m/s, respectively. El Aynaoui's moving period is the shortest one (0.30 s).

3.3 *Sroke phase*

Stroke stage means the period from both feet of player landing on the ground after the skip step is over to one foot of player landing after the stroke.

Table 4 shows that: moving distances of left foot of Sampras and Moya are greater (0.892 m and 0.667 m respectively); when the moving time is equal, the moving speed of their foot is faster, and the follow speed of center of gravity of their bodies is faster accordingly and they can hit the ball in one step while El Aynaoui needs two steps. This shows that a player shall know well the follow adjustment action of center of gravity of received action, so that he/she can prepare in advance to ensure center of gravity of body and the movement of step is

Table 3. Comparison of player's step characteristics in the skip step stage.

players	Maximum jumping height in skip step	Distance between both feet after skip step	Horizontal moving distance of left foot	Horizontal moving distance of right foot	Moving period	Average moving speed of the left foot	Average moving speed of the left foot
Sampras	0.101 m	0.832 m	0.501 m	0.237 m	0.44 s	1.139 m/s	0.539 m/s
Aynaoui	0.175 m	1.001 m	0.400 m	0.382 m	0.30 s	1.333 m/s	1.273 m/s
Moya	0.091 m	0.746 m	0.437 m	0.314 m	0.89 s	0.491 m/s	0.353 m/s

Table 4. Comparison of step characteristics of players in the stroke stage.

Players	Stance	Horizontal movement distance of left foot	Moving period	Average moving speed of the human body	Step amount of moving
Sampras	open type	0.892 m	0.40 s	2.23 m/s	1
Aynaoui	open type	0.141 m	0.38 s	0.37 m/s	2
Moya	open type	0.667 m	0.34 s	1.96 m/s	1

accurate. When moving step, the distance from the body to the ball shall be appropriate, and this is easier to swing the racket to hit the tennis ball effectively. The player can improve the effect of return of service only by dosing so. The moving speed of Sampras and Moya are also faster, being 2.23 m/s and 1.96 m/s respectively.

4 CONCLUSIONS AND RECOMMENDATIONS

There are no special requirements on the body posture when doing stance in the preparation stage, as long as the stroke posture can be formed rapidly. The average horizontal angles between the two feet of Sampras, El Aynaoui and Moya are 12.1 deg., 2.2 deg. and 45.6 deg. respectively, and the distances between the two feet are 0.465 m, 0.874 m and 0.583 m respectively.

In the skip step stage, the maximum jumping heights of Sampras, El Aynaoui and Moya are 0.101 m, 0.175 m and 0.091 m respectively, the horizontal moving distances of their left feet are 0.501 m, 0.400 m and 0.437 m respectively, and the average moving speeds of their left feet are 1.139 m/s, 1.333 m/s and 0.491 m/s respectively. The skip step jumping heights of Sampras and Moya are reasonable while El Aynaoui is a little higher which will affect the effect of his return of service.

In the stroke stage, the moving distances of left feet of Sampras and Moya are greater (0.892 m and 0.667 m respectively) and the moving speeds of Sampras and Moya are faster, being 2.23 m/s and 1.96 m/s respectively.

REFERENCES

Bao Qin. (2005). On the Basic Play of Return of Service of Tennis. *Journal of Nanjing Institute of Physical Education (Natural Science)*. 03.

Li Guanghai & Sun Lei. (2008). Analysis of the Application of Technical and Tactical of Return of Service of Pingpong—Take Ma Lin's Skill of Return of Service as the Example. *New West*. 07.

Peng Guoxiong. (2002). Analysis of the Influence of the Step Moving to the Skill Mastery. *Journal of Wuhan Institute of Physical Education*, 04.

Sun Yanming & Chang Chunfang. (2010). Comparison and Analysis of Application of Open and Closed Stroke Skills in Baseline. *Journal of Beijing University*, 28(11), 126–128.

Wang Qining. (1997). Analysis of the Structure of the Player's Start-up Action of Return of Service. *Journal of Nanjing Institute of Physical Education*. 02.

Yu Jiangwei & Cui Lifang. (2007). The Basic Play of Return of Service of Tennis. *Liaoning Sport Science and Technology*. 02.

Zhou Lancui & Zhang Chengren. (1997). On the Technical and Tactical of Return of Service of Tennis. *Tennis World*. 04.

Modeling and Computation in Engineering II – Xie (ed)
© 2013 Taylor & Francis Group, London, ISBN 978-1-138-00058-2

Research and practice on grouting technology with new cement-based/polymer composite

Zhen Li & Jing Ma
Road and Bridge Engineering College with Hunan Communication Polytechnic, Changsha, Hunan province, China

Hang Yuan
Hunan Tenda Geotechnical Engineering Co., Ltd., Changsha, Hunan province, China

ABSTRACT: Cement concrete pavement becomes one of the two kinds of road surfaces in our country for its high strength and excellent durability. With the development of national economy and intense increase of traffic, cement concrete road surface has undergone worsening damage, among which cracks are the most common phenomena. The new grouting material of cement-based/polymer composite is a method of comprehensive grouting. Its stone body is a network formed by intertwining of high polymer and inorganic material. Its rigidity is between common cement concrete and asphalt concrete. Its pressive strength can reach 3–5 Mpa in days. As a semi-rigid material, it has low temperature sensibility, excellent ability of crack resistance and impermeability. It matches basically the elasticity modulus of the roadbed material, and thus is suitable to be used as grouting material to prevent the void slabs from permeation as well as reinforce them.

1 INTRODUCTION

Cement concrete road surface is widely used in construction of roads of all levels as a result of its large stiffness, high flexural strength, strong bearing capacity and water-resistance, excellent weatherability and high temperature resistance, extensive sources of material, long fatigue life, convenience when driving at night and other merits. However, lots of investigations and researches show that after 5 year's usage, cement concrete roads will endure damage of various severities, such as plate angle fault and corner flaking (see figure 1). This is called "five-year phenomenon". Among all damage, cracks are the most common, serious and complicated, causing the safety degradation and loss of some functions of the concrete roads (Meng Fanhong, Zhang Zengke & Cui Sumin. 2003).

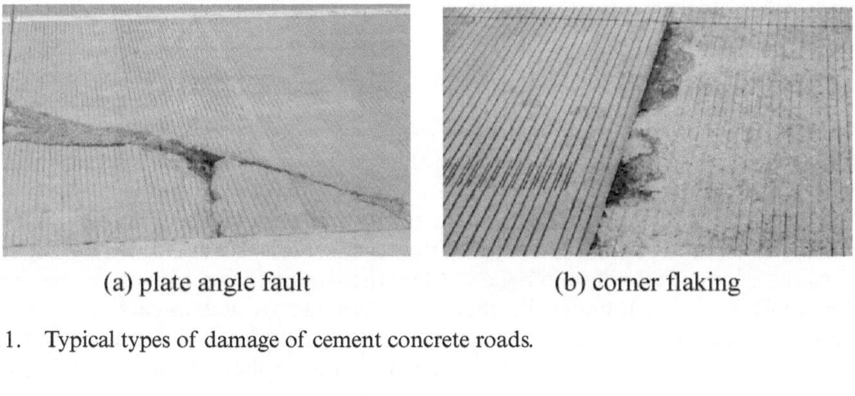

(a) plate angle fault (b) corner flaking

Figure 1. Typical types of damage of cement concrete roads.

Grouting technology has always been the most efficient way to deal with such disease as pumping, faulting of slab ends and void beneath slab. And the best way is to apply interlayer processing to disease sections through grouting. But currently, the mainly adopted method is cement grouting. The requirement is focused on the structure strength of grouting materials. As for concrete pavements, its surface rigidity is larger than that of the foundation, while as for asphalt pavements, its base modulus is larger than surface modulus. Cement is rigid material and its rigidity in the base and surface are not matched, which is harmful for structural force. Besides, the water proof performance of cement stone bodies is poorer, a far cry from the setting functions of original prime coating, adhesion coating and seal coating. Therefore, this kind of control is passive and with less durability and its maintenance period is about only 1year. Given this control status, new grouting material is badly needed to be researched and developed (Xiao Yimin & Tang Lin. 2001).

2 CAUSE ANALYSIS OF ROADBED AND PAVEMENT DISEASE

The damage of cement concrete pavements is generally classified into 4 categories: crack, deformation, pump, and other type of disease. Most of the diseases detected in Tan-Lei cement concrete expressway are void beneath slab, pumping and faulting of slab ends. And the two main reasons for these diseases are repeated effect of free water and heavy load. For this reason, this paper analyzes causes for common structural diseases in cement concrete pavements, with the typical diseases in Tan-Lei expressway (Si Guoliang. 2003).

1. After the operation of expressways, under the repeated effect of traffic load, joint load transfer capacity of cement concrete pavements are decreasing, and finally result in the failure of caulking material. Pumping, faulting of slab ends and void beneath slab of different severities appear around joints. Sides and angles of accelerated plates are cracked and damaged.
2. In the rainy, freezing cold areas, on one side, with the change of groundwater level, the roadbed will alter between continuous consolidating and softening. On the other side, after underground water melted in spring is evaporated, the water gas will be absorbed by the roadbed, which then increases water content in the roadbed and thus reduces its strength. Therefore, this period symbols the peak during which cracks in roadbed and pavement develop.
3. Long-lasting erosion from underground capillary water and upper precipitation, the inner structure of pavement flakes away day by day. In addition, gathered water in the surface directly washes and erodes the pavement structure with the substantial dynamic water pressure produced under the effect of high-speed traffic loads.
4. As the contact between the roadbed and the surface cannot be completely continuous, and as rainwater will continue eroding the roadbed, therefore, diseases will occur under the repeated effect of heavy loads (Li Jinyou. 2004) (Peng Fuqiang & Yuan Hang (ed.). 2011).

3 IDEAS OF DISEASES CONTROL

The STRAINED condition of roadbed and pavement is directly related to the contact state of the two, which has significant influence on the strained condition of the pavement. The occurrence of void beneath slab is generally a process of first even bearing of the roadbed, second uneven bearing of the roadbed, third partial loss of support (become void), and finally slab breakage and faulting of slab ends under the co-effect of the weight of the cement slabs and traffic loads. In the state of void beneath slab, the strained condition of the pavement is very unfavorable. As a result, under the effect of external loads such as cars, the angles near the pavement slab seams is in the work state of similarly cantilever girder and will produce great flexural stress, finally causing cracks in cement concrete slabs. As shown in figure 2:

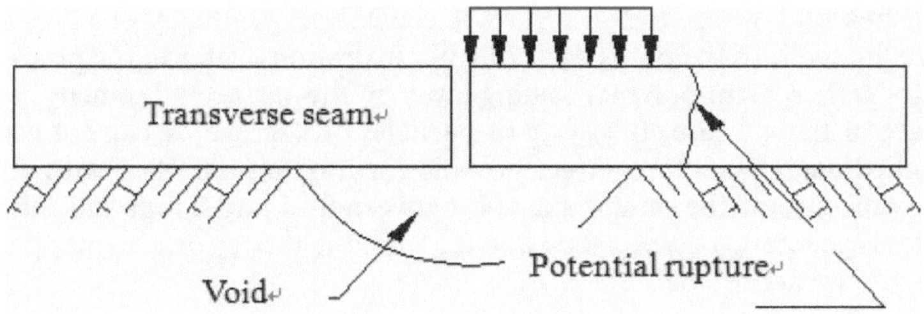

Figure 2. Rupture caused by the void of cement concrete board.

Practice has proved that the occurrence of diseases in the roadbed and pavement is directly related to interlayer processing techniques. Good interlayer processing measures can effectively prevent semi-rigid roadbed shrinking crack reflection as well as stop infiltration of surface water and rise of roadbed water, thus greatly improving the performance and duration of roadbed and pavement. For this reason, the basic requirements cement grouting technology must meet are: ①good pre-processing performance for availability; ②no negative effect on cement hydration; ③high resistance to Ca2+, Al3+ and other ions produced in cement hydration; ④excellent mechanical stability to avoid demulsification phenomenon occurred under the effect of high shearing in the process of storage, transporting and mixing; ⑤the ability to produce films having good adhesive force with hydration products and aggregate after the hardening of cement and the temperature of production should be low; ⑥excellent abilities of water and alkali resistance and weatherability after the hardening of cement (Jiang Shuozhong & Wang Zaiqing (ed.). 2008).

4 STUDY OF NEW GROUTING MATERIAL, I.E., CEMENT-BASED/POLYMER COMPOSITE

As mentioned above, the research and development of new grouting material, i.e., cement-based/polymer composite is to realize the building of an adhesive waterproof layer between the roadbed and the pavement. Therefore two basic conditions must be met: first the material, as adhesive waterproof layer, should possess excellent mechanical and adhesive properties, impermeability and durability; second, as grouting slurry, the material should have high stability and mobility, be permeable and meet required setting time and viscosity.

4.1 *Material composition*

The mixture ratio of new cement-based/polymer composite is: in component A, common Portland cement: fly ash = 4:6 (mass ratio); mixing the ash with water, water-cement ratio = 0.8:1; before grouting adding component B (high polymer) into it, to meet the requirement of reinforcing the road, even bearing of concrete pavement slabs and design requirement of expressway, the ratio of A to B is 3:7; finally, adding 1% surface active agent.

4.2 *Working principles of the material*

Ohama model divides the formation process of polymer grouting material into 3 stages (Yoshihiko Ohama. 1998).

1. After adding Polymer emulsion into the mixture of water and cement, polymer particles spread around evenly in the cement paste, forming polymer cement paste. With the hydration of cement, cement gel gradually forms in the mixture and Ca $(OH)_2$ in the solution reaches saturation state. Meanwhile, polymer particles gather on the surface of cement gel particles, which resembles the process of producing calcium silicate gel after the reaction of Ca $(OH)_2$ in aqueous phase and silicate on the surface of minerals.

2. With the decrease of water, the structure of cement gel gradually changes and polymer particles are limited into the pores. With further hydration, water in the pores losses; polymer particles gather, form polymer sealing layer on the surface of cement gel and adhere to the surface of the mixture consisting of particles of aggregates, cement gel and cement particles unhydrated. Therefore, larger poles in the mixture are filled with adhesive polymer. As the pore size of the cement paste is between 0.25 and 2 mm and the size of polymer particles is generally between 0.04 and 0.5 mm, the theory of polymer particles filling up in the pores are acceptable.
3. With the ongoing process of cement hydration, water between gathered polymer particles gradually participates in the action of cement hydration and forms chemical hydration water. In the end, polymer particles congeal together and form continuous network. The polymer network combines cement hydrates, i.e., hydrates and polymer mingle, and thus improve the structure of cement paste.

4.3 *Characteristics of the material*

The new grouting material of cement-based/polymer composite has good mobility and stability. Its stone body is a network formed by intertwining of high polymer and inorganic material, possessing excellent mechanical and waterproofing properties. Its strength is between cement stone body and asphalt stone body. As a semi-rigid material, it has the following characteristics (Yuan Hang, Wu Yi & Liu Ming. 2008).

1. The using of polymer provides the grouting slurry better stability.
2. This material combines the good waterproofing property of the polymer with the high strength of the cement. The slurry is semi-rigid, and solves the problem of only improving the roadbed strength and the resulting unfavorable force bearing of the pavement, which is caused by applying purely cement.
3. The slurry has low viscosity and high mobility, and can be effectively spread into the small pores in roadbed and the pavement under the effect of pressure. It can adhere to the roadbed and the pavement in a desired way, functioning as prime layer, adhesive layer as well as sealing layer.
4. Cement is typical hydraulic material. Using it to build framework can avoid new void and human destruction caused after the syneresis of common slurry.
5. By adjusting the ratio of polymer, the rigidity of the mixture can be adjustable and the mixture can be applied to structure types of various pavements.
6. The performance price ratio of this material is high. Also it is nontoxic and environmentally-friendly. Convenient construction process and short maintenance period are also its features. Only 4 to 6 hours after grouting, the road is ready to be put in operation.

5 ENGINEERING APPLICATION

In October, 2011, entrusted by Hunan Modern Investment Company, we did a disease treatment to the roadbed and pavement of Tan-Lei expressway.

This section is built with cement concrete pavement. According to field observation, lots of cracks on the pavement has been spotted, the widest of which reaches 2 cm. Part of the section has serious problem of void beneath slab and local chuck holes. The surface has been covered by yellow mud. In order to treat effectively these different diseases on the roadbed and the pavement, we adopted the method of combining human field investigation and falling weight deflectometer testing. We investigated the types of diseases, their respective seriousness, as well as drainage and water accumulation condition on the pavement. In line with the testing results, the degrees of void can be divided into two kinds: light and severe ones and be treated respectively.

As for light void sections, the treatment is to apply directly new cement-based/polymer composite to carry out steel flora tube grouting. Grouting pressure is generally 0.2–0.4 MPa, and the space between holes is 2.5 × 2.5 m (can be adjusted according to specific construction conditions), arranged in the pattern of plum blossom. The depth of the grout hole should be the same as the width of the pavement slab. The mixture ratio of cement and polymer should be: in component A, common Portland cement: fly ash = 4:6 (mass ratio); mixing the ash with water, water-cement ratio = 0.8:1; before grouting adding component B (high polymer) into it, to meet the requirement of reinforcing the road, even bearing of concrete pavement slabs and design requirement of expressway, the ratio of A to B is 3:7; finally, adding 1% surface active agent. For serious void sections, as the roadbed is greatly eroded, cement grouting should first be applied to reinforce the roadbed. After that finished, cement-based/polymer composite grouting should immediately be applied through static press, extruding diluted slurry from pavement cracks and filling the bedding voids with modified new composite (Jiang Shuozhong (ed.). 2006) (Research Institute of East China Investigation and Design Institute of the Ministry of Power Industry. 1984) (Geotechnical Grouting Theory and Engineering Examples Collaborative Group (compiled). 2001).

After the construction is completed, we did repeated field observations to examine the effect of treatment. The pavement no long endures the problem of pumping. Besides, after comparing the tested data of deflection value after grouting (see table 1), we can conclude that strained condition of the roadbed and pavement has been greatly improved.

After applying LTD-2100 ground penetrating radar test to a section of 500 m on the site, we found that fracturing grouting stone bodies were in the pattern of vein-shape cross grid in the roadbed, improving roadbed strength. The new cement-based/polymer composite forms a dense waterproofing layer in interlayer gaps (see figure 3).

Table 1. The deflection value contrast of disease sections and non-disease sections before and after treatment.

Pile Number	Lane	Gagging point quantity	Distribution of deflection value	Mean deflection value	Representative deflection value	Notes
K484+720 ~K485+000	Left Slow lane	20	0~5.0	1.8	3.1	Non-disease sections, untreated
K483+000 ~K486+000	Left lane	150	3.0~20.0	5.5	9.1	Disease sections, before treatment
K483+000 ~K486+000	Left lane	150	0~5.0	2.7	4.7	Disease sections, after treatment

*Note: The unit of deflection value is 0.01 mm

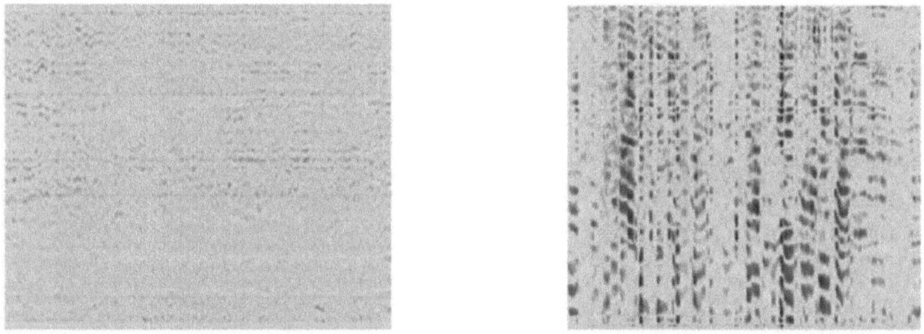

(a) detection picture before grouting (b) detection picture after grouting

Figure 3. Distributions of interlayer gaps of concrete slabs before and after grouting.

6 CONCLUSION

The new grouting material of cement-based/polymer composite is a method of comprehensive grouting. Its stone body is a network formed by intertwining of high polymer and inorganic material. Its rigidity is between common cement concrete and asphalt concrete. Its pressive strength can reach 3–5 Mpa in days. As a semi-rigid material, it has low temperature sensibility, excellent ability of crack resistance and impermeability. It matches basically the elasticity modulus of the roadbed material, and thus is suitable to be used as grouting material to prevent the void slabs from permeation as well as reinforce them.

In practical engineering, the test and treatment of trial section in Tan-Lei expressway acquired good effect. Relevant effective treatment has been made to disease sections. This kind of treatment features low cost, quick effect, simple operation, fast opening, and little impact on vehicles as well as is seldom affected by natural factors, having substantial economic and social benefits in the project of road construction and maintenance.

REFERENCES

Geotechnical Grouting Theory and Engineering Examples Collaborative Group (compiled). 2001. *Geotechnical Grouting theory and engineering examples*. Beijing: Science Press.

Jiang Shuozhong (ed.). 2006. *Green Chemical Grouting Technology*. Wuhan: Changjiang Press.

Jiang Shuozhong & Wang Zaiqing (ed.). 2008. *Technology Innovation and Chemical Grouting*. Wuhan: Changjiang Press.

Li Jinyou. 2004. The Application of Grouting Technology in Dealing with the Void beneath Concrete Pavment. *Transportation Standardization* 7: 46–48.

Meng Fanhong, Zhang Zengke & Cui Sumin. 2003. Research on Cause of Road Longitudinal Crack. *Journal of Highway and Transportation Research and Development* 20(6): 35–38.

Peng Fuqiang & Yuan Hang (ed.). 2011. *Highway Engineering Grouting Technology and Practice*. Beijing: People's Communications Press.

Research Institute of East China Investigation and Design Institute of the Ministry of Power Industry. 1984. *Chemical Grouting Technology*. Beijing: Water Power Press.

Si Guoliang. 2003. Causes and Prevention of Diseases of Cement Concrete Pavement. *Northeast Highway* 3: 38–42.

Xiao Yimin & Tang Lin. 2001. Reunderstanding of Pumping and Grouting Technology of Cement Pavement. *East China Highway* 6: 21–23.

Yoshihiko Ohama. 1998. Polymer-based Admixtures. *cement & Concrete Composites* 20: 189–212.

Yuan Hang, Wu Yi & Liu Ming. 2008. *Study on Grouting Technology of Modified emulsified asphalt/cement composite*. Wuhan: Changjiang Press.

Modeling and Computation in Engineering II – Xie (ed)
 ISBN 978-1-138-00058-2

Development law of accumulation landslide in LUE YANG of Shanxi province in China

Da-wei Lv
Guangdong Road and Bridge Construction Development Co, Guangzhou, Guangdong, China

Chao Xu, Nan Geng & Guan-jun Xu
Chang'an University, Xi'an, Shanxi, China

ABSTRACT: Accumulation landslide is one of the most Geological disasters in LUE YANG of Shanxi Province in China. On the basis of Accumulation landslide in LUE YANG, its development laws are studied by Site investigation and Theoretical analysis, and the conclusions are as follows: (1) The slope gradient lead to develop accumulation landslide is 8~60°, with average gradient of 35° and concentrate in 8°~40°; it often exist in the slope which the height is 16~60 m. Most of them are mainly distributed in the low-erosion mountains and the middle-erosion mountains. (2) Accumulation landslide occurred in cycles. Cyclical performance in two ways, one is it performance in different years, with a significant increase in frequency of geological disaster every few years. Secondly, accumulation landslide occurred in May to August each year. It mainly concentrated in the north area in LUE YANG city. (3) Accumulation landslide were concentrated and developed in the sunny slopes, also a small amount of development in shady slopes of LUE YANG city; (4) Accumulation landslide disasters happened with a chain nature; always with relevant to the intensity of human engineering activities; so they have characters of high-dimensionality, complexity, randomness and "immunity".

1 INTRODUCTION

LUE YANG County is located in Qin-Ba Mountain Area in South-Shanxi province. Due to special reasons such as the complicated geological structure, rock crushing, changeable landform, weather conditions etc (Liu, 2000), it became an area where geological hazards occurred frequently (Zhao, 2010). Loose accumulation horizon landslide is widely distributed there which outbreaks frequently and has sustainable harmfulness to become one of the largest geological disasters. It is one kind of landslide happened in the fourth system and modern loose accumulation horizon, the slope body structure is unconsolidated with largely void ratio, strongly permeability, variability and the lower layer of rock is obvious. Cause the slope of the combination is particular and its physical and mechanical properties have a big change, the groundwater effect rule is complex etc, determined this landslide is different from other types of landslide (such as rock landslide) rule. HE Guo-qiang researched and summarized the groundwater dynamic action of accumulation landslide stability in the evolutionary procession, dynamic response rules and its characteristics to find water induced accumulation of landslide displacement and instability was directly controlled by groundwater variation, which also existing corresponding relationships between the displacement law and underground water level variation (He, 2008). ZHANG Zuo-chen pointed out that the direct induction factor of most landslides is rainfall infiltration leading to groundwater state changes (Zhang, 1996). LIU Cai-hua deduced the calculation formula of reservoir slope stability which took the influence of groundwater into consideration, and pointed out measures which can effectively reduce the negative influence of groundwater to slope stability (Liu, 2005).

XIANG Xian-chao put forward the method of combining optimization technique with geological investigation to search the most dangerous sliding (Xiang, 2009). This method can be well proceeding slope sliding and searching, also it can provide guarantees for the reliable evaluation of accumulation landslide stability. These researches mentioned above provided meaningful exploration and laid a solid foundation in specific accumulation characteristics of landslide stability (Xiao, 2004), prevention and controlling measures (Urciuoli, 2002). This article aims to study the distribution law of landslide disaster in LUE YANG city, it has important theoretical significance and realistic economic benefit and social benefit in the regional accumulation landslide disaster prevention and mitigation, also the control of prediction.

2 THE DEVELOPMENT LAW OF ACCUMULATION LANDSLIDE IN LUE YANG

2.1 *The along topography distribution law*

Table 1 was the slope gradient survey about the internal accumulation horizon of LUE YANG city. It indicates the slope angle that can develop accumulation landslide is 8~60°, average 35°, concentrate on 8°~40°, which means the landslide happened easily when the slope angle in 8°~40°. Survey of 340 landslide showed 226 of them happened in the steep slope which occupied 84.3% of the total investigation, 114 of them happened in the gentle slopes which occupied 15.7%. The gradient of the slope between 25°~40° was high easy-happening area of landslides. The gradient of the slope between 8°~25° was medium easy-happening area; The gradient of the slope between 41°~50°, 51°~60° was less likely to happen landslide.

Table 2 is the slope height distribution statistics of accumulation landslide in LUE YANG city. It can be seen from the table 1: There are obvious controlling relationship between the slope height of accumulation landslide and the occurrence of landslide, when the slope height of landslide distributes under 240 m, average height of 40 m. Accumulation landslide generally occurs in the slope height ranges from16~60 m, 71 of accumulation landslides happened in this area, which occupied 48.63% of the total investigation, the slope which height under 16 m or over 60 m was less likely to develop landslide.

2.2 *The along geomorphology distribution law for accumulation landslide*

Accumulation landslide disaster mainly distributed in the low-erosion mountains and the middle-erosion mountains. The height of middle-erosion mountains ranged from 1000 m to 1600 m, mainly located in the western region of LUE YANG city, river cutting depth was between 300 m and 1000 m, the ridge was gently with smaller fluctuation. The population of this area was intensive, human action caused damage to vegetation, randomness excavation,

Table 1. The slope gradient survey points.

Slope interval		Accumulation landslides in LUE YANG county
Platform	≤8°	5
Gentle slope	8~25°	22
Steep slope	25~30°	33
	31~40°	31
	41~50°	17
	51~60°	14
Cliff	61~70°	16
	71~80°	6
	81~90°	2
Total		146

Table 2. The disaster point where slopes slope height segment tables.

Height interval	Accumulation landslides in LUE YANG county
0~15 m	3
16~30 m	16
31~45 m	31
46~60 m	24
61~75 m	15
76~90 m	13
91~105 m	11
106~120 m	6
121~135 m	9
136~150 m	11
150~200 m	6
<240 m	1
Total	146

and unreasonable irrigation lead to the development of accumulation landslide. Accumulation landslide disaster occupied 52.9% of all disasters.

2.3 *The regional distribution law for accumulation landslide*

By reasons of the complicated geological conditions, landform conditions, meteorological hydrological conditions, the strong new tectonic activities and human engineering activities, all of these factors mentioned above lead to enhance the development of geological disasters in this area. Almost 22 villages and towns have accumulation landslides. But the distributions of geological disasters are uneven with the geological conditions, hydrogeology conditions and geotechnical characteristics differ. Even in the same town, the development of geological disasters is different in different location. For instance, 6 of them locate in BAI QUE SI village comparing to 30 of them in the town of LUE YANG city. Accumulation landslide is widely distributed in the towns which have developed economy and large population density, with actions relevant to destroy the original slope body form such as artificial road and the infrastructure building etc. For instance, towns in LUE YANG city like JIE GUAN TING, CHENG GUAN, HENG XIAN HE, XIA KOU YI and XI HUAI BA are in this case.

Accumulation landslides are mainly distributed in the north of LUE YANG city, for example there are 30 of them in CHENG GUAN town, 11 of them in BAI SHUI town and 24 of them in JIE GUAN TING town.

2.4 *The time law for happening accumulation landslide*

Accumulation landslide happened cyclically. Cyclical performance in two ways, one is it performance in different years, with a significant increase in frequency of geological disaster every few years. Secondly, accumulation landslide occurred in May to August each year.

It can be seen from table 3 and table 4 that there is a corresponding relationship between the frequency of landslide disaster and precipitation. The frequency of landslide disaster increased with the annual precipitation increasing. For instance, the intensity of rainfall was heavy in the year of 1981, 1984, 1990, 1992 and 1998 (annual precipitation more than 870 mm), the disaster happened more frequently in these years, the intensity of rainfall was heavy in the July, August and September, the disaster happened frequently as well. The cyclical for disaster happening in this area is decided by the cyclical of rainfall intensity. The answer is rainfall; especially heavy rain fall is the main trigger factor for accumulation landslide. Its happening has involvement with rainfall, but the time for happening accumulation landslide is lag behind the duration of rainfall. Because a rainfall usually lasts for hours or

Table 3. The accumulation landslide disasters occurring year frequency tables.

Year	Annual precipitation mm	Landslide frequency	Proportion of total %	Year	Annual precipitation mm	Landslide frequency	Proportion of total %
1964	1085.1	1	0.5	1992	874.9	6	2.8
1970	767.7	1	0.5	1995	672.9	2	1
1973	842.7	1	0.5	1997	555.2	1	0.5
1978	1014.2	2	1	1998	880.7	7	3.3
1981	1353.3	102	48.6	1999	716.5	2	1
1983	1062.0	5	2.3	2000	629.5	4	1.9
1984	1055.0	9	4.3	2001	666.2	14	6.7
1986	682.0	3	1.4	2004	639.4	1	0.5
1987	621.7	2	1	2008	701.5	28	13.3
1990	1001.1	9	4.3	2009	678.6	8	3.8
1991	532.5	2	1				

Table 4. Month frequency of accumulation landslides.

Month	February	May	June	July	April	September
Average monthly precipitation (mm)	7	81	101	179	145	129
Landslide frequency	3	28	2	30	123	17
Proportion of total (%)	1.5	13.8	1	14.7	60.6	8.4

days, part of the rainfall erodes along the slope, others infiltrate through the slope. Rainfall infiltrate through the slope body and move in the aeration zone, the same time rainfall infiltrates in different sizes of the pore and at different speed through the aeration zone. When the fastest water slips to the sliding surface of the slope, slope body water level begins to rise. Then the largest proportion of water reach to the surface of groundwater, underground water level rises to a peak, after the slowest water reach to the surface of groundwater, the influence of rainfall comes to an end. Corresponding to the rainfall pulse, the groundwater which as the response of the rainfall pulse rises its level and behaves as a wave. Comparing with the pulse, the emergence of the wave shows time-delay effect, and continues delaying. The delaying wave leads to delay the development of accumulation landslide. There are also some accumulation landslide producing in the rainfall process, because of the two adjacent or more rainfall process are closely, the waveform of groundwater uplifted by each rainfall are mutually superimposing. When each wave superimposed to some degree, the wave became higher and appearing resonance and amplification phenomenon. When the accumulation of slope deform to its critical point, landslide disaster happened suddenly. For instance, the SHI YAO WAN landslide happened on 21:00 pm, August 20, 1981 in SHI GOU ZHUANG village of HENG XIAN HE town of LUE YANG city, which had buried a family of three people and six houses, without any deformation sign before land sliding.

2.5 *The sunward distribution law of landslide point*

The development of accumulation landslide concentrated in the sunny slope, also a small amount of them located in the shady slope. There are 93 of them located in the sunny slope, which occupied 63.6% of the total unstable slopes; 53 of them located in the shady slope, which occupied 36.4% of the total unstable slopes. Moreover, the aspect of slope exposure ranged from 45°~180° was the advantage development range. (Table 5)

Both of accumulation landslides in LUE YANG County dominate in the sunny slope development, also a small amount of them located in shady slope. Reasons for this phenomenon are longer sunshine time in the sunny slope, bigger temperature difference between

Table 5. Statistics of proportion to the number of different slope aspect.

Original slope aspect range	0~45°	45~90°	90~135°	135~180°	180~225°	225~270°	270~315°	315~360°	Total
Landslides in LUE YANG county	24	33	18	12	7	14	22	16	146

day and night. In addition, human activities are mainly centered in sunny slope, as human living habits are facing south which mean facing south, intensive human activities are the main factors for the development of unstable slope and collapsing. The weather in HAN ZHONG area is monsoon climate, most of wind directions are northwest along with intensive physical weathering; although the vegetation coverage is increasing, most places of the area are covering less which lead to water and soil erosion. All these factors mentioned above enhance the advantage development of accumulation landslide in the shady slope.

2.6 *Accumulation landslide has a character of chain*

Chain nature of landslide geological disaster is mainly reflected in it can be induced or associated with other geological disasters, most of the disaster points are sliding along with collapsing which called slump phenomenon; loose deposits produced by landslide is an important material source to form debris flow when in rainstorm. For example, the debris flow happened on April 12, 1992 in JI JIA GOU village of GUAN YING town had many landslides in the groove shore slope of forming region and communication area.

2.7 *Correlation between accumulation landslide and human engineering activities intensity*

Accumulation landslide has closely relations with human engineering activities. In recent years, with the increasing social economy development, human engineering activities such as randomness highway building, slope cutting for house building, our ecology environment is becoming from bad to worse, present a rising tendency. Human engineering activities destroy the structure of the slope as well as the slope height and gradient, leading to change the original slope stress. Geological disasters are happening because of no reasonable protection.

The influences of human engineering activities are mainly reflected in changing the topography, destroying the slope structure type, changing the structure of the slope as well as the slope height and gradient, leading to change the slope stress and reducing the stability of the slope.

All of these factors mentioned above contribute to slope instability which lead to land sliding disaster and collapsing disaster.

2.8 *Accumulation landslide has characters of high-dimensionality, complexity, randomness and "immunity"*

The system of accumulation landslide has multilayer structure, multiple time scale; various control parameters and a variety of action process. It is not only the development of a dynamic, non-linear and open system of disaster, but also the complex system with the characteristics of uncertainty and the social economy.

Accumulation landslide disasters such as sudden geological disasters phenomenon is a result of geological processes of gravity, and its breeding process inevitably along with the accumulation of the potential energy. After long process of development, landslide once happened, then accumulation of potential energy dies, slope immediately enter into the next round development of slope evolution stage. Most of the time, compared with accumulation landslide disaster prevention of concern time scale, the evolution process is quite long, often can only use landscape time scale, and even the geological time scale to measure. And in many cases, the same place next landslide in formation mechanism, movement characteristics has been different with the round in many aspects. Recognizing the characteristics of the landslide, the dangerousness for evaluation of landslide is beneficial. Slope of a place, the meaning which evaluates geological disaster risk of it by historical disaster frequency and intensity is not big. The landslide happened in the history of the place must not happen again in the future, the site of landslide which not happened in the history does not necessarily don't happen in the future. In this way, relying on historical disaster data to forecast disaster frequency and strength is very difficult.

3 CONCLUSIONS

(1) The slope gradient lead to develop accumulation landslide in LUE YANG County is 8~60°, average gradient of 35°, concentrate in 8°~40°, and often exist in the slope which the height is 16~60 m. They are mainly distributed in the low-erosion mountains and the middle-erosion mountains, from the viewpoint of regional it mainly concentrated in the north of LUE YANG County; the development of accumulation landslide in the sunny slope is more than those in the shady slope.

(2) Accumulation landslide occurred in cycles. Cyclical performance in two ways, one is it performance in different years, with a significant increase in frequency of geological disaster every few years. Secondly, accumulation landslide occurred in May to August each year.

(3) Accumulation landslides disasters happened with a chain nature; with relevant to the intensity of human engineering activities; they have characters of high-dimensionality, complexity, randomness and "immunity".

REFERENCES

He, Keqiang, Wang, Ronglu, and Li, Xinzhi etc. 2008. Searching for Slip Surface and Stability Analysis of Colluvial Slope. *Research of Soil and Water Conservation* 27(8): 1644–1650.

Liu, Xingchang. 2000. A Preliminary Study on Regional Regularities of Debris Flow in Western Qinling Mountains of Shanxi Province. *Bulletin of Soil and Water Conservation* 20(1): 17–20.

Liu, Caihua, Cheng, Congxin, and Feng, Xiating etc. 2005. Effect of groundwater on stability of slopes at reservoir bank. *Rock and Soil Mechanics* 26(3): 419–422.

Urciuoli, G. 2002. Strain preceding failure in infinite slopes. *International Journal of Geomechanics* 2(1): 93–112.

Xiang, Xianchao, and Tu, Pengfei. 2009. Searching for Slip Surface and Stability Analysis of Colluvial Slope. *Research of Soil and Water Conservation* 16(4): 60–67.

Xiao, Li, Qiulin Liao, and Jianming He. 2004. In-situ Tests and Stochastic Structural Model of Rock and Soil Aggregate in the Three Gorges Reservoir Area. *International Journal of Rock, Mechanics and Mining Sciences* 41(3): 494–499.

Zhao, Zhiqiang, Lie, Shouzhi, and He, Wei. 2010. Analysis on Stability and Influencing Factors' Sensitivity of Badong Landslide. *Journal of Water Resources and Architectural Engineering* 8(1): 98–100.

Zhang, Zuochen. 1996. Mechanism of Groundwater Effect Landslide Stability and Control Construction. *Journal of EngineeringGeology* 4(4): 80–85.

Modeling and Computation in Engineering II – Xie (ed)
© 2013 Taylor & Francis Group, London, ISBN 978-1-138-00058-2

Flexural strength of corroded C-shape steel members

Shanhua Xu, Jingbo Wang & Bin Qiu
Department of Civil Engineering, Xi'an University of Architecture & Technology, Xi'an, China

ABSTRACT The objective of this article is to provide data to engineers on the structural behavior of corroded steel C-shape members under two concentrated force on one-third of specimens. 7 C-shape members were hot rolled, from a decommissioned steel frame structure with severe corrosion damage. The influence, of corrosion ratio of compression flange and tensile flange, thickness loss and residual thickness, on flexural capacity was measured. Recommendations are drawn from this research to provide guidance to engineers on how to evaluate flexural capacity of corroded members. Needs for future work are also highlighted.

1 INTRODUCTION

Existing steel structures exposed to the environment are subject to corrosion. Even galvanized steel can experience corrosion after galvanic protection is consumed, and their capacity is reduced accordingly. In practice, when a member is found corroded during inspection, it is necessary to estimate the residual capacity of corroded members in order to decide whether to change the member, repair it or just remove corrosion and re-protect the member.

It is very difficult to assess the residual capacity of an existing structure and determine when it is no longer safe after corrosion has started.

Chapkis (1967) made a close analyses of the effects of pit corrosion on ultimate strength of steel plate. Paik JK et al. (2004) evaluated compressive ultimate strength of corroded steel plate with pit corrosion by applying the method of equivalent uniform corrosion.

L-V Beaulieu et al. (2010) evaluated the residual capacity of corroded steel angle members according to the average residual thickness. Kayser & Nowak (1989) developed a corrosion model for steel girder bridges, which takes into account the location and rate of corrosion. Yanina Chen, Xin Li et al. (2010) evaluated compressive the flexural capacity of corroded steel pipes. Sarveswaran V & Smith J. W. (1999) proposed a minimum capacity curves which vary with the percentage loss of flange thickness are presented.

Corrosion might also modify steel from a metallurgical point of view and characteristics of remaining steel, such as yield strength, can be altered. The phenomenon of corrosion is well studied, but very little research has been done on how the flexural capacity of steel members is affected once corrosion has developed. There is a lack of experimental data on the relationship between weight loss due to corrosion and residual strength.

The objective of this study is to provide experimental data on flexural capacity of corroded steel C-shape members that could be used by practical engineers.

2 EXPERIMENTAL PROCEDURE

2.1 *Specimen*

The material we used here is Q345. The specimens for flexural capacity studies were 7 specimens, All specimens were hot rolled C-shaped steel ($50\,mm \times 37\,mm \times 4.5\,mm \times 7.0\,mm$), from a decommissioned steel frame structure with severe corrosion damage, and 600 mm in length.

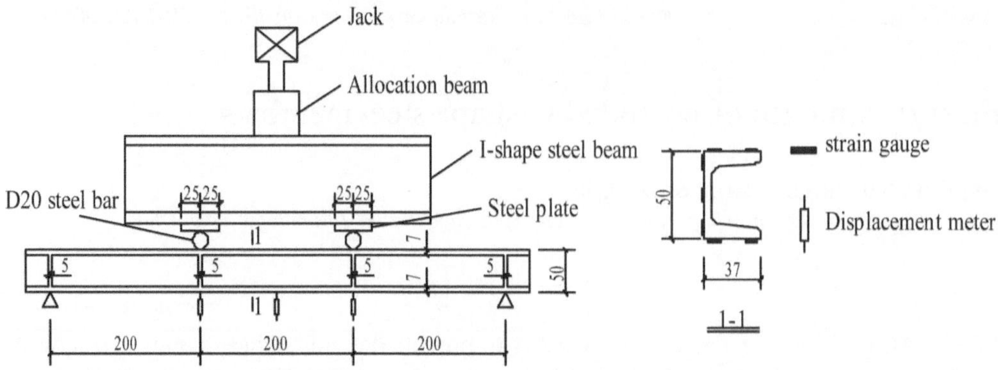

Figure 1. The experimental set-up (unit (mm)).

For decreasing error, both ends and one-third of the specimens for flexural capacity studies had a ribbed stiffener that is 5 mm in thickness welded on. As is shown in Figure 1.

The plate welded on both ends and one-third of specimens should be of sufficient thickness so that the welding deformations can be avoid. The covered electrode we used in manual welding is E43, weld seam is orthogonal fillet weld, size of the fillet weld is 10 mm. All the fillet weld must be of sufficient strength.

2.2 *Experimental set-up*

The experimental set-up developed at Xi'an University of Architecture and technology in China and the hydraulic operating gear was used here to assess the flexural strength of the C-shape corroded members. The maximum load of the machine is 50 ton.

The displacement meter and 7 pieces of strain gauge located in each flange was placed in the middle span of the specimen in the plane of bending. Figure 1 shows the experimental set-up.

The number of strain gauge we used here is BX120-5 AA, its rated value of resistance is 118.5Ω ± 0.1%, coefficient of sensitivity is 2.11 ± 0.52%, and size is 5 mm × 3 mm. The location placed strain gauge should be artificial buffing and cleaned with ACETONE, then sticked strain gauge with glue named 502 in china.

The displacement measured by the displacement meter (range is 30 mm), the compressive strain and the tensile strain measured by strain gauge, in plane of bending, were measured during the tests to observe the ultimate strength and the failure modes of the corroded steel C-shape members and all the datum collected by TDS-303 static data acquisition logging system.

3 RESULTS AND DISCUSSION

3.1 *Corrosion ratio measuring*

The corrosion level was measured in terms of thickness loss, can be calculated by Eqs. (1). The thickness on compression flange and tensile flange, for all the specimens, was measured respectively for corrosion level measuring, after corrosion.

$$\eta_s = \frac{t_0 - t_c}{t_0} \times 100\% = \frac{\Delta t}{t_0} \times 100\% \tag{1}$$

here: η_s = Corrosion ratio;
t_0 = the thickness of specimens before corrosion (g);
t_c = the thickness of specimens after corrosion (g);

The corrosion ratio in terms of thickness loss for each specimen are shown in Table 1.

It is observed from the surface of the specimens that corrosion on the specimens was mainly uniform. Generally, after corrosion, the corrosion pit on surface is much shallow and the distribution of the corrosion pit on the surface is well-distributed. For some particular location, the corrosion pit is overlapped.

3.2 *Effects of corrosion on flexural strength*

Before testing, all the instruments including the hydraulic operating gear, the displacement meter and static data acquisition logging system, should be regulated. When testing, the load, imposed on C-shape steel beam, should be continuous, monotonous, static and without impact action.

Table **1**. shows the results of flexural resistance tests.

It is evident from Table 1 that thickness loss and the corrosion ratio in terms of thickness loss on flange is significantly large. The original thickness is 7 *mm*, and the thickness loss range from 1.23 *mm* to 2.52 *mm* of compression flange, 1.29 *mm* to 1.77 *mm*. The corrosion ratio rang from 17.57% to 36% of compression flange, 22.59% to 33.84% of tension flange.

It is interesting to note that thickness loss and the corrosion ratio on flange is significantly large, but the range of flexural strength of steel C-shape members is significantly narrower, it just range from 28.8 *KN* to 34.36 *KN*.

Figure 2 shows the results of the tests in terms of residual thickness and flexural strength. It is evident from Figure 2 that the flexural strength of steel C-shape members tends to increase with the residual thickness increased. The relationship between residual thickness, of compression flange and tensile flange, and flexural strength is almost linear.

Table 1. The results of flexural resistance tests.

	Compression flange			Tensile flange			
Specimens	$t_{c(mm)}$	$\Delta t_{(mm)}$	η_s	$t_{c(mm)}$	$\Delta t_{(mm)}$	η_s	$P(KN)$
1	5.58	1.42	20.29%	5.62	1.38	24.56%	33.34
2	5.55	1.45	20.71%	5.59	1.41	25.22%	33.41
3	5.77	1.23	17.57%	5.71	1.29	22.59%	35.41
4	4.48	2.52	36.00%	5.23	1.77	33.84%	28.88
5	5.66	1.34	19.14%	5.68	1.32	23.24%	34.36
6	5.52	1.48	21.14%	5.58	1.42	25.45%	33.12
7	5.65	1.35	19.29%	5.64	1.36	24.11%	34.01

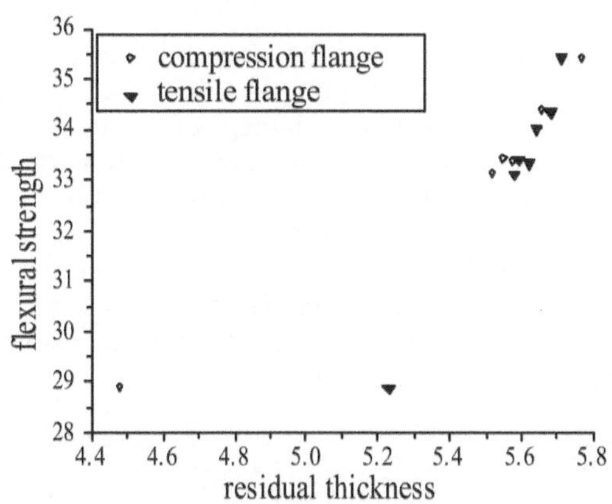

Figure 2. The relationship between residual thickness and flexural strength.

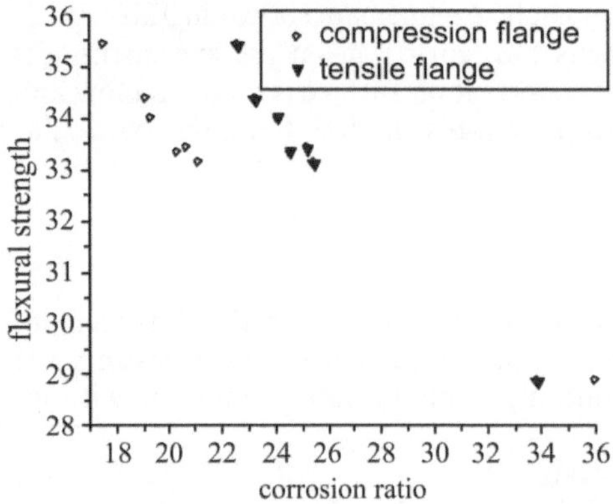

Figure 3. The relationship between corrosion ratio and flexural strength.

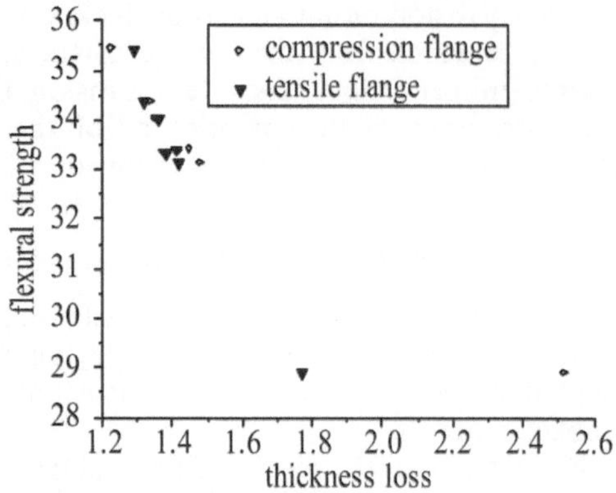

Figure 4. The relationship between thickness loss and flexural strength.

Figure 3 shows the results of the tests in terms of corrosion ratio and flexural strength. It is observed in Figure 3 that the flexural strength of steel C-shape members tends to decrease with the level of corrosion ratio increased. The relationship between corrosion ratio and flexural strength is almost linear, both on compression flange and tensile flange.

Figure 4 shows the results of the tests in terms of thickness loss and flexural strength. Evident from Figure 4 that the larger the thickness loss is, the more the flexural strength of steel C-shape members is smaller. The relationship between thickness loss, of compression flange and tensile flange, and flexural strength is almost linear.

Generally, the relationship between residual thickness and flexural strength, thickness loss and flexural strength is almost linear, the effects of residual thickness and thickness loss on flexural strength is almost the same. But for corrosion ratio, when a comparison was made between compression flange and tensile flange, which had the same flexural strength, it could be made that corrosion of compression flange is more critical than corrosion of tensile flange.

4 CONCLUSIONS AND RECOMMENDATIONS

4.1 *Conclusions*

This article presents the flexural resistance test of 7 corroded steel C-shape members, which corroded at a level corresponding to thickness loss ranging from 17.57% to 36% of compression flange, 22.59% to 33.84% of tension flange and the thickness loss ranging from 1.23 *mm* to 2.52 *mm* of compression flange, 1.29 *mm* to 1.77 *mm*.

(1) With the corrosion damage ratio increase, the flexural strength decrease, corrosion damage of flange is more critical than corrosion of web, and corrosion damage of compression flange is more critical than corrosion damage of tensile flange.
(2) The larger the thickness loss of flange is, the smaller the flexural strength is, the relationship between thickness loss of flange and flexural is almost linear.
(3) The large the residual thickness is, the larger the flexural strength is, the relationship between residual thickness and flexural strength is almost linear.
(4) The effects of residual thickness and thickness loss on flexural strength is almost the same. But for corrosion ratio, it could be made that the corrosion of compression flange is more critical than corrosion of tensile flange.

4.2 *Recommendations*

(1) To develop empirical charts that can be used in practice to evaluate directly the residual strength, it is believed that additional experimental data would be required, in particular

for members with high level of corrosion. Additional data for other steel C-shape members and different slenderness ratio would also be interesting.

(2) For utilities, residual strength curves as a function of time or the capacity loss curves as a function of corrosion ratio are also necessary to plan asset repairs and replacements. For this purpose, additional research is required to transform corrosion level into remaining life time, depending on environment aggressiveness, on the other hand, a better method for assessing the level of corrosion.

REFERENCES

Chapkis, D.T. Simulation of pitting corrosion of hull plating under static loading [J]. Trudy TSNIIMF, 1967, 82: 34–50.

Chen, Y.F., Li, X., Chai, Y.H. & Zhou, J. Assessment of the flexural capacity of corroded steel pipes [J]. International Journal of Pressure Vessels and Piping, 2010, 87: 100–110. (in china)

Kayser, J.R. & Nowak, A.S. Capacity loss due to corrosion in steel-girder bridges [J]. Journal of structural engineering, ASCE, 1989, 115: 1525–1537.

Beaulieu, L-V., Legeron, F. & Langlois, S. Compression strength of corroded steel angle members [J]. Journal of Constructional Steel Research, 2010, 66: 1366–1373.

Paik, J.K., Lee, J.M. & Ko, M.J. Ultimate compressive strength of plate elements with pit corrosion wastage [J]. Journal of Engineering for the Maritime Environment, 2004, 217(4): 185–200.

Sarveswaran, V. & Smith, J.W. Structural assessment of corrosion-damaged steel beams using minimum capacity curves [J]. Struct Eng, 1999, 77: 17–23.

Modeling and Computation in Engineering II – Xie (ed)
© 2013 Taylor & Francis Group, London, ISBN 978-1-138-00058-2

Simulation prediction model for payback period of industrial construction project

Yunxiu Sai
Xi'an Technological University, Xi'an, China

Weiran Wang
Xi'an University of Architecture and Technology, Xi'an, China

Xing Fang
Xi'an Shiyou University, Xi'an, China

ABSTRACT: According to the Characteristics of industrial construction projects, this paper establishes the system model and economic model of the investment recovery period, analyses the random factors that influence the investment recovery period of industrial construction projects, confirms the probability distribution of random factors statistical data with the hypothetical test method of statistics, and then forms the simulation model of payback period combined with system simulation theory. By using the matlab language, it achieves the simulation model, then analyses the precision of simulation results according to central limit theorem. Finally, combined with practical examples, further demonstrates the model's feasibility, the results show that the simulation predicting model can be used as the important basis for the industrial construction projects' investment decision making.

1 INTRODUCTION

Industrial construction project is a kind of investment project, and requires a lot of manpower, material and financial resources, the construction period is relatively long. In addition, construction project has "one trial" characteristics, once finding for the decision-making and construction reasons, resulting in investment out of control during construction, will cause enterprise and national economy great losses. This requires feasibility study must be done early in the project construction. One of the economic indicators of the project feasibility study is the payback period. So the engineering construction project payback period simulation predictions are of great significance, and also provide a basis for investment decisions to the project.

Payback period indicators in "The construction project evaluation methods and parameters", a file published by National Development and Reform Commission and Ministry of Construction, are explicitly identified as one of the important parameters of the construction project financial evaluation. It is provisioned as one of the indicators that can inspect the project investment earnings level of financial profitability analysis, and made a brief summary of its concepts, formulas and evaluating method. Particularly important is to release the parameter "industry benchmark payback period", make the payback period of this evaluation have evaluation standard. Since then, the payback period as one of the important financial evaluation indicator, play a significant role in the construction projects investment feasibility study (WANG and ZHANG, 2004).

2 INDUSTRIAL CONSTRUCTION PROJECT PAYBACK PERIOD SIMULATION MODEL

2.1 *The payback period system model*

Static payback period does not consider the time value of money under the conditions of the time needed to cover all investment projects' net income. Static investment payback period P_t (fang et al., 2003) is defined as follows:

$$\sum_{t=1}^{P_t}(CI-CO)_t=0 \tag{1}$$

where P_t is static payback period; CI_t is cash inflows of the year t; CO_t is the cash outflows of the year t; $(CI\text{-}CO)_t$ is the net cash flows of the year t.

The actual calculation of static payback period (P_t) can be calculated by the available financial statement of cash flows and cumulative net cash flows, the formula is:

P_t = (*Cumulative net cash flow of positive number years-1*)
+ *Last year the absolute value of the cumulative net cash flow/This year net cash flow* (2)

The dynamic payback period is considered the time value of money let project net income to cover all investments need time. Dynamic payback period (P_t') is defined as (Zhang et al., 2012):

$$\sum_{t=1}^{P'_t}(CI-CO)_t(1+i)^{-1}=0 \tag{3}$$

where P_t' is dynamic payback period; i is fixed discount rate.

The actual calculation of dynamic payback period can use financial statement of cash flows and the present value of the cumulative net cash flow. Its formula is:

P_t'= (*Net cash flows discounted cumulative value of the positive number of years -1*) + *Last year net cash flow discounted value of the accumulated absolute value/ This year net discounted cash flow value* (4)

Seen from the formula (1) and (3), to calculate the payback period, the key is to determine the amount of cash inflows and cash outflows. The essence is to determine the annual net cash flow.

2.2 *Payback period economic model*

Total investment (TZ) = Fixed asset investment (GN) + Circulating fund (LN)

Circulating fund (LN) = GN × Fixed asset value of the funds rate fixed asset investment (taking 8%)

Cash inflows (CI) = Annual sales income (SS)

Cash outflows (CO) = Annual total cost (TC) + Annual product sales tax and additional (ST) + income tax (IT)

Annual output (PQ) = Production scale (GM) × Annual equipment loading rate (LR)

Annual sales (SQ) = Annual output (PQ) × Qualified products rate x_1

SS = Annual sales revenue (SQ) × Product price (KP)

TC = Operating costs (JC) + Depreciation rate (ZJ) + Amortization expenses (TA)

JC = Raw material usage (TQ) × Average unit cost of raw material consumption (LP)

ZJ = GN × Depreciation rate n_1 (taking 5%)

TA = other charges (T) × Amortization rate n_2 (taking 5%)

TQ = PQ/finished product rate x_2

ST = Value-added tax (ZT) + Urban construction and maintenance tax (CT) + Education surcharge (JT)
ZT = SS/[1+Value-added tax rate(t_1)] × t_1-JC × t_1 (taking 17%)
CT = ZT × Tax rate t_2 (taking 7%)
JT = ZT × Tax rate t_3 (taking 3%)
IT = (SS–TC–ST) × Income tax rate (taking 33%)

2.3 *Random variables generation*

From the economic model of industrial construction project payback period, it can be seen the main factors that affect the payback period are product rate of qualified products x_1, equipment loading rate LR, finished product rate x_2 three random parameters (Tang et al., 2012). In order to determine the probability distribution, first using moment estimation method construct the estimated amount, then χ^2 testing the estimated amount to further determine its probability distribution, using MATLAB software to generate the random variable value. The specific steps of χ^2 test (hao, 2001) are as follows:

If H_0: $F(x) = F_0(x)$, where $F(x)$ is the general theory of distribution, $F_0(x)$ is the estimated distribution need to be tested.

1) According to the sample values, divide the real axis ($-\infty, +\infty$) into K subintervals, namely $-\infty = a_0 < a_1 < a_2 < \ldots < a_k = +\infty$
2) Statistics sample value $x_1, x_2, \ldots, x_n$ falling on the respective sub-interval (a_{i-1}, a_i) the actual frequency m_i, $i = 1,2, \ldots, K$.
3) When H_0: $F(x) = F_0(x)$ is true, calculate the overall X falls of each sub-interval (a_{i-1}, a_i) the theoretical probability of $P_1 = P(X \le a_1) = F_0(a_1), \ldots, P_i = P(a_{i-1} < X \le a_i) = F_0(a_i) - F_0(a_{i-1})$, $i = 1,2, \ldots, K$. which can be drawn from a sample $X_1, X_2, \ldots, X_n$ falling on the respective sub-interval (a_{i-1}, a_i) the theoretical frequency number nP_i, $i = 1,2, \ldots, K$.
4) For statistic $x^2 = \sum_{i=1}^{k} \frac{(m_i - nP_i)^2}{nP_i}$. Can prove that, no matter what the distribution $F_0(x)$ is, when H_0 is true, as long as n is sufficiently large, the statistical amount χ^2 to approximately follow the χ^2 distribution of freedom degrees K-r-1. Where K is the number of sub-interval, r is the number of parameters to be estimated.
5) Given the inspection level α to check the χ^2 distribution table, find out the freedom degree K-r-1, for a significant level α threshold $\chi^2_\alpha(K-r-1)$.
6) By the sample values $x_1, x_2, \ldots, x_n$ and obtained m_i and nP_i ($i = 1,2, \ldots, K$) substituting $x^2 = \sum_{i=1}^{k} \frac{(m_i - nP_i)^2}{nP_i}$ calculated statistics χ^2.
7) To make judgments, if $\chi^2 > \chi^2_\alpha(K-r-1)$ this small probability event, then refused H_0, and consider the overall distribution $F(x)$ and the estimated distribution $F_0(x)$ does not match, if $\chi^2 \le \chi^2_\alpha(K-r-1)$, accept H_0, consider the overall distribution and judgments distribution $F(x)$ consistent.

2.4 *The payback period simulation model*

According to the economic model, to generate random number of the equipment loading rate, the rate of qualified products, finished product rate, get the annual sales income (SS), is the change years of cash inflows CI (N). And then get the total cost (TC), sales taxes and surcharges (ST), income tax (IT), calculated annual cash outflows CO (N) = TC + ST + IT. Thus was the annual net income, is the net cash flow. Then calculate the cumulative net cash flows and the present value of the cumulative net cash flow, static payback period and a sample value of the dynamic payback period are based on the formula (2) and (4). Repeated simulation many times, gets the static payback period and the mean dynamic payback period, is the prediction of the static and dynamic payback period. Engineering construction project payback period forecast simulation model block diagram is shown in Figure 1.

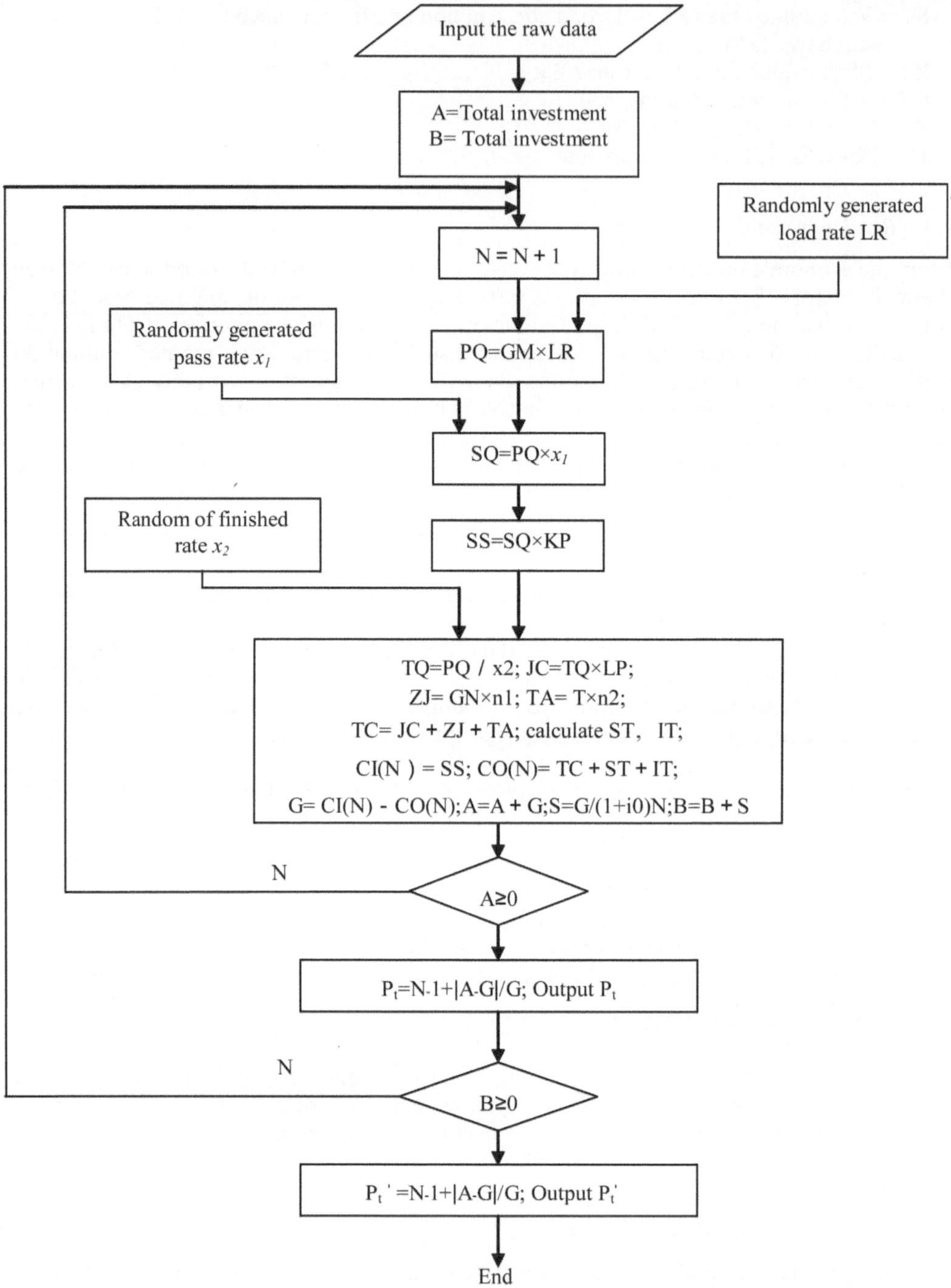

Figure1. Industrial construction projects payback period forecast simulation model.

3 PROJECT EXAMPLES

A hot rolling factory's production scale is 4.5 million tons annually, fixed asset investment is 6.374 billion Yuan, taking 8% of the investment in fixed assets as liquidity of 510 million yuan. Try to predict the static and dynamic payback period of the project (Wu et al., 2012).

First survey on the technical and economic indicators for the country's large and medium-sized hot rolling factories, selected from representative data of several large and medium-sized hot rolling factories statistics, statistical data processing to determine the probability distribution. Processing results is the normal distribution of the products rate x_1 of qualified products meet $(0.981, 0.016^2)$ of the normal distribution, equipment loading rate (LR) according to $(0.88, 0.22^2)$ of the normal distribution, the finished product rate x_2 in accordance with $(0.86, 0.036^2)$ of the normal distribution, for the sake of simplicity, the cost of the product taken as 3100 Yuan/ton, the price is taken as 3900 Yuan/ton. Finally, run the simulation model using MATLAB programming.

4 SIMULATION RESULTS AND THE ACCURACY OF ESTIMATED

After 100 times simulation running, the annual output of 4.5 million tons were 6.374 billion Yuan, fixed assets investment and liquidity of 5.1 billion Yuan hot rolling mills, static payback period are average of 7.1693 years, standard deviation of 1.5052 years; dynamic payback period are mean of 13.8207 years, standard deviation of 7.0040 years.

Estimation method based on the Central Limit Theorem, the error $\varepsilon = U\frac{\sigma}{\sqrt{N}}$ is the payback period of the sample mean instead of the overall expectations. Now N = 100, $\sigma_{static} = 1.5052$, $\sigma_{dynamic} = 7.0040$. With fetch $\beta = 0.99$, from $\Phi(U) = (\beta+1)/2 = (0.99+1)/2 = 0.995$. Check N (0, 1) table have U = 2.57. So the error is $\varepsilon_{static} = U\frac{\sigma_{static}}{\sqrt{N}} = \frac{2.57 \times 1.5052}{\sqrt{100}} = 1.799$

The simulation results is the evaluation of static payback period of 7.1693 years to absolute error is less than 0.386 to 0.99 probability; the average dynamic payback period of 13.8207 years absolute error is less than 1.799 to 0.99 probability.

5 CONCLUSION

By running the simulation and prediction model analysis of actual cases, can prove that the model of industrial construction project investment decision in the field of industrial construction project investment is feasible.

REFERENCES

FANG, Z.M.-., LIANG, H. & HAO, X.X.-. 2003. The Computer Simulation Forecast of the Markov Chain for Construction Project Investment Risk. *Journal of WUT (information & management engineering)*, 25, 55–58.

HAO, X.X.-. 2001. *Construction Engineering System Simulation (In Chinese)*, Science Press.

TANG, S., WANG, D. & DING, F.-Y. 2012. A new process-based cost estimation and pricing model considering the influences of indirect consumption relationships and quality factors. *Computers & Industrial Engineering*.

WANG, J. & ZHANG, X. 2004. Analysis of A Index for Evaluation of Construction Project Finance——Investment Recovery Term *Construction Management Modernization*, 2.

WU, J., XIA, B. & MENG, M. 2012. Research of Popularizing Solar Heating in Countryside. *Information Computing and Applications*, 827–832.

ZHANG, M.F., HUANG, L. & CAI, Y.P. 2012. Forecast on Payback Period of Restaurant Projects Investment by Computer Simulation. *Advanced Materials Research*, 446, 3782–3786.

Modeling and Computation in Engineering II – Xie (ed)
 ISBN 978-1-138-00058-2

Experimental study on factors influence on foundation-pit bursting in soft soil

Yu-yong Sun
Sinosteel Maanshan Institute of Mining Research Co., Ltd., Maanshan Anhui, China
School of Civil Engineering, Heifei University of Technology, Hefei Anhui, China
College of Civil Engineering, Tongling University, Tongling Anhui, China

Yun-min Wang
Sinosteel Maanshan Institute of Mining Research Co., Ltd., Maanshan Anhui, China

ABSTRACT: According to results of centrifugal model tests and model tests, the influence of the soil properties, pile types and layout, foundation size on the anti-bursting stability of foundation pit were studied. It was shown that the mainly failure modes of impermeable layers in Shanghai were overall heaving and surface boiling. Water and sand inrushing in contact between soils and underground structures was mainly occurred when the existence of pile in foundation pit. Foundation size had a significant effect on the anti-bursting stability of foundation pit.

1 INTRODUCTION

Foundation-pit bursting was the focus of underground engineering research and the main research tools included indoor model tests (Marsland A, 1953; Terzaghi K, 1943; Mcnamee J, 1949; Zhou Jian et al., 2006; Sun Yu-yong and Zhou Shun-hua, 2010), theoretical analysis (Ma Shi-cheng and Yin Chang-jun, 2004; Zheng Gang and Yang Jian-min, 2011) and numerical simulation (ARTHUR Marsland, SC M, 1953; Benmebarek N, Benmebarek S, Kastner R, 2005; Ding Chun-lin and Wang Dong-fang, 2007; Hu Qi et al., 2007; Sun Yu-yong et al., 2011). Indoor model tests studied the bursting stability, bursting mode and mechanism of specific foundation-pit, theoretical analysis mainly focused on the criteria and calculation model of bursting and numerical simulation was mainly on account of the impact of soil plasticity and plane size of foundation-pit. Only (Zhou Jian et al., 2006) using self-developed test device, the effect of pile in foundation-pit on bursting stability in homogeneous sand and silt soils was researched.

The system research on factors influence on foundation-pit bursting stability had not been reported. Based on this, using centrifuge model tests and model tests, the factors including the soil properties, pile types and layout, foundation size that affected the bursting stability of foundation-pit were studied.

2 CENTRIFUGE MODEL TESTS

2.1 *Test device*

The centrifuge model tests were completed in L-30 centrifugal model machine in Tongji University, of that the installation capacity was 20 g-t and the effective internal size of the model box was 41.5 cm(L)*22.8(W)*35.5(H).

2.2 *Model rate*

The model rate selected was 100, that the acceleration was 100 g.

2.3 *Model soil preparation*

The soils in testing included ⑤$_2$ micro-confined aquifer and its upper layer of ④ gray silty clay relatively impermeable layer, ⑦$_2$ confined aquifer and its upper layer of ⑥ dark green silty clay impermeable layer. The physical and mechanical parameters of soils were shown in Table 1.

The thickness of confined aquifer and impermeable layer were 7 cm and 5 cm respectively.

According to the characteristics of the soils, the soil samples were prepared by two methods: for the layer of ④, firstly the soil sample would be made into soil extract that the water content was 80%~120% and then to stratification consolidation in the centrifugal model machine. In the consolidation process, the water content, γ and strength indicators were contrasted. As the lower water content and higher strength of ⑤$_2$, ⑥ and ⑦$_2$, the following methods were used: drying → grinding → screening used 1 mm test sieve → determining the water content → preparing the soil extract according to on-site water content → forming the soil sample tamping layered according to on-site density.

2.4 *Preparation of the pile*

Currently, the pile could mainly be divided into two kinds: the precast piles and bored piles, according to construction technology. The material of piles was mainly reinforced concrete. The simulation method of piles was used the following:

1. The prefabricated cement mortar bar was used to simulate the precast piles and the reconsolidation was not implemented when the cement mortar bar driven into soils in order to accord with the actual situation.
2. The field pouring cement mortar pile was used to simulate the bored pile. Firstly, using suitable tube to remove the consolidated soil in model box, then chiseling the inner hole wall using iron wire in order to simulate the roughness of bored pile, finally pouring cement mortar. The simulation pile in testing was shown in Figure 1.

Table 1. Physical and mechanical parameters of soils.

				Shear strength	
No.	γ (kN/m^3)	w (%)	e	C (kPa)	Φ (°)
④	17.5	44.8	1.458	14	11
⑤$_2$	18.1	32.4	0.938	4	29
⑥	19.6	24.3	0.70	42	20
⑦$_2$	18.9	26.7	0.765	0	32.5

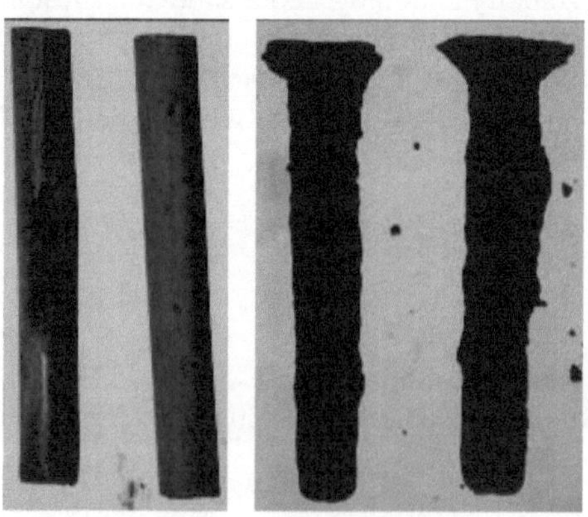

a) Precast pile   b) bored pile

Figure 1. The simulation pile in testing.

a) The distance was 3m b) The distance was 6m

Figure 2. The layout of piles in model test.

The diameter of simulation pile was 8.5 mm and the length of inserting into the confined aquifer was 2 cm that the length of simulation pile was 7 cm. The horizontal distance of pile included 3 m, 6 m and 10 m. The layout of piles was shown in Figure 2.

3 ANALYSIS OF INFLUENCING FACTORS ON BURSTING

3.1 *The soil properties*

The water pressure and failure modes when foundation-pit bursting were shown in Table 2, having no piles. The following conclusions could be got by comparing experimental data.

1. When the water pressure of confined aquifer was equal to the body weight of the overlying soft soil, that achieving the critical state of pressure equilibrium method, the bursting of relatively impermeable layer was not occurred immediately and could continue to withstand a certain water pressure.
2. For the bursting failure mode (Sun Yuyong et al., 2010), ⑥ relatively impermeable layer most likely to occur sand boiling on impermeable layer surface (4 groups occurred in 7 groups), followed by water and sand inrushing in contact surface between soil and underground structures and the probability of occurring overall heaving failure was small. ④ relatively impermeable layer most likely to occur overall heaving failure and water and sand inrushing in contact surface between soil and underground structures and this was confirmed by practical engineering, but the probability of occurring sand boiling on impermeable layer was small.
   Combining the failure mechanism of foundation-pit bursting, the main reasons that caused the difference could be summarized as following: 1) the ④ layer soil had the following properties high water content, flow plastic state and ductile failure, that without cracking even in the event of large deformation; 2) the stiffness of ④ layer soil and the contact strength with the envelope structure were lower and the overall heaving might occurred before sand boiling on impermeable layer occurring.
3. The average confined water pressure ratio η (the ratio between the determined confined water pressure during the experiment and the calculated value using pressure balance method) of ⑥ layer soil was 1.25, while the value of ④ layer soil was 1.17.
4. Through failure mode comparison, it showed that the η of sand boiling on impermeable layer was 1.25~1.32, the value of overall heaving was 1.10~1.20 and the value of sand bursting in contact surface between soil and underground structures was 1.05~1.20. That was to say the order of bursting occurred was sand inrushing in contact surface between soil and underground structures, overall heaving and sand boiling on impermeable layer. So, to prevent foundation-pit bursting, it was necessary to ensure effective contact between the envelope structure and piles and the surrounding soil.

Table 2. Influence of soil properties to bursting in foundation pit.

No.	Soil properties	Thickness of soil/m	Failure mode	Bursting water pressure/kPa		Ratio
				Pressure balance	Experimental value	
6–1	⑥	4.78	Sand inrushing in corner	93.7	105	1.12
6–2	⑥	4.82	Surface boiling	94.5	125	1.32
6–3	⑥	4.83	Sand inrushing in side	94.7	111	1.17
6–4	⑥	4.71	Surface boiling	92.3	118	1.28
6–5	⑥	4.74	Overall heaving	92.9	121	1.30
6–6	⑥	5.56	Surface boiling	110.1	142	1.29
6–7	⑥	5.88	Surface boiling	116.4	149	1.28
4–1	④	4.79	Overall heaving	83.8	102	1.21
4–2	④	4.95	Surface boiling	86.6	114	1.32
4–3	④	5.75	Overall heaving	100.6	112	1.11
4–4	④	5.37	Sand inrushing in side	94.0	99	1.05
4–5	④	4.95	Sand inrushing in corner	86.6	102	1.18

3.2 *The influence of piles*

1. The influence of piles on ⑥ layer

   The bursting mode and water pressure of ⑥ layer soil were shown in Table 3 as the different types of piles and pile distances. Through the analysis the data in Table 3 and compared with the data in Table 2, the following conclusions could be obtained.
   - When the construction quality of bored piles and well consolidation between precast piles and around soils could be ensured, piles had a positive role in against bursting of foundation-pit. For example, the $\eta > 1.40$, when the distance of bored piles was 6 m, while the value about 1.25 without piles. The $\eta = 1.31$ when well consolidation between precast piles and around soils and the distance of precast piles was 6 m, the value also was greater than without piles.
   - The bursting modes of ⑥ layer soil mainly was water and sand inrushing in contact between soil and underground structures and surface boiling was rarely occurred, when the foundation-pit having piles. That was the reinforcing function of piles to relatively impermeable layers.
   - The bored pile distance had certain influence on the foundation-pit against bursting, as when the distance was 10 m, $\eta = 1.32$; when the distance was 6 m, $\eta > 1.40$; when the distance was 3 m, $\eta = 1.43$. So, the distance of bored piles was smaller that having more advantage on against bursting.
   - The bursting mode mainly was water and sand inrushing around piles, when the pile type was precast pile. That was to say that the contact between piles and soils was the weaknesses. So, the different of pile distance had less influence on the bursting stability of foundation-pit, as when the distance was 10 m, $\eta = 1.34$; when the distance was 3 m, $\eta = 1.28$.
2. The influence of piles on ④ layer

   The bursting mode and water pressure of ④ layer soil were shown in Table 4 as the different types of piles and pile distances. As the ④ layer soil in Shanghai was in flow plastic state, it was difficult to form hole without slurry and the piles in ④ layer soil mainly was precast pile. As the shown in Table 4, the bursting mode mainly was water and sand inrushing in the contact between soils and underground structures and partly was overall heaving, this was primarily because the shallow insertion depth of piles (inserting confined aquifer layer 2 m). For the η, precast piles had certain positive on against foundation-pit bursting in ④ layer soil, but the effect was not obvious. For example, $\eta = 1.17$ without piles and $\eta = 1.25$ having piles.

Table 3. Influence of pile on foundation-pit bursting.

No.	Pile type	Pile distance/m	Thickness of soil/m	Bursting water pressure/kPa Pressure balance	Experimental value	Ratio	Failure mode
1	Bored	6	4.78	93.7	132	1.41	Surface boiling
2	Bored	6	4.56	89.4	84	0.94	Water inrushing around pile and in side
3	Bored	6	4.50	88.2	134	1.52	Sand inrushing in corner
4	Bored	6	4.54	89.0	125	1.40	Sand inrushing in corner
5	Precast	6	4.86	95.3	125	1.31	Water inrushing around pile
6	Precast	6	4.69	91.9	97	1.05	Sand bursting around pile
7	Bored	10	4.80	94.1	124	1.32	Surface boiling
8	Bored	3	5.14	100.7	143	1.42	Sand inrushing in side
9	Precast	3	4.89	95.8	123	1.28	Water inrushing around pile and on surface
10	Precast	10	5.05	99.0	133	1.34	Water inrushing around pile and on surface

Table 4. Influence of pile on foundation-pit bursting.

No.	Pile type	Pile distance/m	Thickness of soil/m	Water pressure damaged/kPa Pressure balance	Experimental value	Ratio	Failure mode
1	Bored	6	5.00	87.5	125	1.43	Overall heaving
2	Precast	6	5.00	87.5	102	1.17	Sand bursting around piles
3	Precast	10	4.98	87.2	112	1.28	Sand bursting around piles and in side
4	Precast	3	5.47	95.7	125	1.31	Overall heaving

3.3 *The influence of foundation size*

1. Test introduction

   The test device used the model box of centrifugal testing machine and the effective size was 41.5 cm(L) and 17 cm(W). In the test, the impermeable layer was ④ gray silty clay layer of that thickness 5 cm and the confined aquifer layer was $⑤_2$ layer of that thickness 7 cm.

   The mainly test procedure included: 1) Samples forming. Tamping used in $⑤_2$ layer and consolidation used in ④ layer. 2) Dial indicator installation. Flatting model box, installing four dial indicators on the surface of ④ clay layer, that the distances from the left side of the model box were 5 cm, 15 cm, 25 cm and 35 cm (as shown in Fig. 3) and then recorded the initial data. 3) Applying confined water pressures classification and recording the data of dial indicator. The stability time was 24 hours each applying a confined water pressure.

2. Analysis of experimental results

   The experimental results showed that size effect had a very important role in against bursting of foundation-pit. In this test, bursting had not occurred and the upheaval value was only 4 mm when $\eta = 1.7$ (as shown in Fig. 4). In literature [5], the measured $\eta = 1.58$ in testing when bursting of gray silty clay layer of that the model size was 1200 mm(L)* 1200 mm(W)*1500 mm(H). So, it must consider the size effect of foundation-pit when building the judgment formula in anti-bursting.

Figure 3. Dial gauge installation diagram.

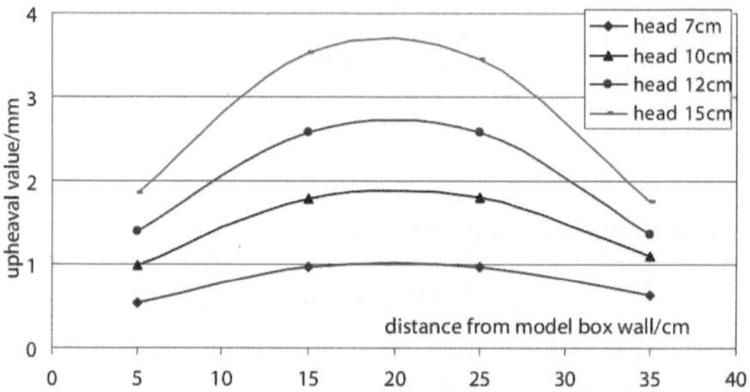

Figure 4. Upheaval & confined water pressure.

4 CONCLUSION

1. It was shown that the mainly failure modes of the ④ and the ⑥ impermeable layers in Shanghai were overall heaving and surface boiling.
2. The mainly bursting mode was water and sand inrushing between soils and underground structures when the existence of piles, especially precast piles.
3. It played an active role in anti-bursting stability when bored piles and precast pile had well contact strength.
4. Foundation size had a significant effect on the anti-bursting stability.

ACKNOWLEDGEMENTS

This study has been financed by Open Foundation of Metal Mine Safety and Health State Key Laboratory (ZDSYS001), Anhui Natural Science Foundation (1208085QE96), Anhui Outstanding Young Talent Foundation (2012SQRL192ZD).

REFERENCES

ARTHUR Marsland, SC M, 1953. Model Experiments to Study the Influence of Seepage on the Stability of a Sheeted Excavation sand[J]. Geotechnique, 6(3): 223–241.

Benmebarek N, Benmebarek S, Kastner R, 2005. Numerical Studies of Seepage Failure of Sand within a Cofferdam[J]. Computers and Geotechnics, (32); 264–273.

Ding Chun-lin, Wang Dong-fang, 2007. A Study on Calculation Model for Piping in Foundation Pit with Confined Underground Water Based on Plastic Failure[J]. Engineering and Mechanics, 24(11); 126–131. (in Chinese)
Hu Qi, Ling Dao-sheng, Zhan Liang-tong, et al., 2007. Model Experiments on Seepage Failure of Deep Foundation Pit Considering Influence of Foundation Piles in Sandy Soil[J]. Chinese Journal of Rock Mechanics and Engineering, 26(10); 2151–2160. (in Chinese)
Ma Shi-cheng, Yin Chang-jun, Zou Yin-sheng, 2004. Analysis and Calculation of the Pit Base Plate Thickness of Bearing Resistance Water Foundation[J]. Engineering Mechanics, 21(2); 204–208. (in Chinese)
Marsland A, 1953. Model Experiments to Study the Influence of Seepage on the Stability of a Sheeted Excavation Sand[J]. Geotechnique, 6(3); 223–235.
Mcnamee J, 1949. Seepage into a Sheeted Excavation[J]. Geotechnique, 4(1); 229–245.
Sun Yu-yong, Zhou Shun-hua, 2010. Mode and Mechanism of Bursting in Foundation Pit Based on Centrifugal Model Test[J]. Chinese Journal of Rock Mechanics and Engineering, 29(12); 2551–2557. (in Chinese)
Sun Yu-yong, Zhou Shun-hua, Zhuang Li, 2011. Calculation of passive earth pressure and shear strength in foundation pits considering residual stress[J]. China Civil Engineering Journal, 44(7); 94–99. (in Chinese)
Terzaghi K, 1943. Theoretical Soil Mechanics[M]. New York; Wiley.
Zhou Jian, Zhang Gang, Hu Zhan-fei, 2006. Model Test Research on Judgment Method of Water Gushing in Pit[J]. Chinese Journal of Rock Mechanics and Engineering, 25(10); 2115–2120. (in Chinese)
Zheng Gang, Yang Jian-min, 2011. Analysis of the Formula for Checking the Heaving Stability of Excavation[J]. China Civil Engineering Journal, 44(2); 123–127. (in Chinese)

Modeling and Computation in Engineering II – Xie (ed)
© 2013 Taylor & Francis Group, London, ISBN 978-1-138-00058-2

Study on the highway rockfall safety risk assessment

C.L. Zhang & S.C. Tang
China Merchants Chongqing Communications Research and Design Institute Co., Ltd., Chongqing, China

W. Yin
School of River and Ocean Engineering, Chongqing Jiaotong University, Chongqing, China

ABSTRACT: Because of the occurred by chance and the randomness of the trajectory of rockfall, to evaluate on-line and high unstable rock by risk assessment is a better suitability method. Combining with specific examples of projects, this paper discusses the rockfall safety risk assessment implementation process in detail. Rockfall safety risk evaluation is divided into the rockfall risk evaluation and highway vulnerability assessment. Influencing factors comprehensive evaluation method can be used for the former while the latter can be used placement probabilistic algorithms.

1 INTRODUCTION

Rockfall and collapse caused a huge threat to the safe operation of the highway, especially in the deep canyon area. It is not uncommon for expressway rockfall and collapse reported at home and abroad, such as Gui-Xin expressway rockfall stabbing incident, Chu-Da highway rockfall blocking road events. At present, the collapse and rockfall mainstream algorithm is using the limit equilibrium analysis method, and this method can provide the basic design parameters for the in-line, low unstable rock and collapse treatment or design. But, for the on-line and high collapse of unstable rock, it is difficult to make a good assessment by the limit equilibrium analysis. Because of the occurred by chance and the randomness of the trajectory of rockfall, to evaluate on-line and high unstable rock by risk assessment is a better suitability method.

2 CONTENTS OF GEOLOGICAL DISASTER RISK ASSESSMENT

Based on risk assessment time, geological disaster risk assessment is divided into pre-disaster pre-evaluation, evaluation of track in the disaster and post-disaster summary evaluation. According to the geological disaster risk assessment range, it is divided into point evaluation, surface evaluation and regional evaluation. The basic contents of the various types of disaster risk assessment for hazard evaluation, vulnerability evaluation, damage loss evaluation and prevention engineering evaluation. The risk assessment system composition by different types and kinds of geological disaster risk assessment is a multi-faceted, three-dimensional evaluation grid (Fig. 1).

According to the above division, the rockfall safety of risk assessment is a pre-evaluation of pre-disaster, so, the main focus of the evaluation is to rockfall hazard evaluation and highway vulnerability assessment.

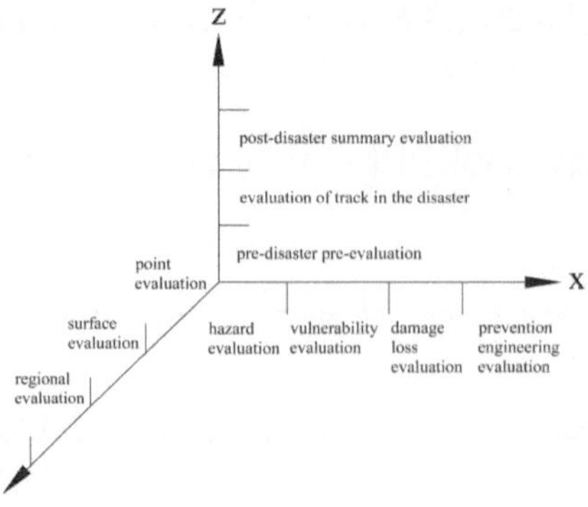

Figure 1. Grid chart of geological disaster risk assessment.

3 THE ROCKFALL SAFETY OF RISK ASSESSMENT

3.1 *Rockfall stability evaluation*

To assess the risk of on-line high rockfall by using comprehensive evaluation method of influencing factors.

Screening to determine the risk of unstable rock stability evaluation index are 7, 3 basic factors as lithology, slope height, slope angle and the other 4 as historical disaster development, rock weathering degree, fissure water development degree and human engineering activities.

3.1.1 *Determining factor set and evaluation set*

According to the evaluation index selected by the previous section, factors set: {lithology, slope height, slope angle, historical disaster development, rock weathering degree, fissure water development degree, human engineering activities}, and evaluation set: {unstable, less stable, more stable, stable} = {I, II, II, IV}.

3.1.2 *Grading evaluation factors*

All the factors grading and criteria listed in the following tables.

3.1.3 *Membership to determine*

The membership (slope angle, slope height) value of quantitative indicators is described by "drop half trapezoidal" distribution of phenomenon function, and the unstable rock stability class is divided into four levels, its expression as follows:

$$\mu_1 = \begin{cases} 1 & x \le S_1 \\ \dfrac{S_2 - x}{S_2 - S_1} & S_1 < x \le S_2, \\ 0 & x > S_2 \end{cases} \qquad \mu_2 = \begin{cases} 0 & x < S_1 \text{ and } x > S_3 \\ \dfrac{x - S_1}{S_2 - S_1} & S_1 < x \le S_2 \\ \dfrac{S_3 - x}{S_3 - S_2} & S_2 < x \le S_3 \end{cases} \tag{1, 2}$$

$$\mu_3 = \begin{cases} 0 & x < S_2 \text{ and } x > S_4 \\ \dfrac{x - S_2}{S_3 - S_2} & S_2 < x \le S_3 \\ \dfrac{S_4 - x}{S_4 - S_3} & S_3 < x \le S_4 \end{cases}, \quad \mu_4 = \begin{cases} 0 & x < S_3 \\ \dfrac{x - S_3}{S_4 - S_3} & S_3 \le x < S_4 \\ 1 & x \ge S_4 \end{cases} \tag{3, 4}$$

Table 1. Unstable rock evaluation factors grading standards table of water-hemp road.

Evaluation index	Grading standards			
	Stable/IV	More stable/III	Less stable/II	Unstable/I
Lithology combination	Full hard rock	Medium hard rock, small joints and fractures	In feeble and white rock, more joints and fractures, small outstanding rock	Loose, more outstanding rock, small fragmentation stone fall, depth development of joints and fractures
Slope angle	≤45°	45°~65°	65°~80°	>80°
Slope height/(m)	≤30	30~60	60~120	>120
Historical disaster development	None	Less frequent occurrence	More frequent occurrence and happen again	Frequent occurrence or cyclical occurrence
Rock weathering degree	Weak	Weak, differential weathering obvious	Stronger, differential weathering significantly	Strong
Fissure water development degree	None	Weaker	Stronger	Strong
The degree of human activity	None	Weaker	Stronger	Strong

Table 2. Fissure water development scores standard.

	None	Weaker	Stronger	Strong
The development degree of fissure water description	Dry rock, no running water and seepage traces	Water flow from the rock mass surface with a small threat to the stability of rock mass	Damp rock, part of the rock mass damage surface with a small amount of water seepage	Wet rock, damaged surface crack is evident in groundwater seepage, and large capacity

Table 3. Human activities influence grading standards.

	None	Weaker	Stronger	Strong
Human activities influence description	No or a small amount of human activity, and highway far away from the dangerous rock mass, dangerous rock mass occurs collapse rockfall disaster without human activities	A small amount of human activity, less impact on the unstable rock stability	Have certain human activities, and the highway is close to the dangerous rock mass, all kinds of activities on the stability of the dangerous rock has a serious potential impact	The highway cutting dangerous rock with a mountain to build, serious damage to the structural stability of the dangerous rock mass and seriously affect the stability of unstable rock

where S1, S2, S3, S4 are the evaluation index for the level of danger in unstable rock classification threshold.

Qualitative indicators of membership values using the empirical method, according to their degree of influence on the unstable rock dangerous build qualitative indicators membership.

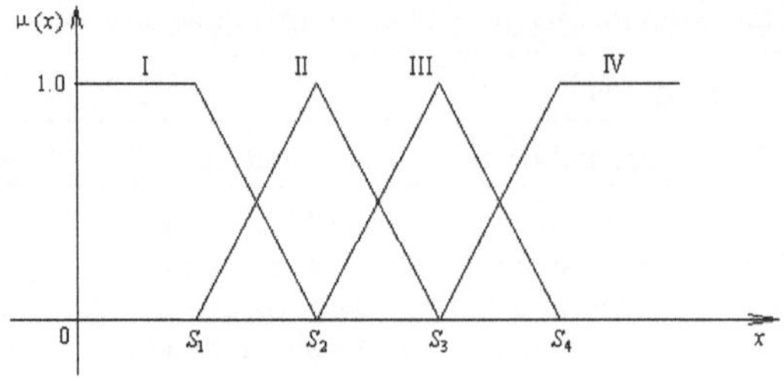

Figure 2. Schematic diagram of the membership function.

Table 4. Unstable rock evaluation quantitative indicators membership value table.

Evaluation index	IV	III	II	I
Slope angle (°)				
x > 80	0	0	0	1
65 < x ≤ 80	0	0	(80 – x)/5	(x – 75)/5
	75 < x ≤ 80	75 < x ≤ 80	75 < x ≤ 80	75 < x ≤ 80
	0	(75 – x)/10	(x – 65)/10	0
	65 < x ≤ 75	65 < x ≤ 75	65 < x ≤ 75	65 < x ≤ 75
45 < x ≤ 65	(65 – x)/20	(x – 45)/20	0	0
x ≤ 45	1	0	0	0
Slope height (m)				
x > 120	0	0	0	1
60 < x ≤ 120	0	0	(120 – x)/30	(x – 90)/30
	90 < x ≤ 120	90 < x ≤ 120	90 < x ≤ 120	90 < x ≤ 120
	0	(90 – x)/30	(x – 60)/30	0
	60 < x ≤ 90	60 < x ≤ 90	60 < x ≤ 90	60 < x ≤ 90
30 < x ≤ 60	(60 – x)/30	(x – 30)/30	0	0
x ≤ 30	1	0	0	0

Table 5. Unstable rock evaluation qualitative indicators membership value table.

Evaluation	I	II	III	IV
Class I evaluation	0.5	0.35	0.1	0.05
Class II evaluation	0.25	0.45	0.25	0.05
Class III evaluation	0.05	0.25	0.45	0.25
Class IV evaluation	0.05	0.1	0.35	0.5

3.1.4 *Weight determination*

For the above established seven evaluation index, calculated using the analytic hierarchy, and the weight of each parameter listed in Table 6.

3.2 *Highway unstable rock on vulnerability assessment*

The unstable rock instability and failure in the form of sporadic small-scale rockfall as the main form, and for the destruction of highway itself and its repair costs in the acceptable range of expressway, the harm of the main object for the traveling vehicles and the value of the members is difficult to use money to measure. On the one hand, the expressway traffic volume varies with the date, weather and travel speed, it is difficult to predict its trajectory.

Table 6. The weight of each evaluation index.

Index	Lithology	Slope angle	Slope height	Historical disaster	Rock weathering	Fissure water	Human activities
Weights	0.231	0.128	0.046	0.121	0.365	0.070	0.039

Table 7. Water-hemp road unstable rock hazard evaluation division table.

Unstable rock dangerous level	High	Higher	Medium	Low
Description	Stability level: stable; vulnerability level: high	Stability and vulnerability of the evaluation results, the lowest is less stable or higher	Stability and vulnerability of the evaluation results, the lowest is more stable or medium	Stability level: stable; vulnerability level: low

On the other hand, once produced casualty accidents, the social impact and economic losses are difficult to accurately estimate. Therefore, this study calculated by using the unstable rock instability in the roadbed placement within the scope of the probability to judge its vulnerability. Generally speaking, the more the unstable rock instability point of fall is located in the roadbed within the scope of the possibility, the larger the chance of loss caused by roads, vehicles and personnel will be, both showed a positive correlation. Calculated the point of fall in the roadbed within the scope of the possibility by using the rockfall analysis software Rockfall of Canadian Rocscience Co.

In the process of simulation, using the linear seeds and random variable input.

3.3 *The rockfall safety risk assessment*

Safety risk assessment system is divided into two aspects of the stability evaluation and vulnerability assessment. On the basis of the unstable rock stability risk assessment and vulnerability assessment results, a certain way to be both comprehensive evaluation of the safety risk assessment.

Safety window theory suggests that the occurrence of disasters is a combination of numerous adverse factors simultaneously, such as highway unstable rock generated disasters need to meet the following criteria at the same time: the one is the avalanche of rock into the roadbed range; the other one is the rock avalanche hit the cars on the road directly resulting in the loss of personnel and vehicles, or a high-speed vehicle failed to discover or to avoid the rock in the road caused by the loss of personnel and vehicles.

Based on the above analysis, this paper puts forward rockfall safety risk assessment follows the principle of "lower but higher". For example, the stability of somewhere unstable rock (zone) is poor (unstable), but its vulnerability is low (the probability of entering into the embankment is less than 30%), the risk comprehensive evaluation of premises unstable rock (zone) is low.

4 CONCLUSIONS

For the on-line and high collapse of unstable rock, because of the occurred by chance and the randomness of the trajectory of rockfall, to evaluate on-line and high unstable rock by risk assessment is a better suitability method.

Combining with specific examples of projects, this paper discusses the rockfall safety risk assessment implementation process in detail. Rockfall safety risk evaluation is divided into the rockfall risk evaluation and highway vulnerability assessment. Influencing factors comprehensive evaluation method can be used for the former while the latter can be used placement probabilistic algorithms.

REFERENCES

Chapman, C.B. & Cooper, D.F. 1987. *Risk Analysis For Large Projects: Models, Methods and Cases*. John Wiley & Sons Publishers.

Chapman, C.B. 1990. A Risk Engineering Approach to Project Risk Management. *Project Management* 18(1).

Ellis D. 1989. Environment at risk, case histories of impact assessment. Berlin: Springer-Verlag.

Fell R. 1994. Landslide risk assessment and acceptable risk. *Canadian Geotechnical Journal* 31:261–272.

Hearn G J. 1995. Landslide and erosion hazard mapping at Ok Tedi copper mine, Papua New Guinea. *Quarterly Journal of Engineering Geology*, 28:47–60.

Mejia N.M., Wohl E.E. & Oaks S.D. 1994. Geological hazards, vulnerability, and risk assessment using GIS model for Glenwood Springs. Colorado. *Geomorphology* 10:331–354.

Shachter, R.D. 1986. Evaluating Influence Diagrams. *Operational Research* 34(6):871–882.

Modeling and Computation in Engineering II – Xie (ed)
 ISBN 978-1-138-00058-2

Study on reinforcement depth of wind-blown sand foundation reinforced by compaction

W. Wang, L. Liu, L. Wu & C. Li
Hebei Construct Rconnaissance Graduate Co., Ltd., Shijiazhuang, Hebei, China

ABSTRACT: Considering the engineering examples involving wind-blown sand foundation reinforced by dynamic compaction, this article focuses upon the effective reinforcement depth of the wind-blown sand foundation reinforced by such method through in-situ tests and proposes the corresponding formula that can be used to calculate such depth of the wind-blown sand foundation materialized by such method.

1 INTRODUCTION

First paragraph. Featuring a wide range of application, effective reinforcement effect, simple construction tools, relatively low cost and quick construction etc., the dynamic compaction technique has been adopted rapidly and extensively since being introduced in 1975. However, the theoretical study on such technique, in comparison with its application, is obviously left behind. In addition, the study primarily concentrates upon qualitative analysis rather than objective quantitative analysis (Wang Tiehong, 2005). Therefore, it is quite essential to quantitatively analyze the laws in connection with soil stratum settlement and change along the direction of depth, the relation between various compacting energy, the reinforcement depth made by dynamic compaction and the bearing capacity of soil foundation, the proper criteria concerning tamping termination during the process of dynamic compaction as well as the influences upon surroundings exerted by such method and so on under different soil foundation and stratum circumstances. So far it has been a great concern while reinforcing foundations through such method, especially for special soil stratum.

As the Open up the West Program is proceeding successfully, most of the land is rested upon the wind-blown sand foundation and the experience of reinforcing such type of foundation through dynamic compaction is directly originated from the engineering practice of handling the filled soil foundation via such method. Either at home or abroad no study on the wind-blown sand foundation reinforced by dynamic compaction has been reported. To sum up, quantitatively analyzing the reinforcement effect of the wind-blown sand foundation by means of dynamic compaction in relation to the wind-blown sand foundation means a lot in terms of setting up, developing and perfecting the dynamic compaction theories and related design method. Considering the related engineering practices, this article mainly researches the effective reinforcement result of the wind-blown sand foundation achieved by dynamic compaction through in-situ tests.

2 FIELD TEST PROPOSAL

2.1 *Engineering geological conditions*

As per the related survey information, the soil foundation on site is divided as below: Unit Layer 1: Artificial Plain Fill (Q^{ml}); Unit Layer 2: Fine Sand of the Holocene Series in the Quaternary System (Q_4^{eol}); Unit Layer 3: Fine Sand of the Holocene Series in the Quaternary System (Q_4^{dl}); Unit Layer 4: Powder Sand ~ Fine Sand of the Upper Pleistocene Series in the

Quaternary System (Q_3^{al-pl}); Unit Layer 5: Fine Sand of the Upper Pleistocene Series in the Quaternary System (Q_3^{al-pl}); Unit Layer 6: Gravel of the Upper Pleistocene Series in the Quaternary System (Q_3^{al-pl}); Unit Layer 7: Powder Sand Mudstone and Fine Sandstone Interbed of the Yanan Formation in the Lower Series of the Jurassic System (J_1^y), and Unit Layer 8: Powder Sand Mudstone and Fine Sandstone Interbed of the Yanan Formation in the Lower Series of the Jurassic System (J_1^y). For details, please refer to the following Table 1.

2.2 *Dynamic compaction construction parameter design*

The compacting energy adopted in the field test is as follows: 1000 kN·m, 2000 kN·m, 3000 kN·m, 4000 kN·m, 6000 kN·m and 8000 kN·m and as for the parameters designed for each compacting energy level, please see below:

1. Test Area with 1000 kN·m as Compacting Energy
   This said test area is filled with a layer of soil not more than 3 m in thickness. The depth required to be treated via dynamic compaction is 6 m and the energy exerted during spot compaction is 1000 kN·m. The hammer weighs 14.7 tons with a diameter of 2.3 m at the bottom. The dropping distance of such hammer is 7 m. Tamp twice as per the intervals between different compacting spots of D × D = 5 m × 5 m in grid mesh arrangement as shown in Figure 1. Thoroughly tamp once with the energy of 800 kN·m and repeat tamping twice in each point.

Table 1. Soil stratum physical mechanical properties indicators.

Soil stratum	Thickness (m)	Water content ω (%)	Special gravity (Gs)	Natural gravity r (kN/m³)	Dry unit weight r_d (kN/m³)	Porosity (e)	Modulus of deformation E_0 (MPa)	Poisson's ratio (μ)
①	4.2	4.4	–	(13.5)	–	–	(1.0)	(0.45)
②	1.8	4.4	–	(14.0)	–	–	(3.0)	(0.40)
③	5.4	6.6	2.66	15.3	14.2	0.873	(6.5)	(0.35)
④	3.5	7.2	2.67	15.4	14.3	0.866	(7.0)	(0.35)
⑤	23.5	11.8	2.66	16.5	14.7	0.816	(16.0)	(0.30)

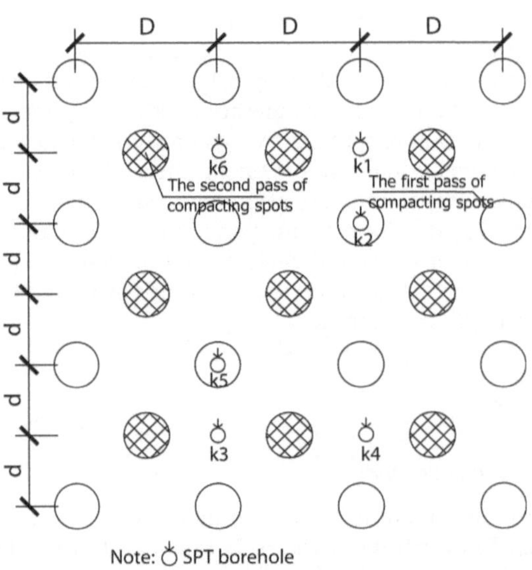

Figure 1. Compacting spots and test point layout.

2. Test Area with 2,000 kN·m as Compacting Energy
   This said test area is filled with a layer of soil ranging from 3 m to 5 m in thickness. The depth required to be treated via dynamic compaction is 8 m and the energy exerted during spot compaction is 2,000 kN·m. The hammer weighs 15 tons with a diameter of 2.34 m at the bottom. The dropping distance of such hammer is 13.5 m. Tamp twice at random as per the intervals between different compacting spots of D × D = 5.5 m × 5.5 m in grid mesh arrangement as shown in the Figure 1. Thoroughly tamp once with the energy of 1,000 kN·m and repeat tamping twice in each point.
3. Test Area with 3,000 kN·m as Compacting Energy
   This said test area is filled with a layer of soil ranging from 5 m to 7 m in thickness. The depth required to be treated via dynamic compaction is 9 m and the energy exerted during spot compaction is 3,000 kN·m. The hammer weighs 25 tons and has a diameter of 2.35 m at the bottom. The dropping distance of such hammer is 12 m. Tamp twice at random as per the intervals between different compacting spots of D × D = 6 m × 6 m in grid mesh arrangement as shown in the Figure 1. Thoroughly tamp once with the energy of 1,500 kN·m and repeat tamping twice in each point.
4. Test Area with 4,000 kN·m as Compacting Energy
   This said test area is filled with a layer of soil ranging from 5 m to 7 m in thickness. The depth required to be treated via dynamic compaction is 9 m and the energy exerted during spot compaction is 4,000 kN·m. The hammer weighs 25 tons with a diameter of 2.35 m at the bottom. The dropping distance of such hammer is 16 m. Tamp twice at random as per the intervals between different compacting spots of D × D = 6 m × 6 m in grid mesh arrangement as shown in the Figure 1. Thoroughly tamp once with the energy of 2,000 kN·m and repeat tamping twice in each point.
5. Test Area with 6,000 kN·m as Compacting Energy
   This said test area is filled with a layer of soil ranging from 7 m to 10 m in thickness. The depth required to be treated via dynamic compaction is 10 m and the energy exerted during spot compaction is 6,000 kN·m. The hammer weighs 40 tons with a diameter of 2.3 m at the bottom. The dropping distance of such hammer is 15 m. Tamp twice at random as per the intervals between different compacting spots of D × D = 6 m × 6 m in grid mesh arrangement as shown in the Figure 1. Thoroughly tamp once with the energy of 2,000 kN·m and repeat tamping twice in each point.
6. Test Area with 8,000 kN·m as Compacting Energy
   This said test area is filled with a layer of soil at least 10 m in thickness. The depth required to be treated via dynamic compaction is 12 m and tamp thrice at random. The energy exerted during spot compaction as per the intervals between different compacting spots of D × D = 8 m × 8 m in grid mesh arrangement as shown in the Figure 2 for the first and second rounds is 8000 kN while for the third round, 3000 kN·m. Thoroughly tamp once with the energy of 2,000 kN·m and repeat tamping twice in each point.

2.3 *Test proposal*

Before and after dynamic compaction, carry out the Standard Penetration Tests (SPT) in the test area via the drilling machine XY-100 so as to obtain soil stratum's physical mechanical indicators. As shown in the Figures 1 and 2, K1~K3 are the SPT holes before dynamic compaction and K4~K6, the SPT holes after dynamic compaction. There are 3 SPT holes drilled before dynamic compaction and 3 SPT holes drilled again after dynamic compaction. As for each set of 3 SPT holes, 2 are among the compaction spots and 1 is in such spot.

The SPT mentioned herein is implemented via the hammer automatic release and free-fall method in which the hammer, heavy as 63.5 kg, drops from a distance of 76 cm and holes are bored via compressed air. Such car drilling equipment is applicable to any mountainous or desert terrain. Once the penetrator reaches 15 cm below the ground surface, write down the number of tamping for every 10 cm and the number of tamping that makes the penetrator reach 30 cm underground is adopted as the standard number of tamping in SPT N. When the number of tamping is accumulated to 50 but the penetrator is failed to reach the depth of 30 cm, terminate

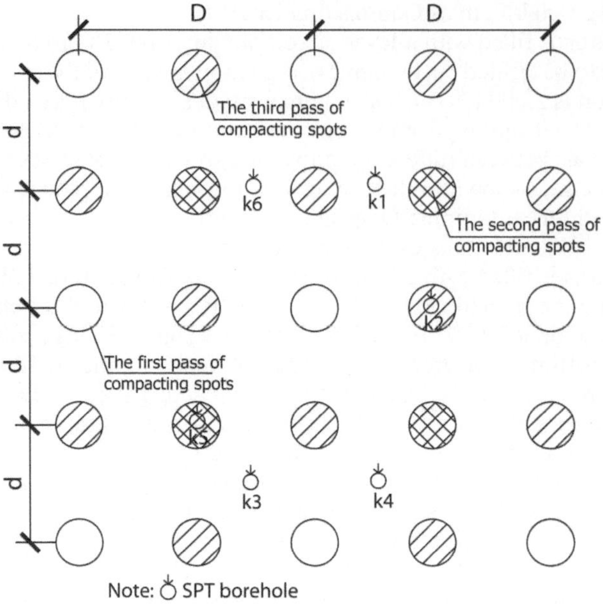

Figure 2. Compacting spots and test point layout.

the test and record the actual penetration distance that is then converted to the number of tamping in SPT for 30 cm. When the pre-penetration distance is failed to reach 15 cm but the number of tamping is already 50, record the actual penetration distance and terminate the test.

$$N = 30 \times \frac{50}{\Delta S} \tag{1}$$

where in ΔS is the penetration distance (cm) at the number of tamping of 50.

3 TEST DATA ANALYSIS AND STUDY

3.1 *Comparative analysis on SPT data*

Figures 3–8 depict the relation between the penetration distance and the average number of tamping in SPT before and after dynamic compaction in each test area. The following figures reveal that the number of tamping in SPT after dynamic compaction is obviously higher than that before dynamic compaction. It is possible to obtain the effective reinforcement depth of the wind-blown sand foundation realized via dynamic compaction from the judging criteria proposed by Lenard (Lenard,1980) that the effective reinforcement depth is achieved by adding 3~5 times of tamping based upon the standard number of tamping in SPT concerning sandy soil.

Figure 3 is related to the dynamic compaction test area of 1,000 kN · m. The judging criteria proposed by Lenard (Lenard, 1980) suggest that the compacting energy of 1,000 kN · m is able to effectively reinforce the wind-blown sand foundation by 7 m.

Figure 4 is related to the dynamic compaction test area of 2,000 kN · m. The judging criteria proposed by Lenard (Lenard, 1980) suggest that the compacting energy of 2,000 kN · m is able to effectively reinforce the wind-blown sand foundation by 8 m.

Figure 5 is related to the dynamic compaction test area of 3,000 kN · m. The judging criteria proposed by Lenard (Lenard, 1980) suggest that the compacting energy of 3,000 kN · m is able to effectively reinforce the wind-blown sand foundation by 9 m.

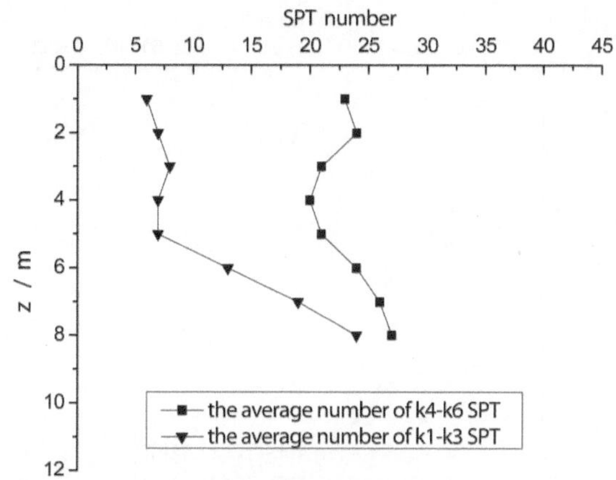

Figure 3. Comparison of the number of tamping in SPT before and after dynamic compaction with the energy of 1,000 kN·m.

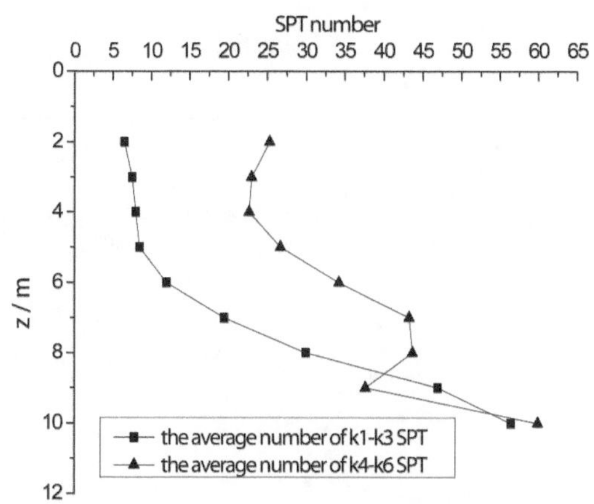

Figure 4. Comparison of the number of tamping in SPT before and after dynamic compaction with the energy of 2,000 kN·m.

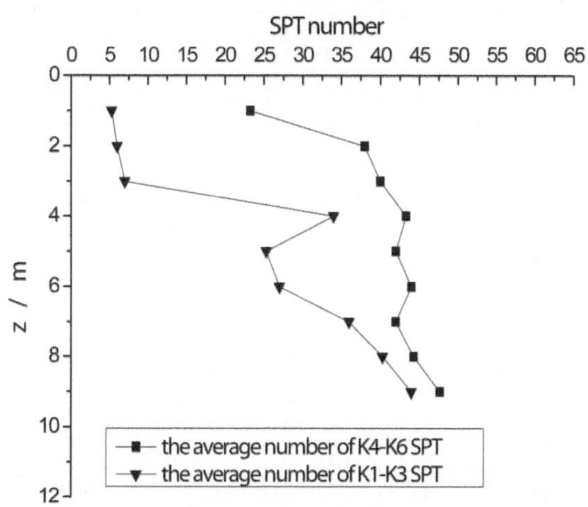

Figure 5. Comparison of the number of tamping in SPT before and after dynamic compaction with the energy of 3,000 kN·m.

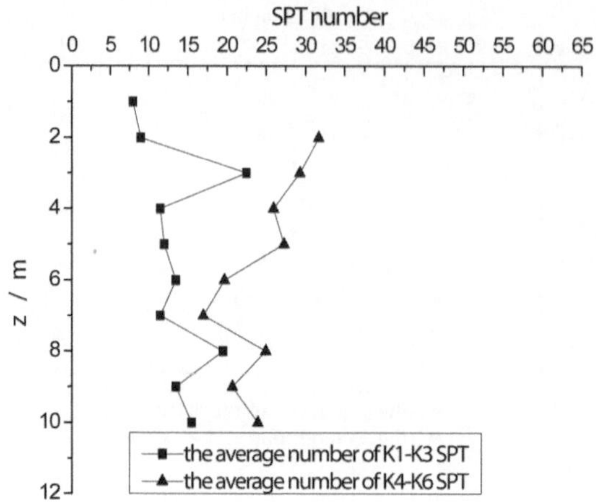

Figure 6. Comparison of the number of tamping in SPT before and after dynamic compaction with the energy of 4,000 kN·m.

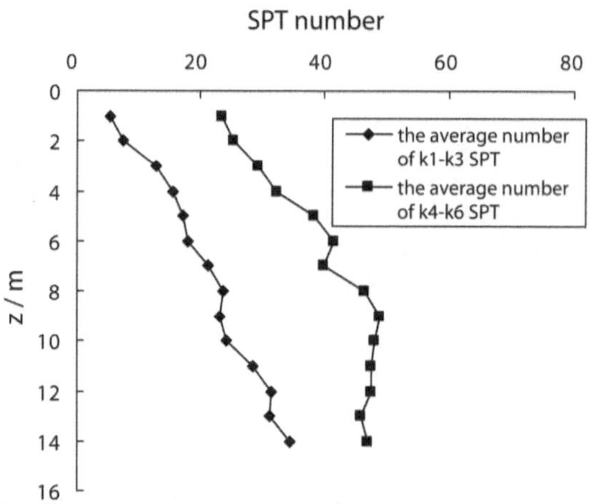

Figure 7. Comparison of the number of tamping in SPT before and after dynamic compaction with the energy of 6,000 kN·m.

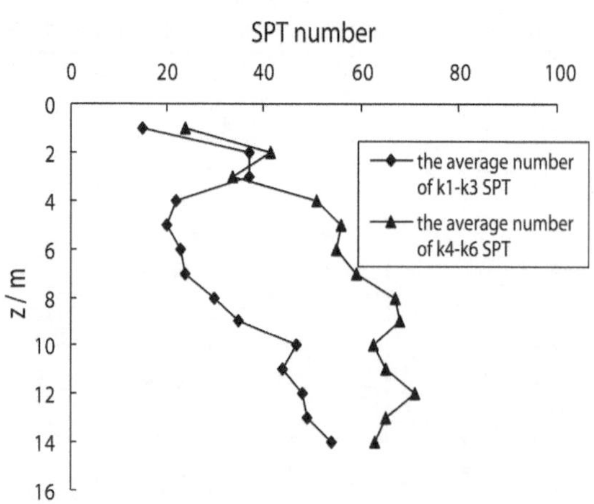

Figure 8. Comparison of the number of tamping in SPT before and after dynamic compaction with the energy of 8,000 kN·m.

Figure 6 is related to the dynamic compaction test area of 4,000 kN·m. The judging criteria proposed by Lenard (Lenard, 1980) suggest that the compacting energy of 4,000 kN·m is able to effectively reinforce the wind-blown sand foundation by 10 m.

Figure 7 is related to the dynamic compaction test area of 6,000 kN·m. The judging criteria proposed by Lenard (Lenard, 1980) suggest that the compacting energy of 6,000 kN·m is able to effectively reinforce the wind-blown sand foundation by 12 m.

Figure 8 is related to the dynamic compaction test area of 8,000 kN·m. The judging criteria proposed by Lenard (Lenard, 1980) suggest that the compacting energy of 8,000 kN·m is able to effectively reinforce the wind-blown sand foundation by 14 m.

3.2 *Analysis upon effective reinforcement depth*

Following comprehensively analyzing the test results of the abovementioned test areas, the relation between the effective reinforcement depth of the wind-blown sand foundation realized by dynamic compaction and the compacting energy at different levels is represented in Table 2 that suggests the former is in a linear relation with the latter. Such relation can be expressed via the linear equation (2) as below:

$$d = 6 + 0.001 \times E \tag{2}$$

where in d is the effective reinforcement depth in the unit of m and E, the dimensionless level of the compacting energy.

Figure 9 shows the relation between the actual values and code values concerning the effective reinforcement depth realized by the compacting energy at different levels. It is revealed by

Table 2. The relation between the actual values and code values concerning the effective reinforcement depth.

Compacting energy level (kN·m)	1000	2000	3000	4000	5000	6000	8000
Results calculated from the formula (2)/m	7	8	9	10	11	12	14
Upper limit/m	6	7	8	9	9.5	10	10.5
Lower limit/m	5	6	7	8	9	9.5	10

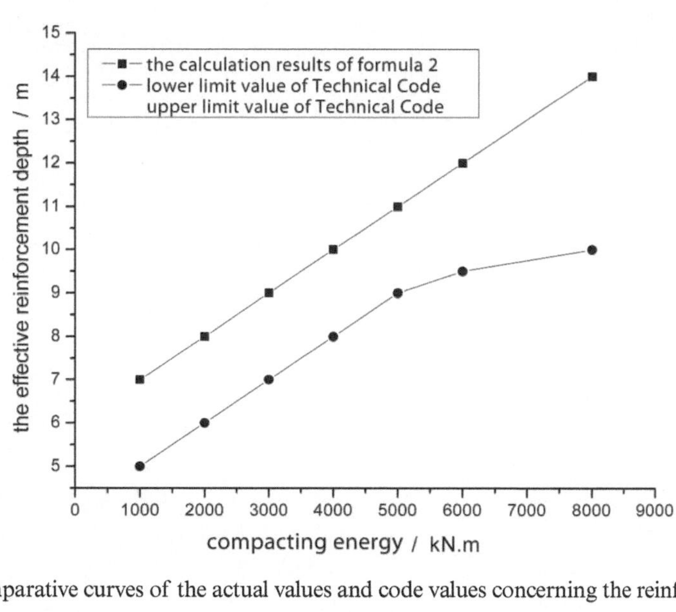

Figure 9. Comparative curves of the actual values and code values concerning the reinforcement depth.

such figure that when the compacting energy is not more than 4,000 kN·m, the code values are in a linear relation with the actual ones and when such energy exceeds 4,000 kN·m, the actual values remain the same but the code values slow down and meanwhile, the actual values are larger than the maximum code values. Therefore, the effective reinforcement depth of the wind-blown sand foundation achieved by dynamic compaction is higher than the upper limit as stipulated. It is due to the characteristics of the wind-blown sand which is formed in single grain structure with coarse particles piled together through no or just a tiny coupling force between grains.

4 CONCLUSION

1. The dynamic compaction method effectively reinforces the wind-blown sand foundation and therefore, constitutes a cost-efficient and speedy approach to handle foundations in the west of China where there is vast land formed by wind-blown sand.
2. Owing to the characteristics of wind-blown sand, it is proved by the tests that the effective reinforcement depth of the wind-blown sand foundation materialized by dynamic compaction in which the compacting energy ranges from 1,000 to 8,000 kN·m is higher than the value prescribed in the Technical Regulations on Treatment of Construction Foundations (JGJ79-2002).
3. The effective reinforcement depth of the wind-blown sand foundation materialized by dynamic compaction can be calculated via the following formula:

   $d = 6 + 0.001\times E$, where in d is the effective reinforcement depth in the unit of m and E is the dimensionless level of the compacting energy.

REFERENCES

Gerald A. Leonard, William A. Cutter, Robert D. Holtz 1980. Dynamic compaction of granularsoils. Journal of The Geotechnical Engineering Division, Vol. 106 (1), PP35–40.

The China Academy of Building Research etc 2002.Technical Regulations on Treatment of Construction Foundations (JGJ79-2002).

Wang Tiehong. 2005. New Records of the Foundation Treatment for Major Domestic Projects. China Construction Industry Press.

Modeling and Computation in Engineering II – Xie (ed)
© 2013 Taylor & Francis Group, London, ISBN 978-1-138-00058-2

Centrifugal model experimental study on natural foundation of immersed tube tunnel

Xia-bing Yue, Yong-li Xie, Zhi-nan Hu & Hong-guang Zhang
College of Highway, Chang'an University, Xi'an Shanxi, China

Guo-ping Xu
CCCC Highway Consultants Company Limited, Beijing, China

ABSTRACT: Base stress distribution characteristics and the deformation characteristics of immersed tube tunnel in excavation and filling process of tunnel engineering occupies an important position. Taking one cross-sectional of the natural foundation in Hong Kong-Zhuhai-Macao immersed tube tunnel as research object, centrifugal model test of the whole construction process with different cushion layer thickness was carried out. When excavation rebound and backfill compression in the immersed tube tunnel, stress distribution and deformation characters of natural foundation, the strain characteristics of immersed tube structure and the influence of cushion layer thickness had been revealed. That provides the design basis for the construction of immersed tube tunnel and scheme of ground treatment. The results show that, excavation unloading makes basement produce arch distribution rebound. Basal stress distribution is the shape of a saddle after backfill, and only the stress in the middle of base is increasing with time and silting. The strain distribution of immersed tube plate is the shape of a saddle and strain has little change after construction. Cushion layer thickness thin lead to smoothing effect on basal stress distribution is poor, and recompression quantity and foundation settlement after construction is large.

1 INTRODUCTION

With the development of the underwater tunnel, immersed tube tunnel already more and more be used in underwater engineering field. As immersed tube tunnel in the formation with rich groundwater, the foundation condition is particularly important. Liu Yan, Liu Goubin (2006) analyzed the measured data of deformation. Zhang Qing-he, Gao Wei-ping (2003) analyzed the treatment method of immersed tube tunnel foundation. The effects of the different methods in solving base trough stability and controling tunnel settlement were revealed. The space stress condition of immersed tube tunnel structure and the influence of water level and overlying soil thickness, etc factors on its internal force were analyzed by Chen Qing-jun, Zhu Hehua and Li Xing (2000). But there is no concrete research on rebound and compression deformation characteristics of immersed tube tunnel foundation. As centrifugal model test can satisfy the key conditions of similarity. So it has been widely used in geotechnical engineering field.

This paper takes one cross-sectional of the natural foundation in Hong Kong-Zhuhai-Macao immersed tube tunnel as research object, centrifugal model test simulate excavation, cushion paving, immersed tube immersing and backfill back silting in the construction process under the conditions of 1 m and 1.5 m cushion layer thickness. Results had been revealed. That provides the design basis for the construction of immersed tube tunnel and scheme of ground treatment.

2 CENTRIFUGAL MODELING TEST

2.1 *Test scheme*

Taking typical cross sectional whose soil distribution from top to bottom were silt, silty, clay, clay sand inclusion, clay, medium sand and coarse sand, centrifugal model test simulate excavation (to 3 m thick clay layer), cushion paving, immersed tube immersing and backfill back silting in the construction process under the conditions of 1 m and 1.5 m cushion layer thickness. Stress distribution and deformation characters of natural foundation, the strain characteristics of immersed tube structure and the influence of cushion layer thickness were analysed. So as to reveal excavation rebound and backfill recompression characteristics of immersed tube tunnel natural foundation.

The clearance size of model box used in test is 70 cm × 40 cm × 50 cm. In the test, model rate n = 100, and Centrifugal acceleration is 100 g. Process of centrifugal test is as follows:

1. Adding water in the basement to soil saturated and fully mixing so as to eliminate soil gas, then it will be made to reshape the soil sample. Choosing organic glass as immersed tube material and width of immersed tube model as the prototype one 0.01 times. According to the similarity theorem (ZUO Bao-cheng et al, 2004). Ensure model floor wall and middle wall stiffness (EI = E*bh3/12) is equal to the prototype, calculating the thickness. To ensure the effect on basal stress equal for the principle, calculate the model height. Processing immersed tube model. As shown in Table 1.
2. In order to speed up the soil consolidation and reduce model box edge wall friction effect, silicone oil is blotted out the walls in the model box. The soil depth is narrow one hundred times and stratification in model box according to section actual foundation soil distribution. Each fill a layer of soil, then open the centrifuge make it preconsolidation in the centrifugal acceleration of 100 g. Four sample were taken out according to the sample standard of undisturbed soil from the model by ring knife that wiped Vaseline. Conducting direct shear test on them. In comparison with the field test results, make the soil in the model box reach initial consolidation condition.
3. Overlying soil cannot be completely in model box because the limit of model box size. According to the similarity theory, high density weight block instead of the overlying soil and it also meet the base stress unchanged. Model box is placed into a centrifuge hanging in blue. Then earth pressure and pore pressure sensor are connected and laser displacement sensor is placed.
4. Construction process simulation is as following: Removing weight block, Simulation excavation to make foundation soil stress release and rebound on cases of the top unloading. After laying hardcore layer, putting immersed tube tunnel model sink and positioning weight block that converted from backfill. The model box is graded loading according to actual working progress. When t = 15 min (about 104 d of the realistic environment) and centrifugal acceleration is up to 100 g, now is equivalent to actual backfill state of prototype. Since then, centrifuge in 100 g continuous running 156 min. That is equivalent to the condition of three years operation after completion of the back silting.
5. Reading and processing the data from each measuring system through the test process.

Table 1. Model and prototype parameter table.

Name	Prototype	Model
Modulus (Gpa)	34.5	50
Frontier walls thickness (m)	1.5	0.02329
Base plate thickness (m)	1.5	0.02329
Middle wall thickness (m)	0.8	0.0035
Model height (m)	11.44	0.126

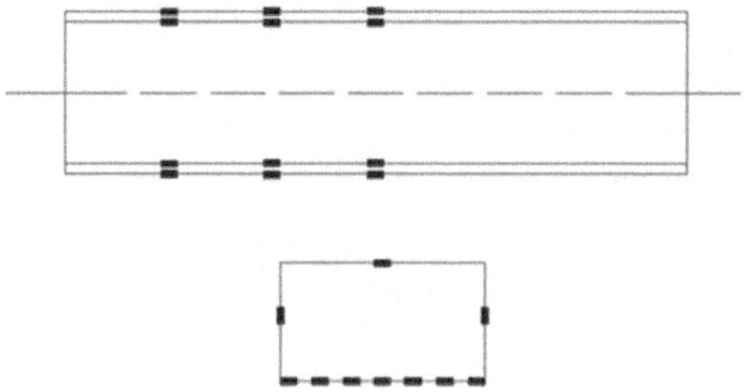

Figure 1. Strain gauge layout diagram.

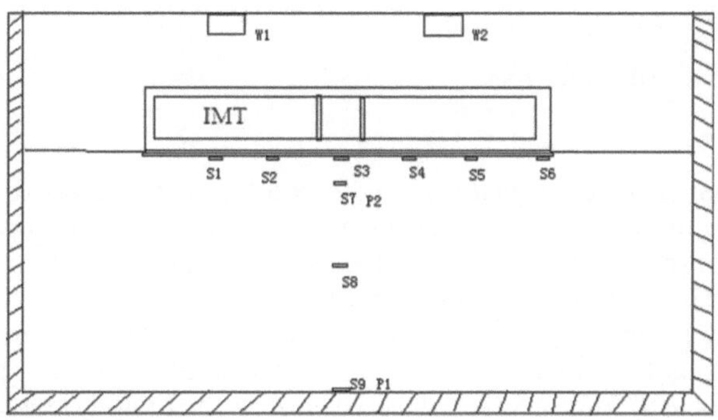

Figure 2. Pressure box layout diagram.

2.2 *Experimental measure*

Earth pressure cell and pore water pressure cell were buried in the foundation soil and laser shift sensor were mounted in the top of model box in order to monitor vertical stress distribution of foundation soil, stress distribution of basement, development process of pore water pressure and vertical rebound recompression displacement of basement during the test procedure. As shown in Figure 2 (S1–S9 are earth pressure cell; P1 and P2 are pore water pressure cell; W1 and W2 are laser displacement sensor). High sensitivity strain gauges were mastered on the bottom and sides of immersed tube in order to measure the strain of immersed tube during the test process. All above tests data was automatically collected by DDS data acquisition system.

3 TEST RESULT

3.1 *Analysis of stress results*

According to the similarity law, Model test data was converted to the numerical value for prototype. Analyzing these conversion results

1. Stress value is increasing with depth along the foundation increases. Stress curve present the transition in different soil boundary. That is because each soil density is different so as to the stress curve in the soil slope is also different.
2. Base stress distribution curve of various conditions section is as Figure 3 shows. As the diagram shows, all curves are in the shape of a saddle distribution. Basement stress value on both sides of the immersed tube maximum, the stress value on the partition of the immersed tube is the second, and basal stress of immersed tube tunnel roadway part minimum. Variation gradient of basement stress curve under cushion layer 1 m condition

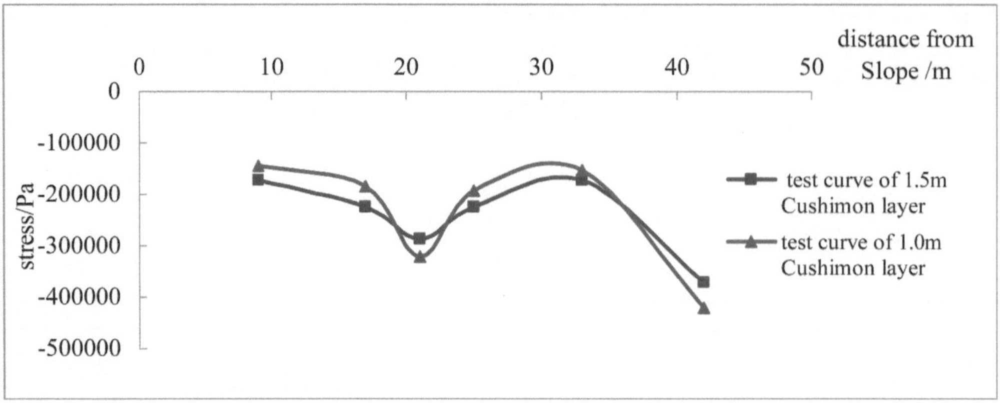

Figure 3. Basal stress distribution curve.

is bigger than that of cushion layer of 1.5 m condition. In other words, the difference between maxima and minima of the basement stress under the condition of cushion layer 1 m is greater. The overall extent change of that curve is larger. Homogenization effect of 1 m cushion layer for foundation stress is poor.

Three years after construction completion, stress value shows attenuation type growth with time increasing. Stress increment is gradually smaller. Maximum increment is about 10 kPa. Eventually tend to be gentle. This is because in the long-term loads, the springback soil of the excavation foundation is gradually consolidation, then pore water discharge. That making the base stress presence increases phenomenon with time increasing. Stress increment on the middle traffic lane part of immersed tube relatively greater than other parts, making the original saddle become gentle and stress have a certain degree of diffusion. The stress on foundation edge increases with the overlying effect increases, so there has a certain plastic deformation in basal soil. Along with the back silting and time increasing, edge stress has no longer increases but intermediate part stress accordingly increases. But as a result of buoyancy effect with immersed tube, stress increment is limited and curve has little change.

3. Pore pressure value in the loading process with time has the obvious growth of the each condition. After achieving steady speed, the pore pressure tends to be gentle, and has a slow dissipation state.

3.2 *Analysis of strain results*

The strain of Immersed tube structure on the backfill plays an important role in the design of immersed tube and foundation treatment (WAN Xiao-yan et al, 1999). Figure 4 is strain distribution with the test on the immersed tube bottom under two kinds of condition after construction is completed. As is shown in the figure, after construction is completed and centrifugal machine achieving steady speed, the strain curve of immersed tube model are in the shape of a saddle distribution. Parts of the sidewall and partition wall are on the underside tension state, while traffic lane part of immersed tube is on the upper tension state. The strain of immersed tube has no obvious change with time increasing.

Comparing the two strain test curves of immersed tube with the two kinds of condition, strain curve has a large distribution under the cushion layer 1 m condition, and its strain of each points are larger than that of cushion layer 1.5 m condition.

3.3 *Analysis of displacement results*

Settlement change value of the foundation after each construction stage completed was measured from the laser displacement sensor. Excavation due to basement

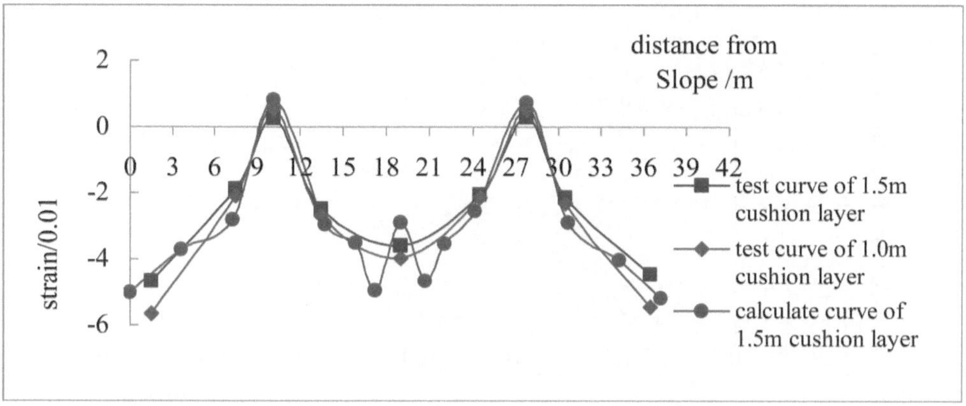

Figure 4. Immersed tube strain curve.

rebound (LUO Zan-you, 2002). The excavation rebound displacement of the basement is 0.0407 m and the compression quantity after backfill is 0.0512 m under the conditions of 1.5 m cushion layer. Because there distribute 3 m thickness of soft clay layer in the mediate section. They are the main factors of springback phenomenon (HU Qi-zhi et al, 2001). When the pressure on the part of that clay represent was excavation removal, original elastic deflection of the soil particle produced unloading rebound. Combined with electric magnetic effect, bound water that was forced out of the soil on the original stress condition is inhaled and adhered in the soil particle surface. That is shown as expansion rebound.

The excavation rebound displacement of the basement is 0.0415 m under the conditions of 1 m cushion layer. Because rebound displacement of the basement has nothing with cushion layer thickness, it has no difference with the conditions of 1 m cushion layer. The compression quantity after backfill is 0.0561 m. This value is larger than that of 1.5 m cushion layer conditions. This is because the cushion thickness decreases, and basal distributing certain thickness of soft clay, so its recompression quantity is slightly bigger.

Settlement change value of the foundation after construction completion three years was measured from the laser displacement sensor. By the data, the compressed displacement changes curve with time increasing can be received, as is shown in Figure 5. Basal recompression quantity change is small after construction is completed. Recompression is mainly produced in the loading process. After achieving steady speed, growth rate of settlement decrease gradually diminishing. Lately, the curve tends to be gentle. The largest settlement is 0.02035 m after three years construction completion under the 1.5 m cushion layer conditions. Under the 1 m cushion layer conditions, the largest settlement is 0.0307 m after three years construction completion. Cushion layer decreasing makes the post-construction settlement bigger.

There has obvious difference of cushion changes after take out immersed tube model between the 1.5 m cushion layer conditions and the 1 m cushion layer conditions. Under the 1.5 m cushion layer conditions, cushion has no obvious change after testing; While under the 1 m cushion layer conditions, there has most gravel infiltration clay layer after test and cushion layer and clay layer part mixed as a whole. This is the main reason of the differences of stress and displacement under two conditions.

3.4 *Analysis of displacement results*

Figure 6 and Figure 7 are displacement curve of basement excavation rebound and backfill recompression. From the diagram, excavation base slot on the rebound displacement curve is roughly arch distribution after excavation. Springback deformation on the center is the largest and to both sides gradually reducing. The recompression quantity is small in the middle of the base and large on both sides after backfill back silting. Recompression quantity on the excavation slope feet is the biggest.

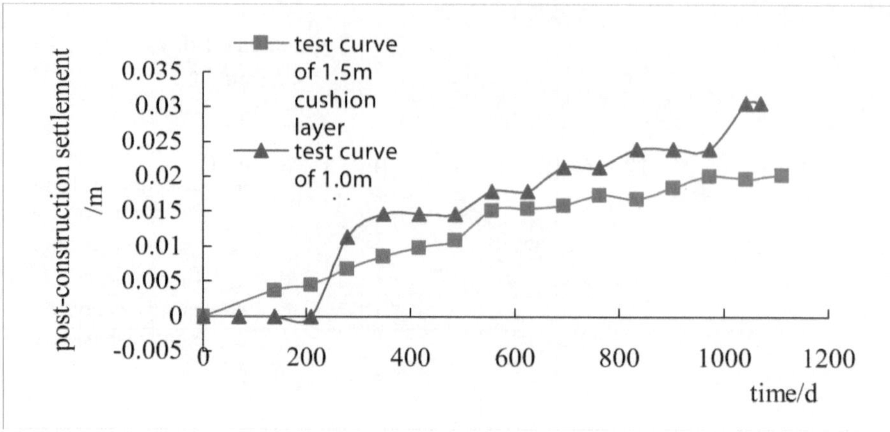

Figure 5. Post-construction settlement curve.

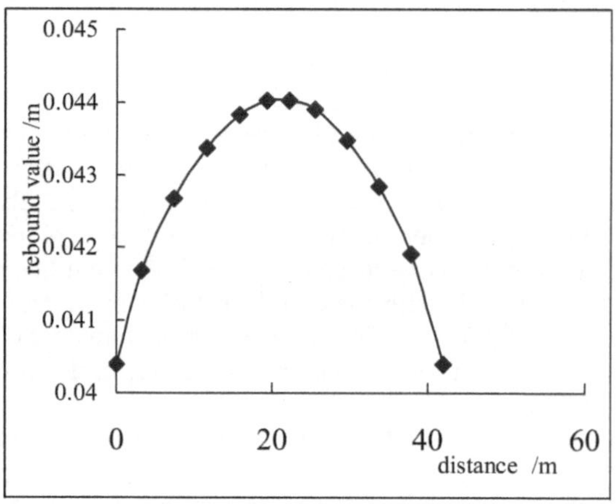

Figure 6. Resilience curve of section base.

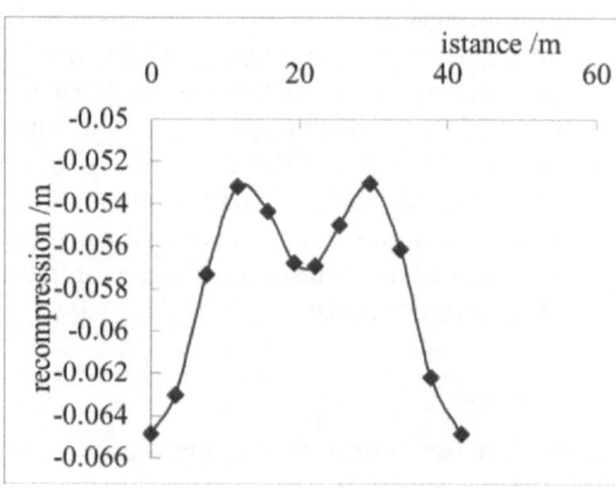

Figure 7. Recompression displacement curve of section base.

4 CONCLUSION

Taking one cross-sectional of the natural foundation in Hong Kong-Zhuhai-Macao immersed tube tunnel as research object, centrifugal model test of the whole construction process with different cushion layer thickness was carried out. When excavation rebound and backfill compression in the immersed tube tunnel, stress distribution and deformation characters of natural foundation, the strain characteristics of immersed tube structure and the influence of cushion layer thickness had been revealed. That provides the design basis for the construction of immersed tube tunnel and scheme of ground treatment. The results are as follows.

1. Excavation unloading cause springback deformation, backfill produce recompression. Standard condition (1.5 m cushion layer thickness) base resilience and recompression quantity were 0.0407 m and 0.0512 m from test measurement. Excavation produces basement rebound and arch up in a curve distribution, the biggest rebound quantity is 0.0443 m in basal center.
2. Basal stress in the shape of a saddle after the completion of backfill. With the back silting and time increasing, stress value of basement intermediate part was increased and change of stress value on both sides was not obvious on the conditions of 1.5 m cushion layer. Cushion layer 1 m condition, stress value of test point were increasing, and increment of the stress value on the middle part and both sides have not quite differ.
3. The strain curve of immersed tube model are in the shape of a saddle distribution. Parts of the sidewall and partition wall are on the underside tension state, while traffic lane part of immersed tube is on the upper tension state. The strain of immersed tube has no obvious change with time increasing.
4. Small cushion layer thickness makes backfill recompression quantity and post-construction settlement quantitative large, and smoothing effect on basal stress distribution is poor.
5. Pore pressure value in the loading process with time has the obvious growth of the each condition. After achieving steady speed, the pore pressure tends to be gentle, and has a slow dissipation state.

REFERENCES

Chen Qing-jun, Zhu He-hua, Li Tong, Yan Hao, Zhao Yun-feng, Jiang Wen-hui. (2000). The 3D Internal Force Analysis of Immersed Tunnel Structure. Chinese Quarterly of Mechanics, (02).

Hu Qi-zhi, He Shi-xiu, Yang Xue-qing. (2001). Estimate on excavation of basement uplift. Soil Eng. and Foundation, 15(2), 29–31.

Liu Yan, Liu Guo-bin, Sun Xiao-ling, Wang Hai-ping. (2006). Analysis of deformation laws by using the rule of time-space effect in soft soil excavation. Chinese Journal of Geotechnical Engineering, (S1).

Luo Zan-you. (2002). Discussion about "the research of deep foundation pit unloading rebound". Chinese Journal of Geotechnical Engineering, (05).

Wan Xiao-yan, Guan Xin-min, Tang Ying. (1999). Calculation and analysis on segment structure of immersed tube tunnel. The World Tunnel, (6), 19–22.

Zhang Qing-he, Gao Wei-ping. (2003). Comparison analysis on treatment methods of pipe-sinking tunnels. Rock and Soil Mechanics, (S2).

Zuo Bao-cheng, Chen Cong-xin, Liu Cai-hua, et al. (2004). Research on similar material of slope simulation experiment. Rock and Soil Mechanics, 25(11), 1805–1808.

Modeling and Computation in Engineering II – Xie (ed)
© 2013 Taylor & Francis Group, London, ISBN 978-1-138-00058-2

Piece-wise control approaches to improving GPS signal reacquisition

Z. Lei
Shanghai Institute of Technical Physics, Chinese Academic of Sciences, Shanghai, China

Yuanfei Wang
Key Laboratory of Geographic Information Science, Ministry of Education, East China Normal University, Shanghai, China

ABSTRACT: Two piece-wise approaches are proposed and tested, including the method of a piece-wise Frequency-locked Loop (FLL) followed by a piece-wise Phase-locked Loop (PLL) and the method of a direct piece-wise PLL. The performance of each approach was assessed using GPS data collected on a vehicle-based mobile platform under mild to moderate operational conditions. Both piece-wise methods were shown to be able to reduce the transition time by a factor of approximately three for the frequency pull-in period and four for the phase pull-in period. Finally, the scheme of piece-wise FLL followed by piece-wise PLL is recommended to reduce the reacquisition time during the carrier phase reacquisition process.

1 INTRODUCTION

How to improve the GNSS (Global Navigation Satellite System) receiver architectures and optimize the algorithms of carrier phase acquisition and tracking is an important research area (Kaplan, E. D. 2006). According to the concept of software radio which is the input signal is digitized as close to the antenna as possible, so that the great flexibility can be achieved (O'Driscoll et al., 2008). Thereby, using hardware to design the front-end sub-system of GNSS receivers and using software to acquire and track the signal are the better approaches to improve performance and efficiency.

During the reacquisition process, this procedure is restarted, albeit often with smaller search spaces. However, given the reduced search space, other approaches may be used to provide a more rapid transition to carrier phase tracking. For example, the relationship between rise time, overshoot, natural frequency, and damping ratio has not been investigated in previous work during the reacquisition process (Kaplan, E. D. 2006). Therefore, if the damping ratio and natural frequency are increased in the loop filters during the carrier phase reacquisition process, a smaller convergence time and overshoot is expected. Based on above observations, a piece-wise control approach is proposed in this article. The approach divides the reacquisition and tracking process into two separate periods, each with different natural frequencies and damping ratios. It is noted that the differences between an FLL and a PLL are filter orders and specific loop filter parameters.

2 FREQUENCY AND PHASE TRANSITION

Generally, frequency lock indicator (FLI) and phase lock indictor (PLI) can be used to evaluate the performance of frequency tracking and phase tracking, respectively.

In the carrier phase reacquisition process, the Doppler and carrier phase transition time have been of particular interests. As mentioned above, FLI and PLI can be used to evaluate the tracking errors (O'Driscoll et al., 2008, LIU Ruopu et al., 2008). However, usually a filter is used in the software receiver to obtain the FLI and PLI. Therefore, a delay is produced, and the use of FLI and PLI is not effective when evaluating the transition time. In order to avoid the delay in the transition time evaluation process, some primary metrics, which are Doppler jitter and phase jitter, are introduced in this article.

The evaluation process for the frequency transition time is conducted by a polynomial fitting and Doppler jitter calculation. The polynomial fitting is employed to obtain the interpolation of the tracking results.

Essentially, the coefficients of a polynomial are determined in a least-squares sense. Generally, a first-order polynomial is employed because the Doppler change during the transition process is small, which means that a first-order fit is accurate enough and will be shown later. However, a second-order polynomial would be preferred for a longer transition process. The tracking errors and Doppler jitter are calculated in the second step.

Doppler jitter is a function of signal power, loop filter noise bandwidth, and integration time. Different from the frequency transition time assessment, the polynomial fitting does not apply to the phase transition time analysis. The reason is that generally the carrier phase increments over integration time are in the order of several cycles. Because of this, high accuracy polynomial fitting cannot be obtained. Fortunately, since the loss of phase lock is manually induced in the phase tracking loop, the carrier phase outputs without a manually induced loss of phase lock can be used as a reference. In so doing, the phase errors can be obtained.

3 POLYNOMIAL FITTING

Polynomial fitting is employed here to evaluate the FLL transition process. Without any signal loss, different GPS satellites with different received signal power have been assessed (Peng Xie 2010). Here we take PRN 31 (C/N0 is 51dB/Hz) for example to illustrate how to obtain the Doppler jitter and the frequency tracking transition time. In this case, the given natural frequency is 10 Hz, damping ratio is 0.707, and a second-order loop filter is employed. Generally, the Doppler outputs without manually induced loss of signal lock can be used to obtain the polynomial fitting results. However, the FLL reaches the steady-state after 1 s; to this end, the tracking results thereafter can also be used to obtain the polynomial fitting for the FLL outputs, as well as the Doppler jitter. As discussed earlier, a first-order polynomial fitting is sufficient to fit the FLL results after reaching steady-state in this case.

Figure 1 shows the scale of y-axis being enlarged to the range of –1 Hz to 1 Hz in order to show the transition time more clearly. The 3σ Doppler jitter used as the threshold to determine the FLL transition time (blue line) is also shown. It is observed that for a 200 Hz initial frequency error, the transition time is 0.6 s (green line).

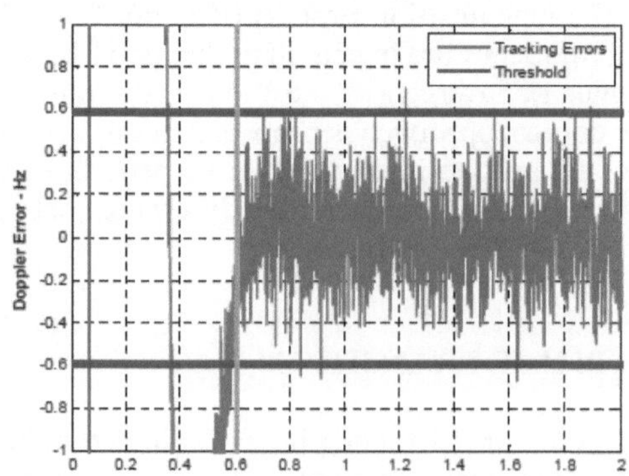

Figure 1. Frequency errors during the FLL transition process for the PRN 31.

4 PARAMETERS TUNING

Different natural frequencies and damping ratios have different transition processes. From this, a problem arises in finding tracking loop parameter values during the reacquisition process in order to meet the fast carrier phase reacquisition requirement. Generally, two types of logic can be employed in the piece-wise method to control the parameter transitions. The first one is time-based logic, and the second one is FLI-based (PLI-based) logic. In the time-based logic, a fix period is used to control the transition between the piece-wise method and general tracking loop (i.e., 10 Hz natural frequency and 0.707 damping ratio for standard FLL). In the FLI-based (PLI-based) logic, the transition is controlled by the FLI (PLI) value since the FLI (PLI) is a function of frequency (phase) error. However, FLI (PLI) is not effective to evaluate the transition time, and the time-based logic is used.

Different tracking loop parameters were assessed during the tuning process. Specifically, the natural frequency range was from 8 Hz to 80 Hz with a step of 4 Hz, and the damping ratio range was from 0.707 to 8.4 with a step of 0.707. Furthermore, since a time-based logic was ultimately adopted, different piece-wise intervals (i.e., intervals with constant parameters) were ranged from 10 ms to 80 ms with a step of 10 ms. Table 1 shows the specific parameter sets of interest, as they give an idea of how to conduct the parameters tuning. The first parameter set is used to show the performance of increased natural frequency compared to the general parameter set; the second parameter set is used to show the performance of larger damping ratio and longer piece-wise control period; the third and fourth parameter sets are used to show the performance of even larger natural frequencies and damping ratios.

Figure 2 shows the transition process for the first parameter set of 200 Hz initial frequency error.

The same threshold (Doppler jitter) is employed in all cases as shown in Figure 1; this is due to the fact that they have the same steady-state behaviors. The transition time for 200 Hz initial frequency error is 0.32 s and is shown in this figure (green line). However, Figure 2 also shows that the overshoot is still relatively large after 20 epochs of the piece-wise period. This should be reduced by further increasing the piece-wise parameter values in the tracking loop.

Table 1. Specific parameter sets of interest.

Parameter sets	Natural frequency	Damping ratio	Piece-wise epochs
1	30 Hz	0.707	20 ms
2	30 Hz	1.414	80 ms
3	60 Hz	5.6	20 ms
4	80 Hz	8.4	20 ms

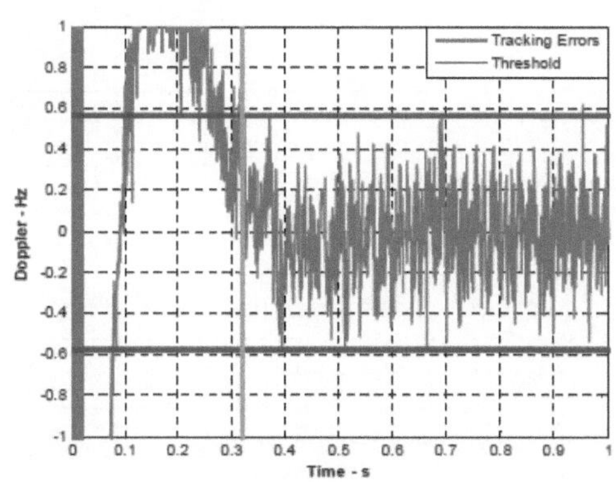

Figure 2. Frequency transition process zoomed in for the first parameter set of 200 Hz initial frequency error.

5 PIECE-WISE PLL

The piece-wise FLL performance shown above suggests that the piece-wise method can reduce the frequency transition time during the reacquisition process (Peng Xie 2010, Borio, D.N. 2009). To this end, the proper parameter set obtained from the piece-wise FLL section can also be used in the second-order piece-wise PLL (since FLL and PLL use the same type of loop filter). Figure 3 and Figure 4 show the transition process of different initial phase errors for the standard PLL and piece-wise PLL respectively.

The phase jitter is used as the threshold to determine the phase transition time where the total phase jitter is smaller than 2 degrees for a signal with a C/N0 of 50 dB-Hz scenarios.

The comparison between standard PLL and piece-wise PLL is summarized in Table 2; again, the improvement with the piece-wise approach is noticeable with a four- to five-fold improvement in all cases.

Given the above, a piece-wise FLL followed by a piece-wise PLL is recommended to reacquire the GNSS signal more quickly. In this way, the piece-wise FLL is used to reduce the frequency error in the tracking loop, and then the carrier phase error is reduced by the piece-wise PLL. For a 200 Hz initial frequency error and 90 degrees initial phase error case, the transition time is reduced from 0.91 s (0.59 s for FLL and 0.32 s for PLL) to 0.25 s (0.21 s for piece-wise FLL and 0.04 s for piece-wise PLL), which means that more carrier phase observations are available after employing piece-wise control methods.

The same as piece-wise FLL method, nine other open-sky loss of signal lock scenarios are also conducted to evaluate the performance of piece-wise PLL method by employing the same proper loop filter parameters, the loss of locks are manually set at every 5 seconds.

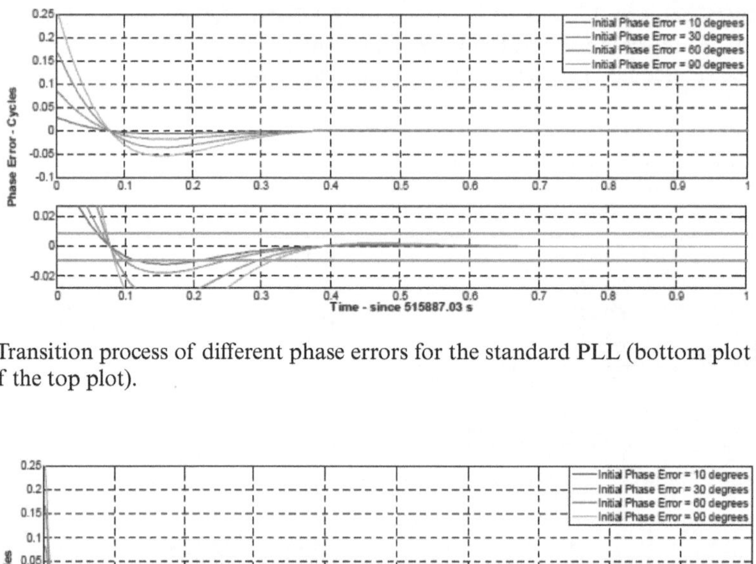

Figure 3. Transition process of different phase errors for the standard PLL (bottom plot is a zoomed in version of the top plot).

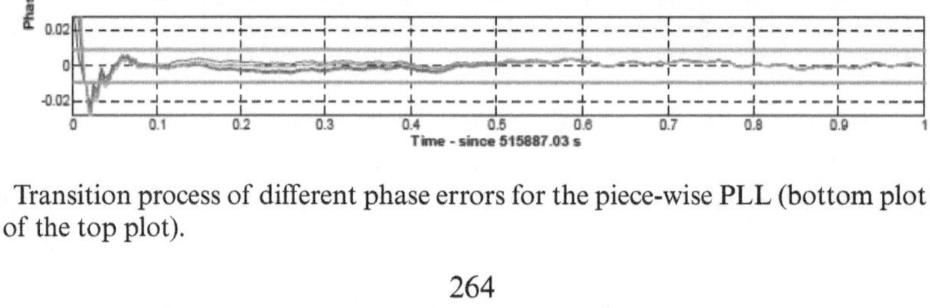

Figure 4. Transition process of different phase errors for the piece-wise PLL (bottom plot is a zoomed in version of the top plot).

Table 2. Standard and piece-wise PLL comparison.

Initial phase error	Transition time standard	Transition time piece-wise
10 degrees	0.21	0.04 s
30 degrees	0.24	0.04 s
60 degrees	0.29	0.04 s
90 degrees	0.32	0.05 s

Table 3. Piece-wise PLL method transition time summary.

Initial phase error	Mean of transition time	STD of transition time
10 degrees	0.03 s	0.01 s
30 degrees	0.04 s	0.01 s
60 degrees	0.03 s	0.01 s
90 degrees	0.04 s	0.01 s

6 CONCLUSIONS

The transition processes of different natural frequencies and damping ratios were evaluated, as well as the logic of how to control the transitions between different loop filter parameter sets. Generally, the piece-wise FLL control divided the reacquisition and tracking process into separate periods, each with different natural frequencies and damping ratios. Proper parameter sets were obtained during the parameter tuning process. At the first period, the natural frequency and damping ratio were 60 Hz and 5.6 respectively, and the piece-wise period was 20 ms. After the piece-wise method, general tracking loop parameters were used. In so doing, the piece-wise method produced a three-fold improvement in transition time. For example, the frequency transition was reduced from 0.59 s to 0.26 s for 200 Hz initial frequency error.

The same tracking loop parameters were employed by the piece-wise PLL, whereas the improvement with the piece-wise approach was noticeable with a four- to five-fold improvement in all initial phase errors. To this end, for a 200 Hz initial frequency error and 90 degrees initial phase error case, the transition time was reduced from 0.91 s to 0.25 s, which means that more carrier phase observations were available after employing piece-wise control methods. Drawbacks of the piece-wise control method include however larger damping ratios that have slow reactions to user dynamics and larger bandwidths that introduce more noise in the tracking loop piece-wise period. Moreover, the time-based logic used in this article sometimes is not long enough to reduce the initial frequency error.

ACKNOWLEDGEMENTS

Funded By Open Research Funding Program of KLGIS (NO. KLGIS2011 A02) and the funding of Changzhou Sci & Tec program (NO. CE20120018).

REFERENCES

Borio, D.N., Sokolova, and Lachapelle, G. Doppler Measurements and Velocity Estimation: a Theoretical Framework with Software Receiver Implementation, in Proceeding of ION GPS GNSS, Savannah, Georgia, 23–25 September, 2009.

Kaplan, E.D. Understanding GPS: Principle and Applications, Second Edition, Artech House, Boston, 2006.

LIU Ruopu, ZHAI Chuanrun and ZHAN Xingqun. A method of GPS positioning based on piecewise and weighted signal-to-noise ratio. Information technology. 2008, 32(9).

O'Driscoll, C., Petovello, M.G. and Lachapelle, G. Software Receiver Strategies for the Acquisition and Re-Acquisition of Weak GPS Signals, in Proceeding of IONNTM, 28–30 January, San Diego CA, 2008.

Peng Xie. Improving Carrier Phase Reacquisition Time Using Advanced Receiver Architectures. Master thesis, University of Calgary, 2010.

Modeling and Computation in Engineering II – Xie (ed)
© 2013 Taylor & Francis Group, London, ISBN 978-1-138-00058-2

Structural optimum design of the hydraulic leakage holes based on ANSYS

Shengyi Zheng & Pengpeng Liu
Hohai University, Nanjing, Jiangsu Province, China

ABSTRACT: Leakage holes, which allow water pour outside when the hydraulic gate is lifting, is usually treated as the secondary design in the hydraulic gate. Small leakage hole would result in a blockage by sundries in the water after long-time gate idle. Consequently, the residual water would cause the corrosion in main girders. Obviously oversized leakage hole would weaken the strength of girders. By using the optimal module in FEM software ANSYS, the reasonable diameter of leakage holes can be obtained. Technically, the diameters are set as design variables and the total weight of the gate is the performance function. Stress and displacement constraints should be also considered.

1 INTRODUCTION

Hydraulic steel gate, which can be divided into emersed gate and DTH gate, is the main water retaining structure of hydraulic structures. No matter what kind of gate type, radial or plain, it may be damaged in the form of anti-corrosion coating spalling, steel corrosion, wear, deformation and so on by the effects of environment and loads in its service period, despite of a variety of anti-corrosion measures taken in the process of construction and management.

During the process of designing a hydraulic gate, the size and thickness of gate panel, main girders, stringers and arm, even the stiffeners are determined by mechanical computation or design standard. However, the size of leakage holes usually is arbitrary chosen because most designers consider it minor importance. The principle function of leakage holes which tend to be set in the main girder web allows the water in the body outflow through it when the gate is open, avoiding the possible corrosion caused by residual water. If the size of leakage holes is set too small, it is very easy to be choked up by mud and dirt. Since water in the gate structure can't be drained successfully, the main girders, front flanges and gate panel downriver side immersed would rust even rust through. Potential safety hazard caused by unreasonable design of leakage holes should be taken into account. As a typical example, the main steel structure of spillway service gate of Peking Yaoqiaoyu reservoir, including its main girder, front flanges, etc, rusted seriously, the average corrosion in some parts even exceeded 3 mm. Oversized leakage hole is not reasonable either because obviously it weakens the strength and stiffness of the main girder. Therefore, it is meaningful for us to study the reasonable size of leakage hole. During the past decades, many scholars have studied the design of hydraulic gate by structure optimization techniques. The optimal layout of truss structure in radial steel gate [Li Huo-Kun, 2007] was studied by Li Huo-Kun, who used the finite element method for structure buckling problems. Zhu Jun-Zuo [Zhu Jun-Zou, 2007] is the first one who use the optimization modulus of FEM software ANSYS to evaluate the static and dynamic response of the hydraulic steel gate such that some optimal structures are obtained. Li Bo [Li Bo, 2006] analyzed the stability of webs with round hole and Cheng Chang-Jun [Cheng Chang-Jun, 1999] did some stability analysis theoretically and applied his results to engineering projects. Depending on the previous researches, we present a viable method to design a reasonable size of leakage hole with which the requirements of structure performance and drainage are satisfied based on the optimization modulus of ANSYS. Combined with the design of gate of diversion tunnel in Hei river reservoir, the basic procedure of leakage holes optimization is introduced. In the following, we first descript the problem of leakage hole size optimization in Section 1 and illustrate the basic method of optimization analysis in ANSYS in Section 2. Subsequently, we take the gate in Hei river reservoir as an example to analysis and show the optimal results in Section 3, followed by some concluding remarks in Section 4.

2 DESCRIPTION OF LEAKAGE HOLES OPTIMIZATION PROBLEM

Leakage holes are set in the position of webs of main girder. In traditional design, usually the number, layout and size of holes are arbitrary chosen without any dependence on evaluation. Theoretical basis and methods for its design are rarely mentioned in the design specification. Designers are usually design it according to actual situation and experience. When excessive leakage hole is provided and the diameter is too large, it will inevitably affect the strength and stiffness of the main girder. Thus, the whole gate is possibly dangerous. On the contrary, too small leakage hole will be choked up by mud and dirt, leading the increase of corrosion. Therefore, for the design of leakage holes, on the premise of satisfying the strength and stiffness requirements, the area of the leakage hole is need to be expanded as large as possible. Moreover, the total weight of the gate weight can be reduced.

In order to achieve the optimal size of leakage holes, structural optimization is applied. In general, the structural optimization consists of three types [Xiao Wei-Hua, 2010], size optimization, shape optimization and topology optimization. Although by using topology optimization, the optimal topology of the shape and position of the holes on the web can be obtained, yet the result is seriously dependent on the mesh density. In addition, the result obtained by using common used OC methods which is already embedded in ANSYS inevitably becomes a form of checkerboard [Bendsoe, 2003]. Moreover, the result is difficult to predict and the final optimized shape is also difficult to process.

Leakage holes set in the main girder, which is usually a welded structure, is generally designed as a circular hole for easy process and manufacture. Thus, the diameters of leakage holes are naturally chosen as design variables. Assume that there are n leakage holes set in the main girders, their diameters, denoted by x_1, x_2, ... x_n, respectively, the design variables set is $X = (x_1, x_2 \ldots x_n)$. For each x_i, taking into account the requirements of the welding process and the minimum aperture requirements, it should be in the range of maximum aperture R_{max} and minimum aperture R_{min}. According to the requirements of *Water Resources and Hydropower Engineering steel gate design specification* [1995], the ratio of maximum deflection f_{max} of the main girder and calculated span should not exceed 1/750, that is, $f_{max} \leq L/750$, where L is calculated span. For strength constraint, the Von Mises stress in the main girder web must be less than the allowable stress. Namely, $\sigma_{r4} \leq [\sigma]$. It is clear that the maximum deflection and stress are functions implicitly with respect to design variables $\boldsymbol{X}$. The total volume of the gate, also is the function of $\boldsymbol{X}$, is taken as the objective function. Therefore, the optimization problem of the gate leakage hole can be described as follows:

$$\begin{aligned} &\min \ V(\mathbf{X}) \\ &\text{s.t.} \ \ R_{\min} \leq x_i \leq R_{\max} \quad x_i \in \mathbf{X} = \{x_1, x_2, \ldots x_n\} \\ &\qquad \sigma_{r4}(\mathbf{X}) \leq [\sigma] \\ &\qquad f_{\max}(\mathbf{X}) \leq \frac{L}{750} \end{aligned}$$

It can be seen that the gate leakage holes design optimization problem is an inequality constrained optimization problems. Such problems can be solved by the penalty function method, the method of least squares, genetic algorithm. Clearly, the response of gate structure, such as stress, displacement, can't be evaluated analytically, whereas the numerical method, FEM should be applied. It is suitable to use FEM solver which has the optimization modulus to solve our problem. Fortunately, gate leakage hole optimization problem is quite simple, and can be easily solved by using optimization modulus [Wu Xin-Miao, 2008] provided in ANSYS.

3 BASIC METHOD OF STRUCTURAL OPTIMIZATION IN ANSYS

ANSYS provides two optimization methods which can solve great number of optimization problems. Zero-order method is a sophisticated method which can efficiently solve most of the engineering problems. First-order method is based on the sensitivity of the objective function with respect to the design variables. Therefore, it is much suitable for the accurate analysis of the optimization. ANSYS provides a series of analysis-assessment-fixed cycle which includes is initial design analysis, results evaluation, and design modification. This loop process is repeated until all of the design requirements are met.

Taken the leakage hole design as an example, the basic steps of structural optimization can be described as follows: Firstly, model the whole gate. Since the hydraulic gate body is welded structure, SHELL63 is applied as main finite element type. Secondly, mesh the gate structure and apply load for static analysis followed by compute the response so that the maximum stress and maximum displacement values are obtained. Theoretically, the mesh is more dense, the evaluated result is more accuracy but with much computational effort. Finally, get into the optimization module, set the design variables, the objective function and optimization algorithm, and start optimizing. The basic procedure is shown as Figure 1.

For the leakage holes design, part of the optimization code in ANSYS is shown as follows:

```
*get,R,30
/prep7
……
/solve
……
/opt
opanl,truss,lgw
opvar,R,dv,30,200
opvar,smax,sv,,160
opvar,vtot,obj,,,2
optype,first
opfrst,16
opexe
```

It must be noted that mirroring, replication, Boolean and some complex operations can't be executed for several times during the optimization modeling process. Therefore, the gate structure would better to be simplified properly. Rear flange of main girder, stringers and stiffeners should be excluded from the model whereas the boundary girder's flange should be retained. With these simplifications, the result should be considered to be safety. First-Order method is taken as the Iterative algorithm.

4 ENGINEERING APPLICATION

4.1 *General design for hydraulic gate*

The gate of diversion tunnel in Hei river reservoir, which belongs to DTH gate, its height, width and design water level are 3.98 *m*, 0.616 *m* and 24 *m*, respectively. It is a common flat

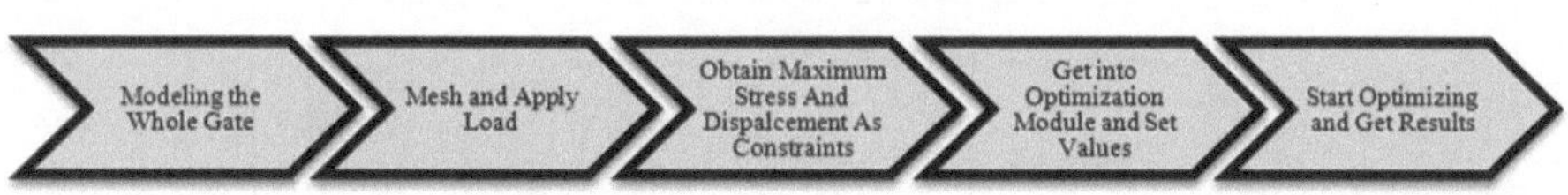

Figure 1. Flow chart of optimum design.

structure, consists of panels, girders, stringers and water seals. The top water seal is set on the flange of top girder while the side one is arranged on both sides of the stringers. Therefore, top and edge girder webs bear stress from water pressure. For convenience, the main girders from top to bottom are successively denoted by 1 to 5, stringers from left to right, denoted by 1 to 5, respectively. Figure 2 shows the basic structure of the gate. The material is A3F steel (Q235). Since the gate panel and stiffening plate thickness should be less than 16*mm*, which belongs to the first group of steel size standard, the yield limit is 235*MPa* and the allowable stress is 160 *MPa*. For the other main component, whose thickness is greater than 16*mm*, its allowable stress is 150 *MPa*. The calculated stress for strength verification should be equal to its Von Mises stress, i.e. based on the fourth strength theory.

4.2 *Modeling and static analysis*

According to the gate structure and the loading, gate panel, main girders, stringers, side column, bottom girders are treated as discrete plate elements. There are 24,702 nodes and 25,425 elements in the generalized FE model of the gate structure. Coordinate system is defined as follows: the x-axis is along the direction of the ground sill, y-axis is upward vertically and the z-axis coincides with the direction of water flow. Natural boundary conditions include the zero vertical displacement of nodes on the sill and all zero displacements of nodes on padding blocks. Due to the irregularity of rusted distribution of the gate, the actual size of each component is not easy to be accurately determined. Therefore, in Finite Element Analysis, the design values of the cross-sectional dimensions are used as the real constants of beam elements. The final FEA results are shown as in Figures 3 and 4.

The result shows that the maximum equivalent stress is approximate 139.3*MPa*, far less than the allowable stress 240*MPa*. For the main girders with four leakages holes whose diameter is set to 30 *mm*, i.e. No. 2 to No. 5 main girder, the maximum equivalent stress appears in the junction of No. 4 girder and padding block. Neglect the stress singular point, the main girder equivalent stress is less than 140*MPa*. In addition, the total volume of the gate is 1.15445m^3, and the weight is 9.005 tons.

4.3 *Size optimization of leakage holes*

It can be seen from the contour of static analysis that the equivalent stress in the middle of the main girder web is very small. This result tells us that if the diameter of the leakage holes

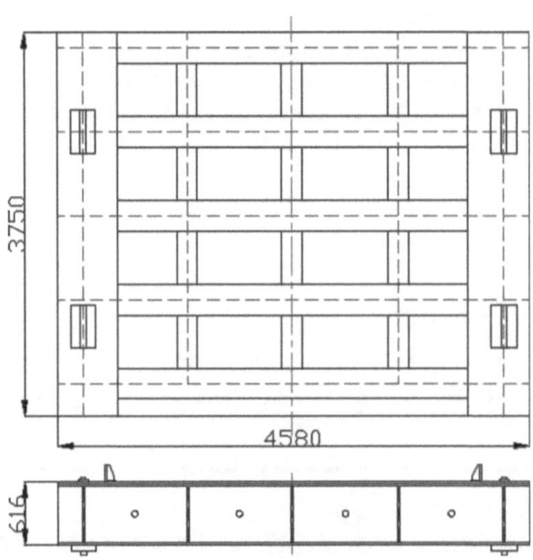

Figure 2. Structure of gate.

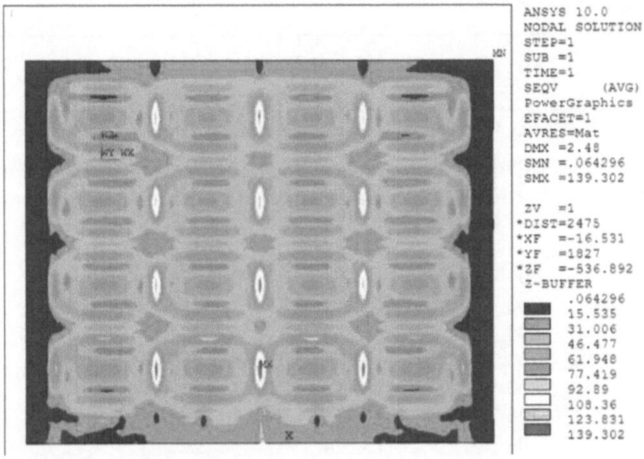

Figure 3. Von stress of gate panel (MPa).

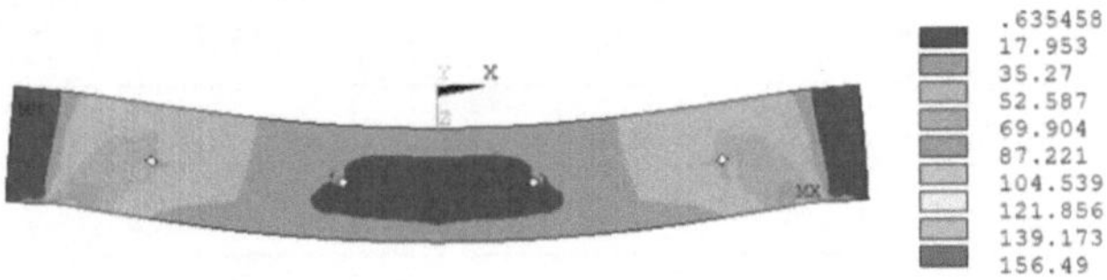

Figure 4. Von stress of 4# beam web (MPa).

is set to 30*mm*, then the bearing capacity of materials is far from being utilized. Considering the symmetry of gate structure and load, we may set two different leakage hole diameters as design variables. The diameter for two leakage holes near the edge of the girder is denoted by R_1 while for the rest, is R_2. Taking account the requirement of welding process, the upper bound of the diameters R_i should be less than 200*mm*. The state variables are the gate overall equivalent stress and deflection. Equivalent stress, σ_{zh} should be less than 160*MPa*. The upper bound of the maximum deflection, also the allowable deflection, which is the value of main girder span (4080*mm*) divided by 750, is 5.44*mm*. The gate overall volume or total weight is treated as objective function. After optimization iterative calculation by ANSYS, the first 16 iterations optimization results are obtained, listed in Table 1. Diameter and maximum equivalent stress curve with the iteration step is shown in Figures 5 and 6.

As shown in Figure 5, we can see that after 16-step iteration, R_1 is equal to 86.504*mm* and R_2 is 200*mm*. If these two values are chosen as design, the stress of main components can be calculated as shown in Figure 7.

The maximum stress is 141.316 *MPa*, far less than allowable equivalent stress. Stress of the leakage holes at both ends of main beam web is relatively high, but the maximum stress appears in the junction of No. 4 main girder and padding blocks. Green colored range shows the largest stress near the leakage hole, is below 100*MPa*. The result indicates that in this design, the gate is adequately safe. Taking account of the factors of manufacture and appropriate safety margin, the final design, R_1 can be chosen as 80*mm* and R_2 is 180*mm*.

4.4 *Remarks to optimization result*

The stress, volume and weight of the main components of gate before and after optimization are listed in Table 2.

From Table 2, we know that the panel stress distribution is almost unchanged after optimization. The maximum stress increase 1.46% while the maximum stress position does not

Table 1. Optimization results.

	R_1 (*mm*)	R_2 (*mm*)	Max stress (*MPa*)	Maximum deflection (*mm*)	Volume (mm^3)
Set 1	30	30	95.955	2.7041	0.98637×10^9
Set 2	56.174	56.174	130.7	2.7189	0.98466×10^9
Set 3	61.81	69.507	132.62	2.725	0.98377×10^9
Set 4	64.068	72.646	151.74	2.7287	0.98343×10^9
Set 5	56.512	73.008	131.18	2.7219	0.9838×10^9
Set 6	51.353	200	129.92	2.7926	0.97096×10^9
Set 7	61.926	200	135.25	2.7992	0.97055×10^9
Set 8	64.722	200	156.34	2.8033	0.97034×10^9
Set 9	56.361	200	133	2.7954	0.97077×10^9
Set 10	62.454	200	131.1	2.7994	0.97053×10^9
Set 11	67.675	200	156.15	2.8056	0.97021×10^9
Set 12	56.515	200	133.24	2.7955	0.97077×10^9
Set 13	62.514	200	133.28	2.7995	0.97052×10^9
Set 14	63.634	200	134.34	2.8003	0.97048×10^9
Set 15	66.05	200	156.89	2.8044	0.97028×10^9
Set 16	86.504	200	158.01	2.8251	0.96916×10^9

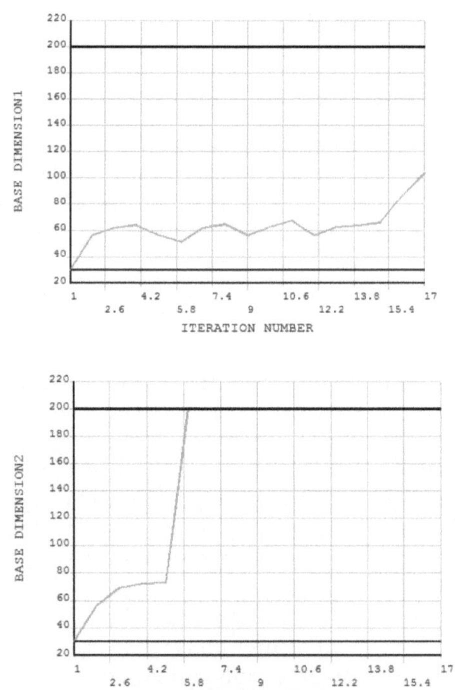

Figure 5. Curve of radius and step.

change. When leakage hole near edge beam varies from 30*mm* to 86.504*mm*, its change has little effect on the size of the gate panel stress. Even the middle of the main beam leakage diameter increases to 200*mm*, a very small affect will be produced to the panel stress. Although the main girder web equivalent stress will be slightly decreased, yet it is concentrated to the hole. It means that setting holes on the web will mainly affect the whole stress distribute of the web, possibly the stress will dramatically increase near the hole. After

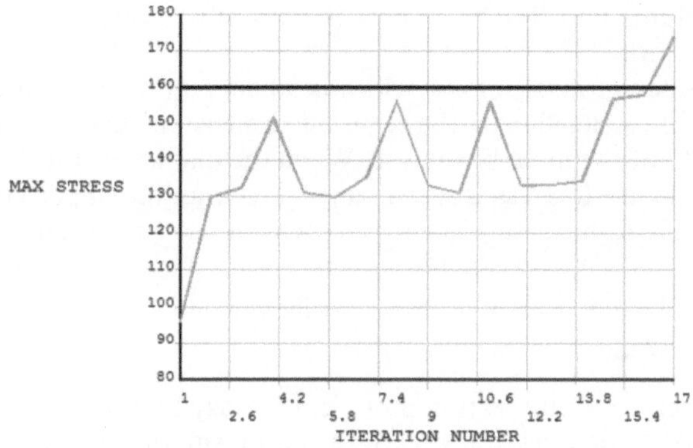

Figure 6. Curve of max stress and step.

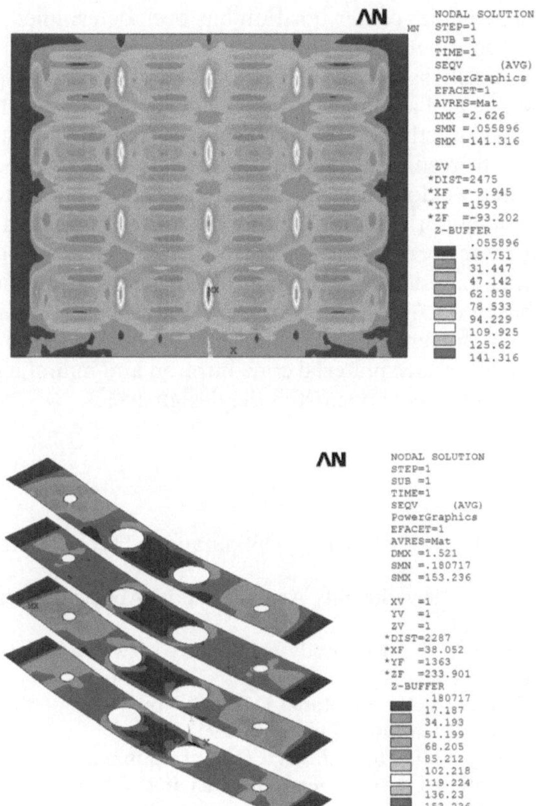

Figure 7. (A) Von stress of gate panel after Optimize (*MPa*); (B) Von stress of beam webs after Optimize (*MPa*).

Table 2. Parameters contrast before and after optimization.

	Plane max stress (*MPa*)	Beam web max stress (*MPa*)	Volume (m^3)	Weight (*t*)
Before	139.302	156.49	1.15445	9.005
After	141.316	153.236	1.13149	8.825

optimization, the volume and weight of gate is naturally decreased, for this specific gate, there is a 2.0% reduction.

For this gate, the dock groove channel width is 900*mm*, and main girder width is 616*mm*. If the design value of R_1 is 86.504*mm*, the sectional dimensions size of leakage hole accounts for 28.1% of the width of the main beam web. When the diameter of leakage hole is less than this value, the web equivalent stress is significantly less than the upper bound. Continue the optimization to step 17, R_1 turns to 173.84*MPa*, exceeds the constraint.

5 CONCLUSION

Based on ANSYS structure optimization module, it is easy to obtain relatively better diameter of leakage holes. Although the result is not the best result of the optimization, the method is simple and the result is also able to satisfy the design requirements. It has not only changed the arbitrary design of the diameter of leakage holes, but also reduces the dead weight of the gate under the premise of the strength and stiffness. However, this design is depended on the assumption that the leakage holes center position has been determined, even the quantity of leakage hole is not being considered and the shape of the hole constrained to be a standard round. Therefore, theoretically speaking, it is just a size optimization problem. In fact, the optimal choice of the shape and arrangement of the leakage holes location can be solved by topology optimization. Because the topology optimization results are dependent on the mesh density [K Maute, 1995], the result is usually difficult to be used for processing, further post-processing is required, here we do not adopt. Further research will be focused on the manufacturability of the topology optimization. Although this optimization problem for leakage hole is relatively simple, there are many designs of other parts in hydraulic gate design similar to the leakage hole. Thus, the method can be easily to be used for some other parts. This optimization method based on ANSYS can be easily accepted by engineering practice [Liu Ji-Liang, 2011]. Further improvement and processing can be applied to similar situations, so as to reduce the gate weight, reduce material consumption and manufacturing costs, improve product quality and performance, and shorten the design cycle.

REFERENCES

Bendsoe, M.P., & Sigmund, O. 2003. Topology Optimization: Theory, Methods and Applications. Berlin: Springer-Verlag.

Cheng Chang-Jun, Lv Xiao-An. 1999. Stability Analysis and Its Application for Perforated Structures. Advances in Mechanics: 85–96.

Li Bo, Wang Zhao-Min, Zhao Hui. 2006. Study on stability of web with circular opening. Sichuan Building Science: 62–65.

Li Huo-Kun & Xu Zhe. 2007. Structural Optimal Design on the Arms of the Radial Gate Based on Dynamic Stability. Journal of Nanchang University: 294–298.

Liu Ji-Liang, Wang Zheng-Zhong, Jia Shi-Kai. 2011. Optimization design of main frame of radial gate with three arms based on rational layout. Journal of Zhejiang University: 1985–1990.

Maute K, E Ramm. 1995. Adaptive topology optimization [J]. Structural optimization: 100–112.

Water Resources and Hydropower Engineering design specifications of the steel gate (DL/T5013-1995).

Wu Xin-Miao, Qie Zhi-Hong, Zhou Zhi-Jun, Zhang Hai-Ru. 2008. Application of Improved PSO to Optimization of Gravity Dam and Sluice Gate. Proceedings of the World Congress on Intelligent Control and Automation: 6178–6182.

Xiao Wei-Hua, Zhang Wen-Ping. 2010. Application of topology optimization theory in two-dimensional plane gate design. Yangtze River: 24–27.

Zhu Zuo-Jun & Wang Zheng-Zhong. 2007. Topology optimization application of optimize the arrangement in a large arc steel gate. Yellow River: 63–64.

Modeling and Computation in Engineering II – Xie (ed)
 ISBN 978-1-138-00058-2

FEM analysis on shape influencing factors

Sheng-li Chen
Metallurgical Engineering Institute of University of Science & Technology Beijing, Beijing, China
Technology Center, Ma Anshan Iron and Steel Co., Ltd., Ma' Anshan, Anhui, China

An-rui He
Metallurgical Engineering Institute of University of Science & Technology Beijing, China
National Engineering Research Center for Advanced Rolling Technology;
Design Research Institute Co., Ltd., Beijing, China

Zhi-rang Xu
Institute of Mechanical Engineering, Anhui University of Technology, Ma' Anshan, Anhui, China

Jian Zhang
Technology Center, Ma Anshan Iron and Steel Co., Ltd., Ma' Anshan, Anhui, China

ABSTRACT: For strip shape quality problem of CSP seven-stand four-high CVC hot rolling mill line in Maanshan Iron and Steel Co, three-dimensional finite element model was established by using finite element software ANSYS based on real mill and technologic parameter in practical production. Major shape influencing factors were analyzed based on loaded roll gap profile. Finally, the model was validated, and results of simulation calculation are basically close to the measured data. Through off-line simulation of rolls, it can play a reference and instruction role for strip profile control.

1 INTRODUCTION

The strip shape is an important quality index of strip product, while the strip shape control is a key technology of strip mill (Wang Guo-dong, 1986). The object of shape control is to change strip exit crown and flatness to satisfy with customer demand through change loaded roll gap profile, so higher loaded roll gap profile accuracy is required in the strip rolling process. This paper studies rolls loaded and deformation based on CSP F7 stand 4-high CVC hot continue rolling mill line in Ma steel by creating static three-dimensional FE body model simulating to real mill using FE software ANSYS, and gets influence relations between strip width, rolling force, bending force, shifting work roll and loaded roll gap profile. Finally the FE model validity is verified through comparing with experiment measuring data.

2 MODEL CREATING

Three-dimensional FE model suits appliance to complex system, and provides certain describes to physical model of real mills, and exactly analyses loaded roll gap profile (Zeng Pan, Lei Liping, Fang Gang, 2011). Three-dimensional entity model is established by using parameter language APDL of ANSYS, and major parameters are shown based on Table 1. SOLID45 is selected and used rolls element; CONTA173 and TARGE170 are selected and used contact parts between work roll and backup roll, respectively. In process of creating model through bottom to top methods from point to surface again to entity, mesh grids are closely divided at contact occurring roll-strip interface, at contact occurring work roll and backup roll interface, and possible contact region for increasing accuracy and decreasing calculation expense.

Table 1. Parameters of finishing mill.

Work roll diameter/mm	Work roll length/mm	Backup roll diameter/mm	Backup roll length/mm	Backup roll bearing distance/mm
Ø540–Ø620	2000	Ø1350–Ø1500	1800	2900

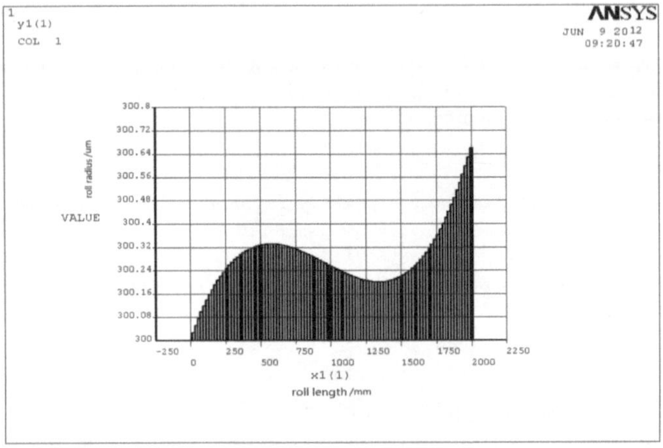

Figure 1. Curve of work roll shape.

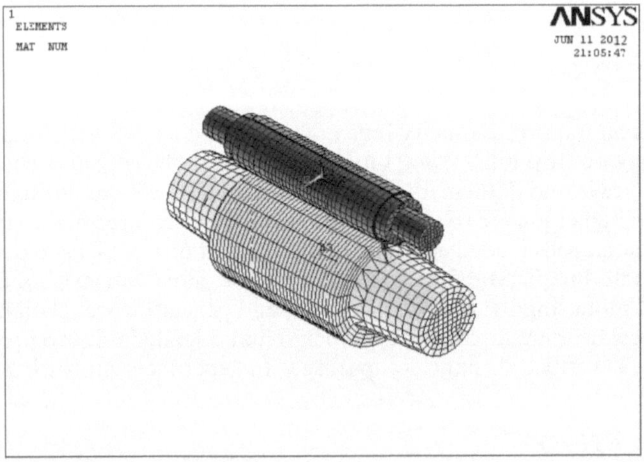

Figure 2. Three dimension entity model.

Grind work rolls contour with five orders curve parameters is generated by using command BSPLIN to fit 101 key points through software ANSYS (Gong Shu-guang, Xie Gui-lan, Huang yun-qing, 2009). Results are shown in Figure 1. The established three-dimensional entity model is shown in Figure 2, which is firstly used command AMAP to generate surface mesh grids then used command VROTAT and VSWEEP to obtain body mesh grids.

3 RESULTS ANALYSIS

Basic on the above FEM, influence relations of strip width, rolling force, bending force and shifting position to loaded roll gap profile are studied by giving an example about rolling varieties SPHC. The initial parameters are shown in table 2. In process of the loaded rolls, rolling force is loaded at contact occurring roll-strip interface, and difference of rolling force between forwards slide and backwards slide is considered. Accordingly R.B. Sims forwards and backwards slide rolling force formulas are loaded, respectively. Positive bending force is loaded at work roll neck position, which left backup roll neck is all restrained by three direction of X axis, Y axis and Z axis, and right backup roll neck is restrained by axial direction and rolling direction. See Figure 3.

3.1 *Strip width effects*

Simply supported beam theory about roll deflection believes that roll deflection bigger along with rolling force increasing causes strip crown increasing, so bigger strip crown is brought in process of rolling width strip (Vladimir B. Ginzburg, 2000). But FE analysis theory indicates that the judgment is right only within the limited width. The relation of strip width and crown is, on the one hand, bigger strip width, greater rolling force, which result in roll elastic deflection deformation increasing and strip crown increasing; on the other hand, result in roll reflection and strip crown decrease because of bad contact section decreases in roll-roll interface. When strip width increase causing bad contact section and bending moment decrease cannot compensate rolling force increasing causing rolls elastic deformation, strip crown decreases. This paper verifies the argument by FE simulation. Relations between strip width and loaded roll gap profile are shown in Figure 4, and expression, $y = -1.89*10^{-10} x^4 + 6.96*10^{-7} x^3 - 0.1*10^{-2} x^2 + 0.71x - 141$, is obtained through software MATLAB fitting simulation data, specified value is critical width at approximately roll barrel length of 75% position.

Table 2. Process parameters of an example.

Width/ mm	Entry gauge/mm	Exit gauge/ mm	Deformation resistance/MPa	Shifting distance/mm	Work roll radius/mm	Backup roll radius/mm	Bending force/kN
1200	4.2	3.5	100	30	300	700	600

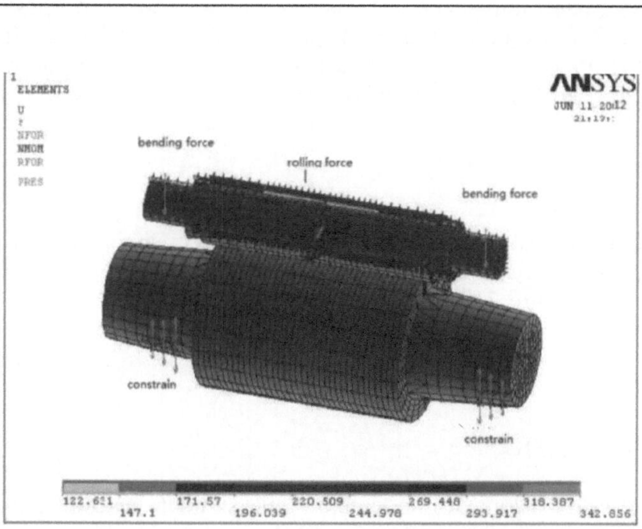

Figure 3. Rolling force and bending force load of CVC rolls. Three dimension model.

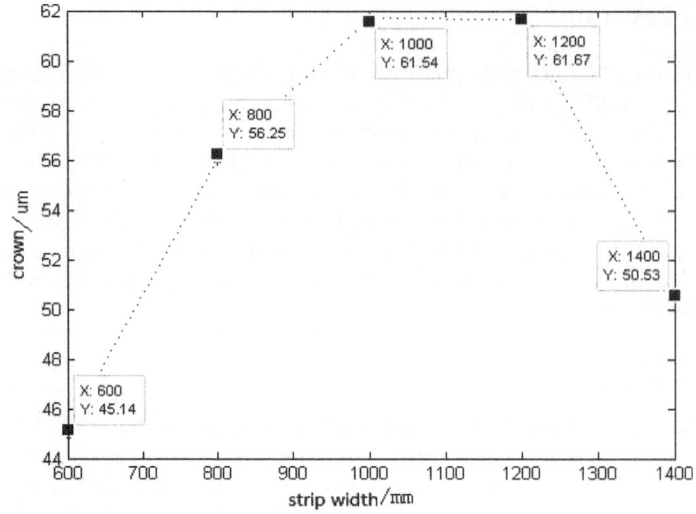

Figure 4. Relation between crown and strip width.

3.2 *Rolling force effects*

Rolling force is one of major effects factors to strip crown, and is affected by deformation resistance, strip width, and friction coefficient and so on in plate rolling process. Assuming rolling force is stable under the specific deformation resistance action, simulation results on rolling force effects to loaded roll gap profile are obtained by using R.B. Sims forwards slide and backwards slide rolling force formula to calculate deformation resistance. The relation curve of rolling force and amount of crown is shown in Figure 5. With excellent line relation, and relation expression formula, $y = 0.03204x - 0.0032041$, is obtained by using software MATLAB fitting simulation data.

3.3 *Bending force effects*

Roll bending is a frequently-used validation on-line control method to control strip crown and flatness during plate rolling. The simulated results with work roll positive bending are illustrated to relation of bending force and loaded roll gap profile in Figure 6. And the relation of bending force and crown is shown being near liner relation, moreover relation expression formula, $y = -0.081x + 114.98$, is obtained by using software MATLAB fitting simulation data.

3.4 *Work roll shifting effects*

Work roll shifting is normal method in the presetting strip shape control model for CVC mill. Studies show that relation of work rolls shifting and loaded roll gap profile is near liner relation illustrated in Figure 7. The relation expression formula between strip crown and work roll shifting position, $y = -1.52x + 181$, is obtained by using software MATLAB fitting simulation data.

4 MODEL VERIFICATION

Verification of three-dimensional finite element model simulation accuracy to loaded roll gap profile is carried out using drafting schedule and site basic technologic parameters shown in Table 3.

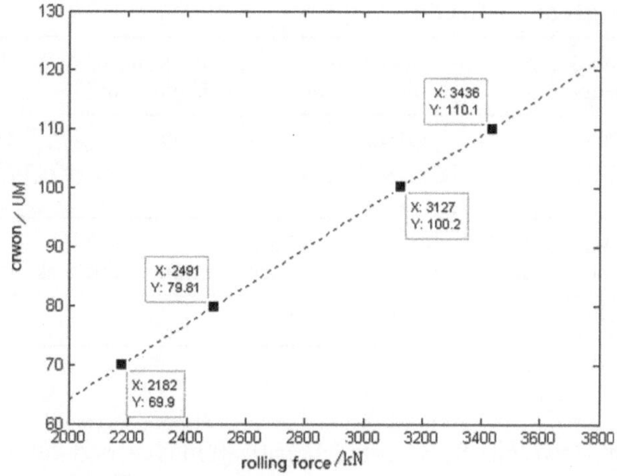

Figure 5. Relation between crown and rolling force.

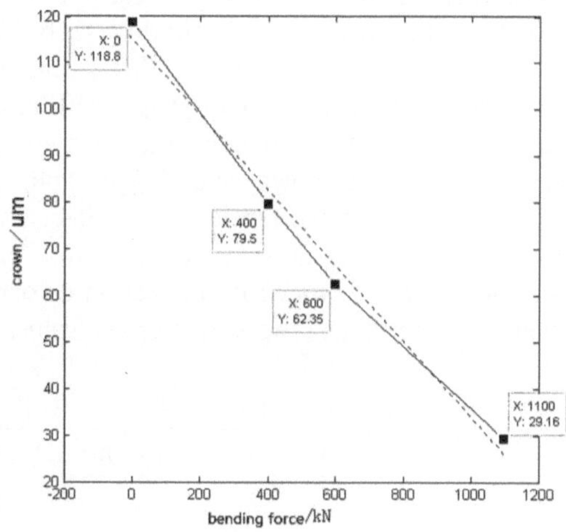

Figure 6. Relation between of crown and bending force.

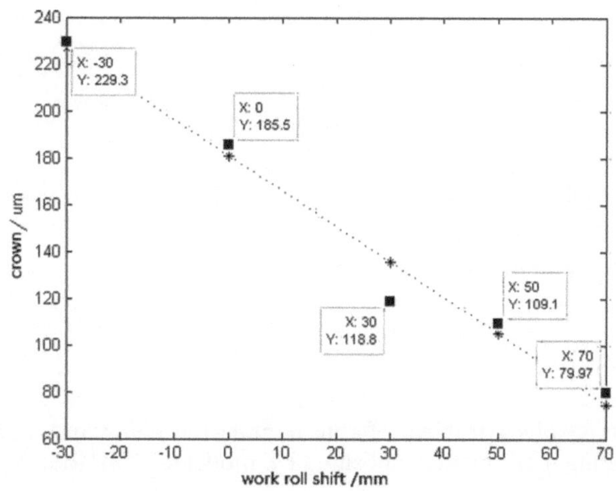

Figure 7. Relation between crown and work roll shifting position.

Table 3. Basic parameters of 1800 model.

Work roll diameter/mm	Work roll length/mm	Backup roll diameter/mm	Backup roll length/mm	Backup roll bearing distance/mm
Ø610	2000	Ø1500	1800	2900

Continue 2

Work roll bending force/kN	Work roll shift value /mm	Strip width/mm	Entry thickness/mm	Exit thickness/mm	Rolling force/ kN
630	25	1225	4.2	3.5	9130

By using for example varieties SPHC, model verification is carried out on basis of Table 3. parameters including actual total rolling force 9130 kN, work roll radius 305 mm, strip width 1225 mm, entry thickness 4.2 mm and exit thickness 3.5 mm, furthermore these parameters are substituted in Equation 1~4 to calculate deformation resistant. Firstly using Equation 3 and Equation 4 work out neutral angle 2.2 rad corresponding strip thickness 3.64 mm, then using Equation 1 figures out deformation resistant 182 MPa. Lastly results are obtained that strip crown is 55.3 um at 40 mm distance edge of strip by FEM simulation calculating. Comparing with experiment measuring strip crown 43.4 um is main agreement, and model validity is verified. Causes for simulation crown value bigger than experiment measuring value are no consideration thermo camber and tension stress effects in practice rolling process. When work rolls temperature raise, its expansion are more; when plate rolling has tension stress, rolling force grows small to lead to work rolls deflection deformation growing small. So simulating value is bigger than experiment measuring crown value.

$$\frac{P}{B\sqrt{R(H-h)}}=\left(\frac{\pi}{2}\sqrt{\frac{1-eps}{eps}}\arctan\sqrt{\frac{eps}{1-eps}}-\frac{\pi}{4}-\sqrt{\frac{1-eps}{eps}}\sqrt{\frac{R}{h}}\ln\frac{h_r}{h}+\frac{1}{2}\sqrt{\frac{1-eps}{eps}}\sqrt{\frac{R}{h}}\ln\frac{1}{1-eps}\right)k \tag{1}$$

$$eps=\frac{H-h}{H} \tag{2}$$

$$h_r=h+Rr^2 \tag{3}$$

$$r=\sqrt{\frac{h}{R}}\tan[\frac{1}{2}\arctan\sqrt{\frac{eps}{1-eps}}+\frac{\pi}{8}\sqrt{\frac{h}{R}}\ln(1-eps)] \tag{4}$$

where P = total rolling force; R = work roll radius; k = deformation resistance;
γ = neutral angle; h_r = neutral angle corresponding thickness
H = entry thickness; and h = exit thickness.

5 CONCLUSIONS

1) This paper properly solves problem of rolls contact flattening and deformation through numerical calculating method of establishing FE model for CSP four-high mill rolling line in Ma steel.

2) Based on real mill and technologic parameters, major strip shape influencing factors to loaded roll gap profile are studied, and simple relation formula is fitted and play reference and instruction role for strip shape level 2 setup control model.
3) CVC mill loaded roll gap profile with special work rolls profile is predicted to accomplish accuracy simulation by using software ANSYS.
4) The established Model can investigate to the difference process conditions, and popularized similar mill to control strip shape (Xu Zhi-rang, Xue Jia-guo, 2000).

ACKNOWLEDGEMENTS

This work was financially supported by Program for New Century Excellent Talents in University (NCET-10-0223) and the Fundamental Research Funds for the Central Universities (FRF-TP-11-003A).

REFERENCES

Gong Shu-guang, Xie Gui-lan & Huang yun-qing. 2009. *Ansys Parameter Design Language-APDL* [M]. Beijing; Metallurgical Industry Press.

Vladimir, B. Ginzburg. 2000. *High-Quality Steel Rolling: Theory and Practice* [M]. Beijing; Metallurgical Industry Press.

Wang Guo-dong. 1986. *Shape Control and Shape Theory* [M]. Beijing; Metallurgical Industry Press.

Xu Zhi-rang & Xue Jia-guo. *2000. Finite element method for the analysis of rolls deflection for 4-high mill* [J]. Mechanics and Engineering, 22(5):22–24.

Zeng Pan, Lei Li-ping & Fang Gang. 2011. *Finite Element Analysis Guide: Modeling and Analysis of Structure* [M]. Beijing; Metallurgical Industry Press.

Modeling and Computation in Engineering II – Xie (ed)
© 2013 Taylor & Francis Group, London, ISBN 978-1-138-00058-2

Analysis of energy-saving and high efficiency for bucket chain continuous ship unloader

Minhui Tong & Yuemin Wang
School of Logistics Engineering, Shanghai Maritime University, Shanghai, China

Huiqing Qiu
School of Mechanical Engineering, Tongji University, Shanghai, China

ABSTRACT: The paper collects the statistics productive data on continuous ship unloader and grab ship unloader in steel work raw material port, compare the advantage and disadvantage of two type of ship unloader on production capacity, equipment working status, energy saving and emission-reduction, corollary equipment support, labor power investment etc; analysis the principle and get the conclusion that the continuous ship unloader is superior to the grab ship unloader, provide decision support for equipment allocation and type selection in the bulk port.

1 INTRODUCTION

1.1 *Background*

At present the domestic and foreign various ports iron ore transport machinery equipment main machine is grab ship unloader. With the development of the port machinery equipment, large-scale and high efficiency, the port machinery operation efficiency, stability, energy conservation and environmental protection put forward higher request. Continuous ship unloader design models to meet the needs of the modern port development (Shen Z. & Shen Y.D. 2010.), but because of its complicated process, use less experience, until 2010, there no successful application cases in china.

1.2 *Research object*

The steelwork in china put the continuous ship unloader into use, from December 7, 2009 began to ore conveyor, as of the end of July 2011, the total transport ore 677.3573 million tons, the total operating time 2632.6967 hours, the total work ships, 85 ships during equipment state very stable, domestic steelwork first continuous ship unloader terminal. Figure 1 is the photo of continuous ship unloader on the port.

Continuous ship unloader is new type equipment of the steelwork raw material. The equipment, production base data accumulation and analysis are good for the future development of the terminal an important significance. In this paper, the continuous ship unloader production efficiency, energy conservation and environmental protection, equipment status, maintenance cost, human resources and the traditional grab ship unloader is compared, for the future development of the bulk handling and to provide the reference (Luo C.Z. & Liu J. 2011.).

Figure 1. Continuous ship unloader in operation.

2 PRINCIPLE COMPARISON OF TWO TYPE EQUIPMENTS

2.1 *Continuous ship unloader operation principle analysis*

Comparison of Continuous ship unloader and grab ship unloader as follows.

For the continuous ship unloader, the weight of bucket is: A, The weight of material is B, the total weight of the empty buckets chain is: ΣA, the total weight of the total weight of the full buckets with material is: ΣA + ΣB, due to the continuous ship unloader is more than the bucket chain cycle operation to handing and hoisting materials, therefore, in the process, half buckets up and the other half bucket down, total buckets weight is balance, cycle theory does not work (friction resistance can be negligible), only hoisting material consume power, as figure shows.

What the continuous ship unloader bucket chain hoisting consumes power P_c is only for hoisting material consumes power.

2.2 *Grab ship unloader operation principle analysis*

For the grab ship unloader, the weight of grab is ΣA, the weight of full grab with material is ΣA + ΣB, due to grab ship unloader (Zheng P. 2004) is up to the individual grab of hoisting materials, therefore, in the operation process, grab and material at the same time to be lifted up, grab weight must at the same time to be hoisted, each work cycle grab always consumed power (Hu Y.N. 2011). Such as following formula shows:

$$P_C = (\Sigma A + \Sigma B)\cdot \vec{V_1} + (\Sigma A)\cdot \vec{V_2} \tag{1}$$

$$\because\ V_1 = -V_2 \tag{2}$$

$$P_C = (\ \Sigma B)\cdot \vec{V_1} \tag{3}$$

What the grab ship unloader hoisting consumes power P_G is NOT only for hoisting material consumes power, also including the hoisting grab self-weight consumes power in each cycle (Zhao H.Y. 2007).

$$P_G = (\ \Sigma A + \Sigma B)\cdot \vec{V_1} \tag{4}$$

Compare the two formulas (3), (4), get:

$$P_C < P_G \tag{5}$$

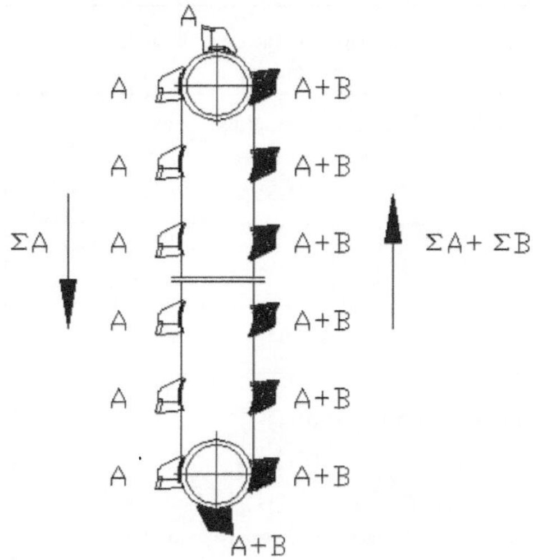

Figure 2. Continuous ship unloader bucket chain operation schematic diagram.

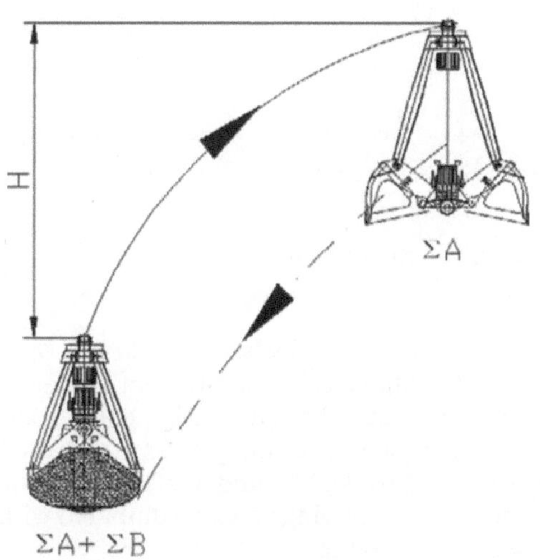

Figure 3. Grab ship unloader operation schematic diagram.

According to above analysis, continuous ship unloader hoisting the operation is consumed power less than grab ship unloader hoisting the operation of the power consumed (Qi Z.Q. 2006). And grab ship unloader is intermittent operation machine, each cycle will consume an invalid load production power, therefore, continuous ship unloader than grab ship unloader low energy consumption, has obvious advantages of energy saving (Peng C.S. & Zheng J.C. 2000).

3 COMPARISON OF ACTUAL PRODUCTIVITY

3.1 *Productivity improvement significantly*

The comparison between continuous ship unloader and grab ship unloader production capacity is as follows: Figure 4 shows the actual productivity; Figure 5 shows the average productivity.

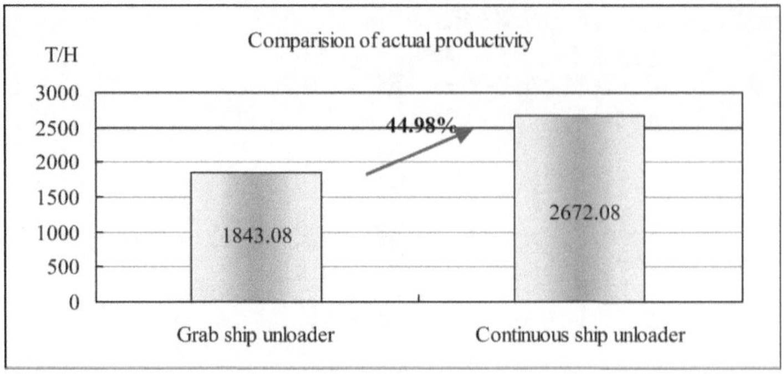

Figure 4. The actual productivity comparison.

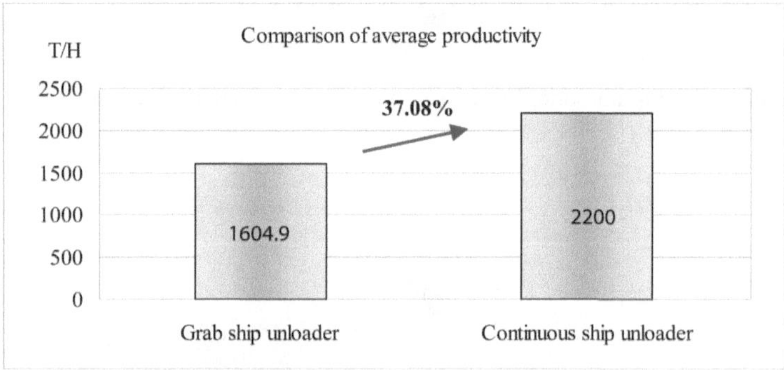

Figure 5. The average productivity comparison.

Note: Figure 4, 5 data get from steel mill wharf, a 3600 t/h continuous ship unloader and two 1800 t/h grab ship unloader. Statistical period is from January 2010 to June 2011.

Analysis the above production data, the same design ability, continuous ship unloader in operation efficiency is obviously superior to the grab ship unloader, actual productivity and average productivity are increased by 44.98% and 37.08%, continuous ship unloader benefit from its highly efficient and stable conveying system, compared with the traditional grab ship unloader operation efficiency advantages.

3.2 *Remarkable energy conservation and emission reduction*

i. Environmental protection effect is remarkable
   On the steel wharf, continuous ship unloader boom belt conveyor and built-in belt dust system, dust control is very effective, compare with grab ship unloader, dust reduce 95% and dropping reduce 100%, continuous ship unloader can be called a real environmental protection equipment.
ii. Power consumption significantly reduced

From Figure 6 can achieved, each ten thousand tons of discharge amount of electric energy consumption, continuous ship unloader than grab ship unloader reduced by 32.02%, and reflects the good energy conservation. Note: the data source as above.

From the chart can see the continuous ship unloader one compartment of the bulldozer oil consumption compared to grab ship unloader both ore fines and lump ore, reduce about 50%.

So from the above material the material handing dust control, power consumption, one compartment bulldozer fuel consumption three aspects, continuous ship unloader has a very

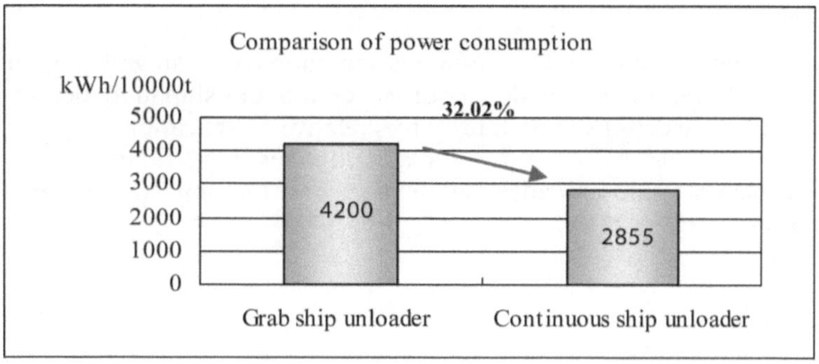

Figure 6. The average productivity comparison.

Table 1. Maintenance cost comparison.

Monthly maintenance cost	Routine cost	Bulldozer cost	Bucket chain	Guide wheel	Wire rope	Total
Continuous ship unloader	14.95	6.95	12.667	1.375	0	35.942
Grab ship unloader	10	24.185	0	0	2.5	36.685

*From statistics of steel work raw material port.

significant reduction for the protection of steel terminal environment, reducing cost and make a contribution, it also embodies the continuous ship unloader for energy conservation and environmental protection equipment superiority (Sun G.Z. 2000.).

3.3 *Equipment state remained stable*

The same design capability, a 3600 t/h continuous ship unloader in equipment status include fault time, failure frequency and every ten thousand tons on the down time is obviously superior to the two 1800 t/h grab ship unloader. Equipment state stability and promote the continuous ship unloader operation efficiency. 2011 1~6 months, continuous ship unloader than grab ship unloader every ten thousand tons of down time reduced by 79.52%, and the fault time reduced by 83.62%, failure frequency decreased by 78.51%.

3.4 *Auxiliary machinery reduce*

Continuous ship unloader compared to grab ship unloader, the auxiliary machinery significantly reduced, this not only reduces the bulldozer and grab the purchase cost and maintenance cost, total daily every year can save material and maintenance cost 2 million, also reduced the push steak machine oil consumption. Due to the continuous operation of the vessel clearance quantity obviously than grab ship unloader operation ship clearance quantity is little, so the corresponding human resources and get the optimization.

3.5 *Maintenance cost comparison*

Comparison is a raw material dock respectively 3600 t/h continuous ship unloader and two 1800 t/h grab ship unloader, data from 2011 1~6 months of the maintenance cost data to the average monthly maintenance cost comparison.

3.6 *Manpower cost comparison*

From the operational driver staffing, continuous ship unloader than grab ship unloader can reduce a operation driver and two bulldozer driver, continuous ship unloader revamping the raw wharf operation the driver's human resources relative optimization.

From the above table can be concluded that a continuous ship unloader maintenance cost basic and two table grab ship unloader maintenance cost the same, no obvious advantages and disadvantages.

4 CONCLUSION

Combined with the above all aspects of contrast and analysis, the same design ability of continuous ship unloader and grab ship unloader compared in production efficiency, energy conservation and environmental protection, equipment state advantage is very obvious, in the human resources advantage, slightly in the maintenance cost on flat, from data on the very good reflected its high efficiency, environmental protection, the advantages of stable.

Continuous ship unloader in steel plant wharf iron ore transportation successful application has the very good reference, the promotion of meaning. Continuous ship unloader efficient, environmental protection and energy saving advantages is the trend of the development of the port machinery.

ACKNOWLEDGMENT

The author would like to acknowledge Baosteel Group Co, Ltd. for their help in data acquiring and collecting production data. This research was also supported by National "Eleventh Five-Year Plan" Science and Technology Support Program Major Funded Project (2007BAF10B01).

REFERENCES

Hu Y.N. 2011. Bridge type grab ship unloader 3 main styles comparation [J]. *Hoisting and Handing machinery*. 2011(6).

Luo C.Z. & Liu J. 2011. Continuous ship unloader integrated shipment technology research and application [j]. *China Water Transport*. 2011(8).

Peng C.S. & Zheng J.C 2000. 2000t/h Bucket type continuous ship unloader research and develop technology question probe [J]. *Journal of Scientific Research Institute of water transportation*. 2000(3).

Qi Z.Q. 2006. Bridge type grab ship unloader different design styles comparison [J]. *Water transportation science research*, 2006(4).

Shen Z. & Shen Y.D. 2010. Lin H., Ship Unloader equitable distribution and optimization design integrated planning on power station port. [J]. *Marine traffic engineering*. 2010 (9).

Sun G.Z. 2000. Aimed at the 21th century Port logistic technology and equipment [J]. *Port handling*. 2000(4).

Zhao H.Y. 2007. Bridge type grab ship unloader capacity analysis [J]. *Heavy industry and lifting technology*, 2007(2).

Zheng P. 2004. Super large grab ship unloader steel structure non-liner dynamic response simulation [D]. *Shanghai Maritime University*. 2004.

Modeling and Computation in Engineering II – Xie (ed)
© 2013 Taylor & Francis Group, London, ISBN 978-1-138-00058-2

Research on measuring and assessing the development of nanotechnology based on co-countries and co-institutions analysis

Xuefeng Wang, Rongrong Li & Shiming Ren
School of Management & Economics, Beijing Institute of Technology, Beijing, China

Qi Zhang
Scientific and Technological Division, Zhejiang Sci-Tech University, Zhejiang, China

ABSTRACT: This paper chooses the SCI (Science Citation Index) data co-published by Chinese author from 1985 to 2011 to research the international technology cooperation of P. R. China resent years. In this paper, we are focused on three main collaboration countries (USA, Japan and Germany).We combine bibliometric analysis and social network analysis method to analyze and compare collaboration between USA, Japan, Germany and China based on the analysis of co-countries, co-institutions together. In addition, we adopted the scientometric analysis and social network method to make further and deeper exploration and analysis of China's international technology collaboration situation. We find that most of the co-countries are from some developed ones, such the United States, Taiwan province, Hong Kong and Japan and the international cooperation between the institutions is not close, just focused on the several American institutions. In addition, the research fields are room term, carbon, magnetic, electron, transmit in nanotechnology.

1 INTRODUCTION

Nano-science and nanotechnology is one of the most important researches in the 21st century. By nano-science we mean that the functions and physical, chemical properties the materials sized from 1 to 100 nanometers owned. So, nano-technology based on nano-science comes down to a technology which manipulates matter on an atomic and molecular scale in order to develop new materials, new devices and new technology. Nanotechnology is a key technology for the future, the American, British, German, Italian, and Japanese governments have invested billions of dollars in this research. China is no exception.

In recent years, economic globalization has increasingly widened. The global flow of scientific and technological (S&T) resources has developed faster, and the major global issues have been more prominent. Owing to the trend, this study shows that strengthening the international S&T collaboration can make an influential use for international S&T resources to promote its economic development, and it can be a general consensus between governments and business community. By collaboration, countries (regions) can share and pass knowledge, set up networks of academic communication and generate new academic thoughts. Since the adoption of the Opening-up and Reform Policy in 1978, China has achieved a spectacular growth in both of economy and S&T development. However, as a developing country, certain gaps have existed between China and the developed countries.

2 LITERATURE REVIEW

In the long history of scientific collaboration, the earliest documented collaborative scientific paper was published in 1665, which was attributed to Hooke, Oldenburg, Cassini, and Boyle

(Beaver and Rosen 1978). It was not until the middle of the 18th century that the growth of scientific collaboration increased dramatically (Luukkonen et al., 1992).

Nowadays, scientific collaboration has become a very important pattern of scientific research. This phenomenon can be measured by co-authorship of published papers. According to Kostoff's study, in terms of research articles, especially in cutting-edge technologies, such as nanotechnology and energetic materials, China has grown significantly and is among the leaders in the world. Moreover, it has been shown that there was a substantial increase in highly cited documents when foreign collaborators, especially from the USA, were included (Kostoff et al., 2007a).

The motivation of international scientific collaboration is complicated. Countries benefit from international collaborations in greater visibility and higher citation impact (Gla¨nzel and De Lange 2002). In addition, the demand for international collaboration is much stronger in countries with low scientific production than in advanced countries (Davidson Frame and Carpenter 1979). Scientific collaborations of China were analyzed both on country level (He2009; Jin and Rousseau 2005) and institution level (Tang and Shapira 2011).

On country level, Jin and Rousseau (2005) observed the exponential growth of internationally co-authored papers of China. Furthermore, Tianwei He's results indicate that international collaboration publication output between China and the G7 countries has grown exponential thanks to the growth of science in China, and notably, the USA is the most important collaborative country for China (He 2009).

On institution level, LiTang and Philip Shapira's research focused on the China—US scientific collaboration in nanotechnology. Through the collaboration analysis of institutions, they concluded that "The pattern of China's nanotechnology R&D collaboration with the US is asymmetrical, with a relatively small number of elite Chinese research organizations and universities working with a wide array of US universities" (Tang and Shapira 2011). However, this is worth discussing, because we need to consider the fact that the biggest institution in China, namely the Chinese Academy of Sciences (CAS), has over 50, 000 researchers, which is much more than any other university in China. In other words, CAS and a few top universities have dominated scientific research in most fields in China. As a result, we consider that collaboration analysis of institutions is not sufficient to reveal the patterns of China—US scientific collaboration.

Every two or more researchers is the fundamental unit of collaboration, because at the most basic level, it is people who collaborate, not institutions. Inter-institutional and international collaboration need not necessarily involve collaboration between every two or more individuals; but it is the individuals that play an important role in the beginning of collaboration (Katz and Martin 1997). Melin (2000) suggested the collaborations are characterized by a high degree of self-organization on individual level. Moreover, regarding collaboration cosmopolitanism, Bozeman and Corley (2004) found that most researchers tend to work with people in their own work group and people within relatively short geographical distance.

In this field, the majority research objects are a certain or a selected number of country or region within an especial scope and most of the research method is patent statistical analysis or on the single level. In our research, different from the existing analyses at the single level, we use the SCI data to go deep into the collaboration between USA, Japan, Germany and China based on the analysis of co-countries, co-institutions together. In addition, we adopted the bibliometric analysis and social network method to make further and deeper exploration and analysis of China's international technology collaboration situation.

3 THE DEVELOPMENT OF NANOTECHNOLOGY IN CHINA

2001, through the joint effort of related departments, national Nano-science and nanotechnology Coordinating Committee was founded and published the Outline of the National Program for Nano-S&T Development. In 2006, the Outline of the National Program for Long and Medium Term Scientific and Technological Development was promulgated by the State Council. Nano-science was considered the most promising field in which we can achieve type development. In the Eleventh Five-Year period, the Ministry of Science spent over 2 billion invest in nano-S&T, there are more than 50 universities, 3000 firms and 3000 researchers in this area. With the help of Chinese government and Opening-up and Reform Policy, China's nano-S&T is advancing rapidly.

In Table 1 and figure l, we can see the top 11 co-countries and co-region with China in nano S&T. This figure shows the collaboration between China and other countries. As one of the most advanced countries in the world, America ranks TOP 1 in the collaboration list with the percent of the share of papers as 6.48%. Then Japan and Germany comes as second and third with their percent of the share of papers as 6.48% and 2.64% separately. The rank indicates that these three countries dominated the development of nano-S&T with its dominated interactivity, influence power and academic level for China. So in our paper, we focus on these TOP 3 countries to perform our comparison and analysis.

Before 2000, merely few countries collaborated with China. But later, this situation changed. The number of international collaboration got a leap and increasing at a high speed. However, the international collaboration countries are focused on the USA, Japan, and Germany. China's international collaboration technology shows a flourishing growth; while compared to the rapid development of independent R&D, especially after 2007, China's international collaboration has gradually fallen behind in nanotechnology. By 2007 China had risen to second place by global nanotechnology publications. And then China contributed 21% of the world's nanotechnology papers identified in our analysis of Science Citation Index (SCI) records.

Table1. Statistical data of International co-country and co-region in nanotechnology (≥500), China, 1985–2011.

No	Co-country/region	The number of papers	The share of papers (%)
1	China-USA	11403	6.48%
2	China-JAPAN	4651	2.64%
3	China-GERMANY	3127	1.78%
4	China-SINGAPORE	2230	1.27%
5	China-AUSTRALIA	2135	1.21%
6	China-ENGLAND	1870	1.06%
7	China-CANADA	1598	0.91%
8	China-SOUTH KOREA	1550	0.88%
9	China-FRANCE	1425	0.81%
10	China-SWEDEN	914	0.52%
11	China-TAIWAN	896	0.51%

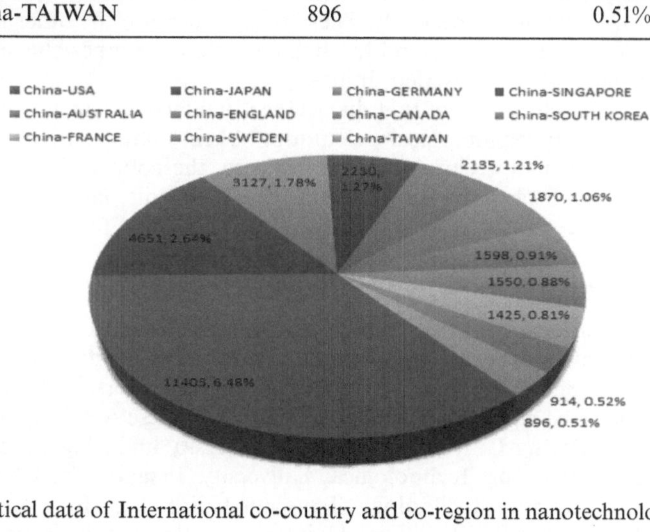

Figure 1. Statistical data of International co-country and co-region in nanotechnology (≥500), China, 1985–2011.

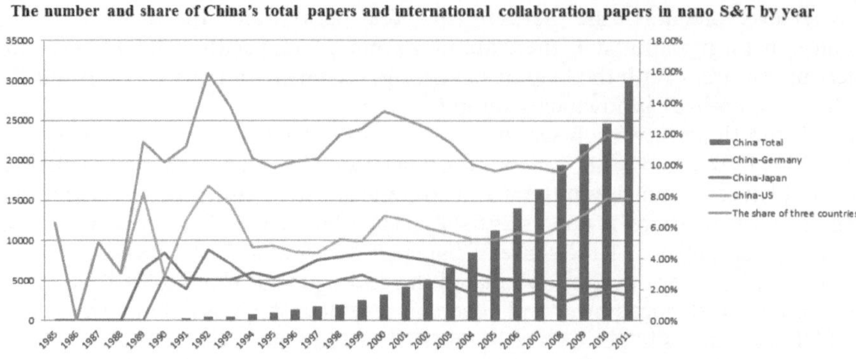

Figure 2. The number and share of China's total papers and international collaboration papers in nano S&T by year.

Figure 2 shows the share of Chinese international collaboration papers in nano S&T by year. We can find out that the percent of China-US collaboration is always in the leading role when compared with other countries. However, in 1990, there is a rapid drop in the share of collaboration. This is because Germany and Japan suddenly collaborated much more frequently with China in nano S&T. And then, Chinese-Foreign countries collaboration comes into a steady trend with America always ranking TOP 1. In Figure 2, we can see this steady trend by the number of collaborated papers more clearly.

4 CHINA'S INTERNATIONAL COLLABORATION RESEARCH

4.1 *Co-countries analysis*

Fig. 3 displays the relationship of the co-inventors' countries and keywords based on different countries. The original time of a big co-countries cluster is also shown by the figure 3. The node size of each cluster time-line represents the number of papers. It is also share the same cluster from inside to outside, which the purple outside ring depth stands for centrality. On the bias of different countries, the main co-inventors exporting countries can be divided into six clusters: one is centered with the United States, surrounding by Taiwan province, Hong Kong, Germany and Japan. The number of SCI papers in this cluster is large and the research emphasis is similarity. The other two clusters are Finland and Korea, and Switzerland, the United Kingdom and Belgium respectively. These five clusters, on the other hand, cover little papers, one co-country is a cluster and the single relationship inner-cluster is weak, only Spain, Italy, Switzerland, Russia, Sweden, India.

From the fig 3, we can also find out that due to the differences of each country or region's innovation capability and technological orientation, their main technology cooperation points are also varied in nanotechnology. For instance, the cooperation with the United States is focused on carbon, electron, transmit, experine, mechanic, Japan is transmits, experine, Germany is transmit, x-ray di, electron, Taiwan region is room term, x-ray di.

4.2 *Co-institutions analysis*

Fig. 2.2 displays the relationship of the co-inventors' institutions and keywords based on SCI. On the bias of different institutions, the main co-institutions exporting institutions can be divided into many clusters: one big cluster and many single clusters, and many institutions are mostly universities. The biggest cluster is centered with Chinese Academy of Sciences with American institutions Peking University, Fu Dan University and Singaporean institutions Natl Univ Singapore, Nanyang Technological University. In addition, Chinese Academy of Sciences, Dalian University Technology, Tsinghua University, Hong Kong University, Zhejiang University and Shanghai Jiao Tong University are the core institutions in China.

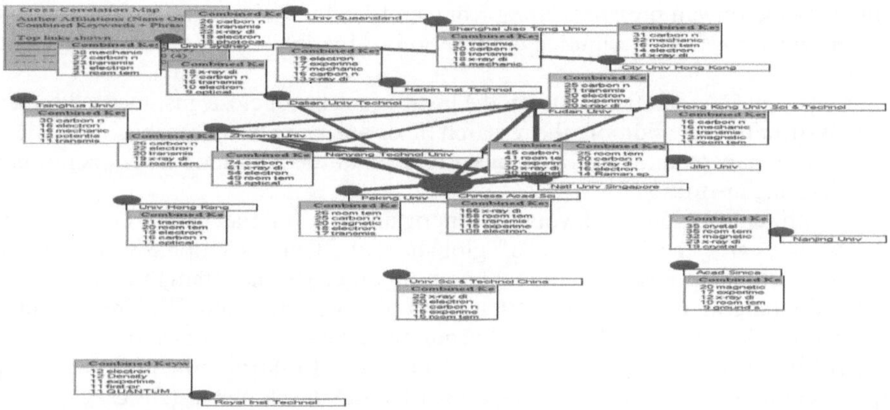

Figure 3. Association map of the co-inventors' countries and keywords in nanotechnology.

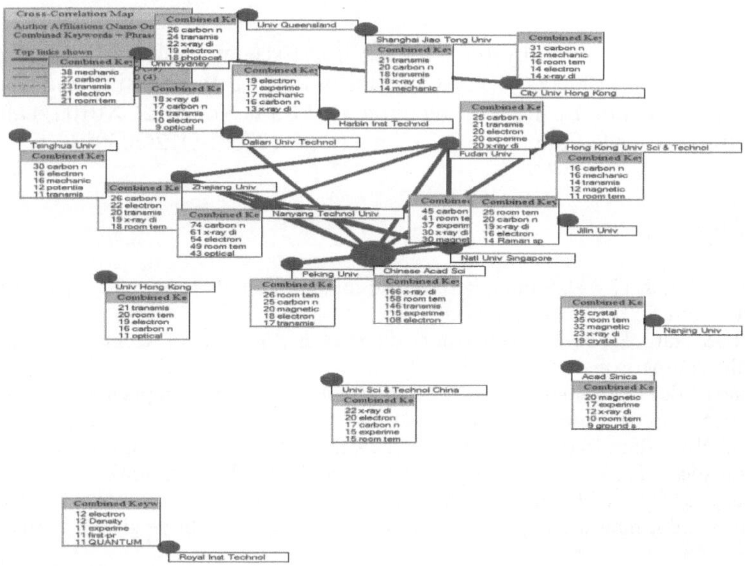

Figure 4. Association map of the co-inventors' institutions and keywords in nanotechnology.

From the fig 4, it is shown that institutions main technology cooperation points are also varied. For instance, the cooperation with the Peking University is focused on room term, carbon, magnetic, electron, transmit.

5 CONCLUSIONS

The collaborative information in paper documents are regarded as an effective tool to study international S&T collaboration development of a country. This article analyzes China's international S&T collaboration from the perspective of paper analysis based on the data to the SCI (Science Citation Index, SCI) from 1985 to 2011. USA, Japan and Germany are the main three collaboration countries with China in nanotechnology. We draw some conclusions as follows:

Even though the international collaboration between China and other countries/regions has stretched around the globe, most of the co-countries are from some developed ones, such the United States, Taiwan province, Hong Kong and Japan. These countries or regions have shown their great power on technology innovation to the world and we are following their steps steadfastly.

The international cooperation between the institutions is not close, just focused on several American institutions. In addition, the research fields are room term, carbon, magnetic, electron, transmit in nanotechnology. So we need to expand the international cooperation with many outstanding institutions.

Overall, although China is playing an important role in the global development of nanotechnology, with the increasing of globalization, China's emphasis of technology innovation development and the national booming on economic strength and comprehensive national strength. We can't ignore China's potential any more. We find outstanding Chinese institutions on a global scale and promote collaborative innovation, develop various forms of international collaboration, make a full use of information, technology, capital and equipment abroad to upgrade its own research and development (R&D) capability to narrow the gaps.

ACKNOWLEDGEMENT

This research is partly funded by the Program for Excellent Talents in Beijing of China (Grant No. 2011D009011000006), Key Cultivation Project of Scientific and Technological Innovation Plan of Beijing Institute of Technology (Grant No. 2012CX01001) and Zhejiang Provincial Natural Science Foundation of China (Grant No. LQ12G03019).

REFERENCES

Beaver, D.B. & Rosen, R. (1978). Studies in scientific collaboration. I. The professional origins of scientific co-authorship. Scientometrics, 1(1), 65–84.

Bhattacharya S. & Nath, P. (2002). Using patent statistics as a measure of 'technological assertiveness': a China-India comparison, Curr. Sci. 83(1), 23–29.

Bozeman, B. & Corley, E. (2004). Scientists' collaboration strategies: Implications for scientific and technical human capital. Research Policy, 33(4), 599–616.

Christian Sternitzke, Adam Bartkowski, Heike Schwanbeck & Reinhard Schramm. (2007). Patent and literature statistics—The case of optoelectronics, World Patent Information 327–338.

Chun-Yao Tseng. (2009). Technological innovation and knowledge network in Asia: Evidence from comparison of information and communication technologies, Technological Forecasting & Social Change, 654–663.

Connie Wu & Yanhuai Liu. (2004). Use of the IPC and various retrieval systems to research patent activities of US organizations in the People's Republic of China, World Patent Information, V26(3), 225–233.

Davidson Frame, J. & Carpenter, M.P. (1979). International research collaboration. Social Studies of Science, 9(4), 481.

Federico Caviggioli. (2011). Foreign applications at the Japan Patent Office—An empirical analysis of selected growth factors, World Patent Information, 157–167.

Gla¨nzel, W. & De Lange, C. (2002). A distributional approach to multinationality measures of internationalscientific collaboration. Scientometrics, 54(1), 75–89.

He, T. (2009). International scientific collaboration of China with the G7 countries. Scientometrics, 80(3), 571–582.

Hullmann, A. & Meyer, M. (2003). Publications and patents in nanotechnology: An overview of previous studies and the state of the art. Scientometrics, 58(3), 507–527.

Ingwersen P. & Larsen B. (Eds.), ISSI: Proceedings of the 10th international conference on scientometrics and informetrics.

Jin, B. & Rousseau, R. (2005). China's quantitative expansion phase: exponential growth but low impact.

Jin, B.H., Rousseau, R. & Sun, X.X. (2006). Key Labs and Open Labs in the Chinese scientific research system: Their role in the national and internationalscientific arena. Scientometrics, 67(1), 3–14.

Kostoff, R.N., Briggs, M., Rushenberg, R., Bowles, C.A., Icenhour, A.S. & Nikodym, K.F., et al. (2007a). Chinese science and technology? Structure and infrastructure. Technological Forecasting and Social-Change, 74(9), 1539–1573.

Li, X., Lin, Y., Chen, H. & Roco, M.C. (2007). Worldwide nanotechnology development: A comparative study of USPTO, EPO, and JPO patents (1976–2004). Journal of Nanoparticle Research, 9(6), 977–1002.

Luukkonen, T., Persson, O. & Sivertsen, G. (1992). Understanding patterns of international scientificcollaboration. Science, Technology and Human Values, 17(1), 101.

Melin, G. (2000). Pragmatism and self-organization: Research collaboration on the individual level. Research Policy, 29(1), 31–40.

Stockholm.Katz, J.S. & Martin, B.R. (1997). What is research collaboration? Research Policy, 26(1), 1–18.

Tang, L. & Shapira, P. (2011). China–US scientific collaboration in nanotechnology: Patterns anddynamics. Scientometrics, 88(1), 1–16.

Wagner-Doebler, R. (2001). Continuity and discontinuity of collaboration behavior since 1800—from a bibliometric point of view. Scientometrics 52, 503–517.

Zhenzhong Ma, Yender Lee & Chien-Fu Patrick Chen. (2009). Booming or emerging? China's technological capability and international collaboration in patent activities, Technological Forecasting and Social Change. 787–796.

Modeling and Computation in Engineering II – Xie (ed)
© 2013 Taylor & Francis Group, London, ISBN 978-1-138-00058-2

Numerical analysis of influence on indoor air distribution by the positions of air inlet and air outlet

G. Li, Z. Li, G.H. Feng & H. Wang
Municipal and Environmental Engineering Institute, Shenyang Jianzhu University, Shenyang, Liaoning, China

H.S. Jin
The Fourth Engineering Limited Company, The Ninth Bureau of China Railway Group, Shenyang, Liaoning, China

ABSTRACT: This paper choose a typical air-conditioned room as a research target, making use of Computational Fluid Dynamics to research the distribution of airflow organization under different positions of air inlet and air outlet. Establish the mathematical models of different vent positions to proceed numerical simulation. The distributions of vent positions are as follows, the upper vent sends air and the lower vent returns air in the same side, the upper vent sends air and the lower vent returns air in the opposite side, the upper vent sends air and the upper vent returns air in the same side, the upper vent sends air and the upper vent returns air in the opposite side. It can obtain the computation results of the indoor airflow velocity field and temperature field under four kinds of vent positions. Through comparative analysis, it finds that the best vent position is the upper vent sends air and the lower vent returns air in the same side.

1 INTRODUCTION

Nowadays, with the improvement of the quality of life, the use of air-conditioning is becoming more and more popular, and the demands for higher indoor air quality and the Thermal Comfort of Human Body is increasing. The Thermal Comfort of Human Body is the condition of mind that expresses satisfaction with the thermal environment, namely thermal neutral sensation (neither hot nor cold). Ventilation and air conditioning measures are important methods to achieve good indoor air quality and satisfy the human thermal comfort, and different airflow patterns for the influence of indoor air quality are different [GBJ 19–87. 2001]. Indoor airflow velocity, temperature and humidity are essential factors of the thermal comfort of human body, and concentration of pollutants is an important indicator of air quality. Therefore, if we want to make movement area becomes a space with suitable temperature, humidity and excellent air quality, which not only should have a reasonable system form, but also should have a reasonable air distribution. Most of the ventilation and air conditioning systems need to send air to the room or the control area and exhaust it [Ma, R.M. & Lian, Z.W. 2000]. Difform rooms, different forms and arrangements of air inlet and air outlet all affect the velocity distribution of indoor air, the distribution of temperature and humidity and the distribution of the concentration of the pollutants. The mode of air distribution depends on the positions of air inlet and air outlet and the forms of air inlet, and so on. Air inlet (its location, form, specifications, export wind speed, etc.) is the main factor which influences the air distribution [Lu, Y.J. & Ma, Z.L. 2007]. In recent years, many researches in domestic and foreign made use of Computational Fluid Dynamics (CFD) to simulate the indoor airflow distribution, mainly to discuss the regular boundary conditions. In this paper, we use computer software to simulate temperature field and velocity field of air distribution under the conditions of different positions of air inlet and air outlet.

Based on a typical air-conditioning room as the research object, we establish the mathematical model, use the Computational Fluid Dynamics (CFD) software to simulate different position of air inlet and air outlet that influence the distribution of indoor airflow organization, contrast the results of indoor airflow organization distribution under different working conditions and determine a reasonable vent position.

2 BASIS OF NUMERICAL SIMULATION AND ESTABLISHMENT OF THE MODEL

2.1 *Computational basis*

We use a typical office of Shenyang Jianzhu University as the object of study. The size of the room is: 6000 mm × 5000 mm × 3000 mm, south wall is the exterior wall, the other three are interior walls, and the adjacent rooms have no air-conditioning. This room uses the aluminium alloy double layer shutter vent as air inlet, the vent of bars grilling as air outlet. There are four distributions of positions to set the vents, those are the upper vent sends air and the lower vent returns air in the same side, the upper vent sends air and the lower vent returns air in the opposite side, the upper vent sends air and the upper vent returns air in the same side, and the upper vent sends air and the upper vent returns air in the opposite side. The models are as follows.

2.2 *Mathematical model*

Because the airflow is basically turbulence in the air-conditioning room, so we adopt the turbulence model to proceed numerical simulation of the incompressible three-dimensional turbulent flow form. According to actual condition, the air movement is unsteady flow that

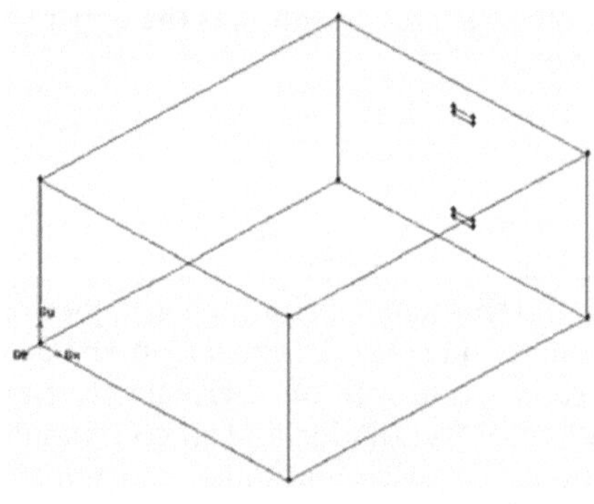

Figure 1. Upside air-supply bottom air-return in the same side.

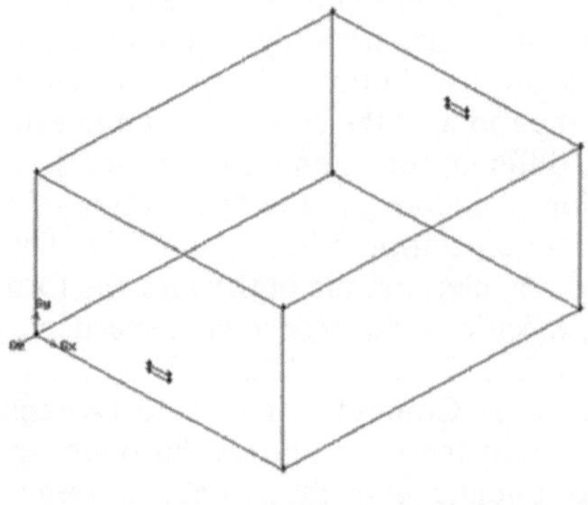

Figure 2. Upside air-supply bottom air-return in the opposite side.

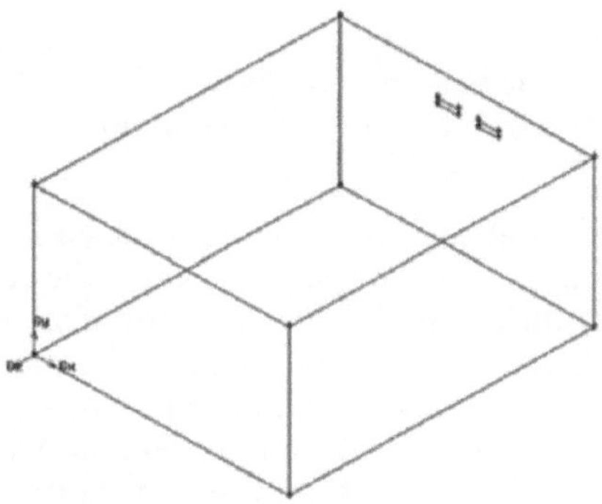

Figure 3. Upside air-supply upside air-return in the same side.

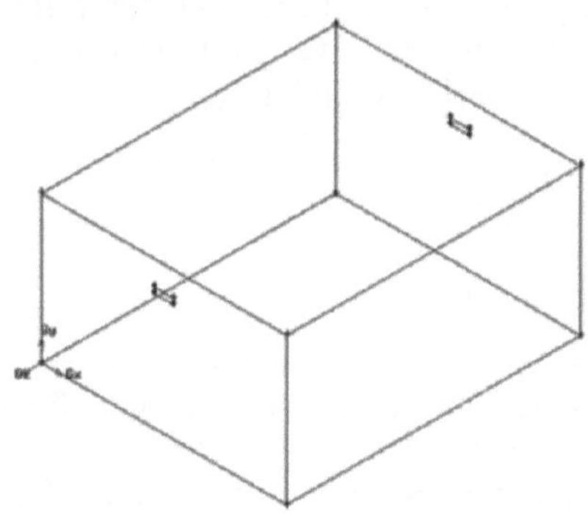

Figure 4. Upside air-supply upside air-return in the opposite side.

is changing over time in the air-conditioning room. The distribution of stable airflow is the most important influence factor. In order to simplify the research, we consider the airflow as the steady flow. The driving force of the flow is natural convection. Turbulence model uses the k-ε model that considers the gravitational influence and combines with the wall function to analog computation [Wen, Z. & Shi, L.C. 2009]. The general form is as follows:

$$\frac{\partial(\rho\phi)}{\partial t} + \nabla \bullet (\rho V \phi) - \nabla \bullet (\Gamma \nabla \phi) = S \tag{1}$$

In the type, $\partial(\rho\phi)/\partial t$ is unsteady state term, $\nabla \bullet (\rho V \phi)$ is convection term, $\nabla \bullet (\Gamma \nabla \phi)$ is diffusion term, and S is source term. The k-ε model of higher Reynolds number is suitable for the turbulent region that departs from the wall a certain distance, but there is a low Reynolds number of viscous laminar flow adjacent to the wall. The influence of the molecular viscosity must be considered in the simulation of air distribution while simulating the air distribution, the influence of the molecular viscosity must be considered and the k-ε equation need to be modified to fit the low Reynolds number in the air-conditioning room.

2.3 *Mesh generation*

We use the GAMBIT to establish the geometry areas that need to be calculated and to divide the grid. In this paper, volume mesh is directly divided and the grid space is set to 0.1 m, the others keep default. The grid division is a very important work, that the quality of the grid often determines the success or failure of the numerical simulation and the speed of convergence [Wen, Z. 2009].

2.4 *Boundary conditions*

We approximately select the parameters of airy physical property under the normal temperature as fluid physical properties in the numerical calculation. The boundary conditions are as follows:

1. Inlet boundary. The double shutter vents are used of the air inlet; the size of the air inlets is 400 mm × 100 mm. The airflow is flatly sent out along the ceiling, the speed of air supply is 3.9 m/s. The velocity-inlet is used of the inlet boundary condition [Wen, Z. 2009].
2. Outlet boundary. The grille vents are used of the air outlet, various variables are disposed according to local unilateralization, and each node parameters have no effect to the previous node. The size of the air outlet is 400 mm × 100 mm. We use the pressure-outlet as boundary condition. When there is backflow, using the boundary condition of pressure-outlet to take the place of the outflow often can have a better speed of convergence.
3. Solid boundary. We adopt the first kind boundary conditions as the measured wall temperature, the south wall temperature is 30°C, the temperature of external windows is 32°C, other wall temperature is 27°C, and roof and floor are disposed for the adiabatic condition. Except the air inlet and air outlet, the fixed boundaries all take the no slip boundary conditions [Luan, R. & Li, R. 2009].

3 NUMERICAL CALCULATION

3.1 *Setting of the calculation model*

1. The definition of solver. We select the pressure based as solver, set implicit in the format of Formulation, set stationary flow as the time, set the three-dimensional space in the space, and set Green—Gauss Node Based in the Gradient Option that can improve the resolution of the gradient, especially for coarse or oblique unit [Wen, Z. & Shi, L.C. 2009].
2. The function of activate gravity. Y is set to −9.81 among the Gravitation Acceleration. We need to activate the gravitational function because the main parts of flowing are driven by the natural convection.
3. The choice of turbulence model. We set the k-epsilon model as equation model in the control panel of turbulence model (k-ε model).

3.2 *Setting of fluid physical properties*

We use air as calculated fluid material in this paper, so we need to modify the characteristic parameter of the air. The density is set as 1.225 kg/m^3 under the condition of 26°C and 1 atm., because such setting is more stable for the problems which contain of natural convection [Ge, F.H. 2008].

3.3 *Setting of solving process*

1. Setting the solving parameter. The option of Under-Relaxation Factor expresses the control equations of the solution and the Relaxation Factor of some variables. We set 0.3 in the Pressure, 0.2 in the Momentum, and 0.9 in the Energy, others Relaxation Factor keep default. The Body Force Weighted is set as the discrete format of Pressure, and the others are set as the discrete format of first-order upwind [Zhao, B. 2003].
2. Initialization. The temperature is set to 20 in the Initial Values.

4 INTERPRETATION OF RESULTS

Figures 5~8 are the simulated results. We select the yz plane where the place is x = 2.5, the calculative results of the velocity vector field and the temperature field are shown as follows.

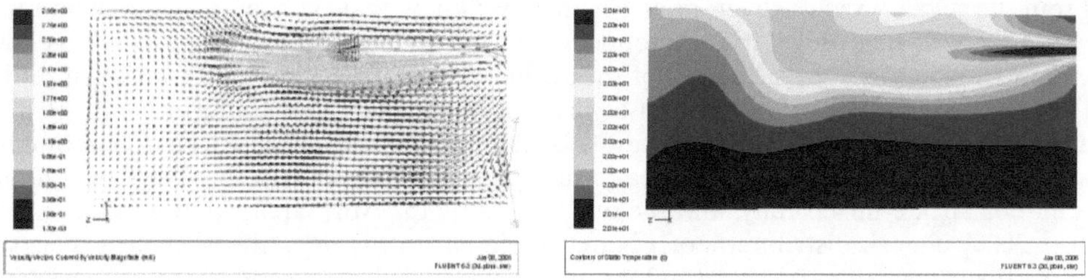

a. The vector diagram of velocity distribution b. The cloud picture of temperature distribution

Figure 5. The air distribution of upside air-supply bottom-side air-return in the same side.

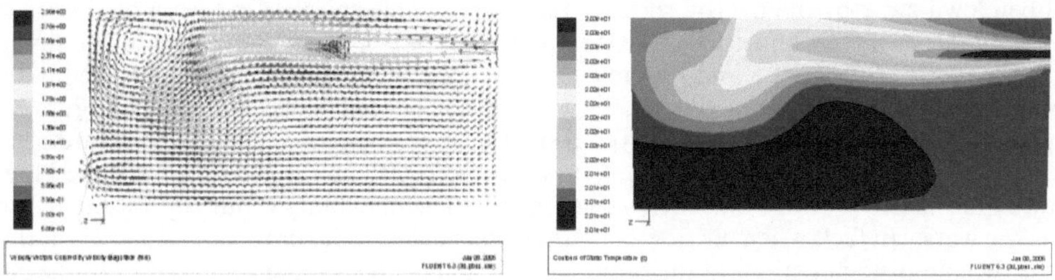

a. The vector diagram of velocity distribution b. The cloud picture of temperature distribution

Figure 6. The air distribution of upside air-supply bottom-side air-return in the opposite side.

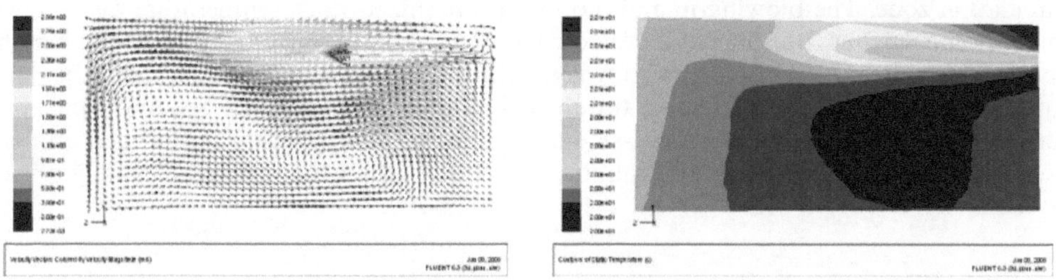

a. The vector diagram of velocity distribution b. The cloud picture of temperature distribution

Figure 7. The air distribution of upside air-supply upside air-return in the same side.

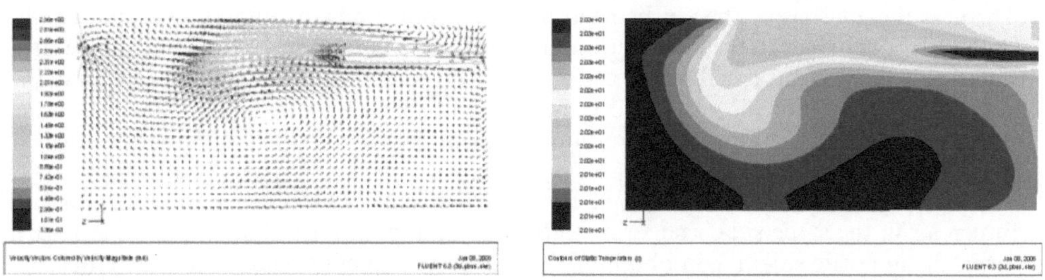

a. The vector diagram of velocity distribution b. The cloud picture of temperature distribution

Figure 8. The air distribution of upside air-supply upside air-return in the opposite side.

From the velocity vector diagram of Figure 5, we know that the incoming air encounters the left side wall then flows to the ground, when the air hits the ground, it will engender backflow from left to right. The most of the airflow is exhausted through the lower air outlet in the process of backflow; a small part of the airflow will climb due to the influence of heat source. The climbing air will be blown away by the nearby jet flow, then forms some spirals. Taking one with another, the work area basically is backflow zone in this room, air-supply and indoor air are mixed fully, and the wind speed of the work area is lower [Wu, X.P. & Zhu, J. 2000]. The air distribution of Figure 7 is similar to Figure 5, the work area is also in the backflow zone, air-supply and indoor air are mixed fully, the wind speed of work area is lower. Through contrasting Figure 5 with Figure 7, it can be seen that the vorticity of Figure 5 is stronger than Figure 7; the ventilation efficiency of Figure 5 is close to Figure 1, higher than Figure 7. Figure 6 shows that the incoming air encounters the left side wall then flows to the ground, the most of airflow is exhausted when arrived at the air outlet, a small part of the airflow returns from left to right, the regurgitant air will climb, the climbing air will be blown away by the nearby jet flow, thus forms some spirals. On the whole, the work area is in the backflow and eddy zone, the mixability of air-supply and indoor air significantly is inferior to Figure 5 and Figure 7, although the wind speed of the work area is also lower, but the air freshness of work area is not high. The incoming airflow of Figure 8 is mostly discharged in the progress of flow, a small part of the airflow flows downward along the left wall. When it hits the ground, it flows to the right. The climbing air will be blown away by the nearby jet flow, thus forms some spirals. The mixability of air supply and indoor air is the worst. In fact, when air flow encounters obstacles in the process of flow, it will flow to the two opposite directions at the same time, thus engenders some small vortex in the corner [Xu, L. & Weng, P.F. 2003].

It can be seen that the temperature of work area is more homogeneous in the airflow organization of supply air and return air in the same side, the distribution of temperature of upside air-supply bottom-side air-return in the same side is the most homogeneous. First of all, the work area of the room which supply air and return air in the same side is basically in recirculation zone. The blowing-in and indoor air are mixed fully, temperature distribution is more homogeneous. But the work area of the room which supply air and return air in the opposite side is in the backflow and eddy zone, the blowing-in and indoor air are mixed not enough, so the uniformity of temperature distribution is a bit weaker. Secondly, the eddy strength of upside air-supply bottom-side air-return in the same side is stronger, and the mixed degree of the supply air and indoor air is higher than the upside air-supply upside air-return.

5 CONCLUSIONS

Through studying the positions of air inlet and air outlet that engender the influence to the air distribution, we reach the following conclusions:

1. Through the analysis of the air distribution under four kinds of different working conditions, it is concluded: to the distribution form of upside air-supply bottom-side air-return, the supply air flow attached to the ceiling, the work area is in the recirculation zone, the supply air and indoor air are mixed fully, the wind speed of work area is lower, and the distribution of temperature is more homogeneous, so this kind of distribution form is the best.
2. Through the simulation analysis, we know that the ventilation pattern of supply air and return air in the same side is superior to the ventilation pattern of supply air and return air in the opposite side; it can obtain higher indoor air quality. The ventilation efficiency of upside air-supply bottom-side air-return in the same side is higher than the ventilation efficiency of upside air-supply upside air-return the same side.

REFERENCES

GBJ 19–87. 2001. *Design Specifications of Heating Ventilation and Air Conditioning*. Beijing: China project press.

Ge, F.H. 2008. Numerical simulation of indoor air distribution and air quality. *Journal of Jilin Architecture Engineering Institute* 24(1): 9–14.

Luan, R. & Li, R. 2009. Numerical simulation of wind speed influences on indoor air distribution. *Journal of Computational Mechanics* 26(1): 17–21.

Lu, Y.J. & Ma, Z.L. 2007. *Heating Ventilation and Air Conditioning.* Beijing: China building industry press.

Ma, R.M. & Lian, Z.W. 2000. The discussion of replacement ventilation problems. *HVAC* (4): 18–22.

Tao, W.Q. 1991. *Computational Fluid Mechanics and Heat Transfer Theory*. Beijing: China building industry press.

Wen, Z. & Shi, L.C. 2009. *FLUENT Fluid Computing Applications Tutorial.* Beijing: Tsinghua university press.

Wu, X.P. & Zhu, J. 2000. The research of indoor airflow and comfort level of low temperature air supply air-conditioning. *East China Electric Power* (8): 11–13.

Xu, L. & Weng, P.F. 2003. The numerical analysis of the indoor air distribution and indoor air quality under three kinds of ventilation patterns. *Journal of Air Dynamics* 21(3): 10–14.

Zhao, B. 2003. *The Vents Model Research and Application of Indoor Air Flow Numerical Simulation.* Beijing: Tsinghua University.

Modeling and Computation in Engineering II – Xie (ed)
© 2013 Taylor & Francis Group, London, ISBN 978-1-138-00058-2

Competitive intelligence, knowledge management, and anthropology

G. Tian, H. Wang & Q. Dai
Shantou University, Shantou City, Guangdong Province, China

ABSTRACT: As Competitive Intelligence (CI) and Knowledge Management (KM) have become important management issues for decisions makers, the contemporary environment requires the CI staff must broaden their ways of collecting, storing, analyzing information to meet the need of business leaders for competitive intelligence. With their unique training and skills, business anthropologists are able to help CI staff for the best solution in their ways of conducting CI programs. This paper will discuss the implementations of anthropology in competitive intelligence and knowledge management.

1 INTRODUCTION

In recent years, due to the faster and faster growing competition among firms in the business world, Competitive Intelligence (CI) has become one of the most important management issues for senior decision-makers. This enhanced phenomenon is being fueled by increased global competitiveness characterized by heightened industry consolidation and fragmentation. More than ever before the top managerial staff entrusted with designing and implementing strategy for their enterprises in such kind of environment have to base on more reliable and valuable competitive intelligence for their decision-making.

Related with CI, another intellectual issue for the contemporary business firms to win the competition is the Knowledge Management (KM). As a new business function, KM comprises a range of practices used in an organization to identify, create, represent, distribute, protect, and enable adoption of insights and experiences. Such insights and experiences comprise knowledge, either embodied in individuals or embedded in organizational processes or practice.

To better satisfy the need from the business leaders for competitive intelligence programs, it is necessary that CI staff to broaden their ways of collecting, storing, analyzing information. Anthropologists will be the best candidates for CI staff to consider for collaboration, and anthropological methods will be the best solution for CI staff to broaden their ways of conducting CI programs. Meanwhile, it is important that business firms implement a well designed KM program to protect their competitive advantages.

2 ANTHROPOLOGICAL APPROACH TO COMPETITIVE INTELLIGENCE

Robert Galvin, former Chairman of Motorola Executive Committee, indicates that a truly thorough understanding of the cultural anthropology of the world can explain the factors that underline a particular nation's conduct, and it is likely that grasping and conveying anthropological knowledge will fall to intelligence department as we expand our awareness of this very complex, multi-faced world. For example, in attempting to break into the Japanese electronic market the US business firms need to realize the extent to which the Japanese respected power, in contrast to the deferential polite mannerisms of the more apparent culture. It takes time to break down the market barriers put up by Japan. Japanese respected power that is an anthropological principle, a very significant piece of intelligence (Galvin & Robert, 2001).

Deeply influenced by the famous anthropologists of business scholars Russell Belk, John Sherry and Melanie Wallendorf, among others, Dr. Walle is a noted social scientist with a strong anthropological background in CI and marketing research. He has moved between consulting in business focusing on marketing and teaching in the fields of business, anthropology, and tourism at college levels. According to Walle, competitive intelligence is the information gathering and information analysis component of building competitive advantage. Although competitive intelligence and contemporary marketing research evolved from different intellectual and quantitative traditions, both are indebted to qualitative methods of research and analysis that anthropologists are well trained for. Nonetheless, the quantitative methodology has dominated in the Competitive Intelligence (CI) and Marketing Research (MR) area.

Walle discusses how the field of CI brings powerful qualitative tools to business research and argues that in an era when many business scholars and practitioners have come to depend upon mostly quantitative techniques, CI professionals have long embraced a qualitative, subjective, and intuitive toolkit that has provided timely meaningful information for many successful cases. But the vital role of anthropological qualitative methods within business strategy, particularly within CI and MR has been long neglected or ignored. He indicates that in the post-World War II era, when business research became more quantitatively oriented, CI started with a qualitative method that drew inferences without formal proof; as a result, the field has gained a special niche within business. CI embraced the intuitive tools of ethnography, and marketing researchers have turned to the techniques of the qualitative social sciences and the humanities. He claims that CI can and should forge linkages between these two qualitative traditions.

Walle presents the current contemporary initiatives within business that link the methods of the social sciences and humanities to business analysis by making a well-reasoned rationale for CI professionals to apply qualitative methods in their business analysis. To Walle CI is qualitative in nature, a discipline that is based on the traditions of espionage, and thus has special toolkit, which is an argument that other CI professionals may disagree. The contemporary business world has elevated marketing theory and methods to a strategic position although much traditional business thought was centered upon management and had dealt with marketing as tactic subordinate activity. The intelligence profession has a strong tradition of embracing relevant aspects of the social sciences and humanities in its toolkit.

CI professionals can enhance their toolkits in relevant ways as CI basically is a qualitative method oriented business field that stems from espionage, which infers information from weak compromised and incomplete data. Although differ from social sciences and humanities in terms of profitability driven, CI nonetheless can benefit from the methodologies used in these disciplines. Walle, by merging intelligence with other qualitative traditions, breaks new ground and offers an expanded vision of applied anthropology that is most relevant to the contemporary business world. Today, many applied anthropologists employ qualitative techniques along with some quantitative approaches (survey, for example) in complex organizations from community centers to large corporations in terms of various anthropological researches (Walle, 2001).

More recently many anthropologists involve themselves into MR although not many are yet involved in the CI field. Meanwhile more and more marketers are using anthropological methods in their marketing practice and research. In practice, business anthropologists almost study everything from marketing strategies to the corporate climate, applying traditional anthropological methods of research and observation to understand and reflect business culture, and thus make their contributions to the business development. It is our brave view of the future that applied anthropologists will become the hottest candidates for business related research jobs given the fact that anthropological methods are becoming more widely acceptable in the business world, especially in the fields of competitive intelligence and marketing research.

Anthropology's main distinguishing method is participant observation. This involves the anthropologist spending a protracted period doing fieldwork in an effort to gain an in-depth understanding of the society under study. By virtue of its eclecticism and experience of facilitating understanding of the processes of change across institutions and other social phenomena, anthropology can make a significant contribution to the implementation of competitive intelligence. Below we will probe what anthropology can offer CI (Hohhof, 2010).

Fundamental to understanding a department is to learn how a group works—its structures and processes (i.e., the culture). In their consultancy experience, Neil Simon and David Grzelak have found many companies that have had separate CI units with their own structures and processes. In other words, the cultures of each group were unique. Upon investigating their formation, each unit was found to have been created in order to fulfill the specific needs of that group. A unique concept of CI emerged in each one of the organizations based on the whole organization's needs.

Simon and Grzelak stress that the organizational theorists tell us that problems arise when differences between two departments become apparent. These differences also can adversely affect the relationship that the CI practitioner needs to develop with the client. This relationship is fundamental in order for CI professionals to be effective for the client. Often, what one department views as different becomes a threat to them, and this can hinder the amount of trust that the two departments establish. These differences can become a point of friction. True 'integration' (acceptance and adoption) becomes impossible. The CI professionals can avoid these potential difficulties by quickly learning about the culture of their client, as well as what the client cherishes. As the anthropologist participates and observes, so should the CI practitioner; the client can understand the results of the work (whether it is the subject of the anthropologist or the department requiring CI).

As Simon and Grzelak explain, to anthropologists "artifacts" represent culture. These are the surface-level manifestations, the appearance, of a culture. As symbols, these artifacts represent the values and beliefs of the culture, usually indirectly presented through their 'myths'. Being able to recognize and understand the meaning of the client's culture allows CI practitioners to integrate some of the client's values into the way business is conducted with the client. Sharing the values and beliefs helps both the client and the CI professional better understand each other and develop trust.

By applying anthropological approach for knowing and understanding, what the CI client values can help CI professionals to recognize what type of information will be important for that client, and to better present their findings. Therefore, the CI staff moves from an 'outsider' to an 'insider', enabling CI to play a more important role in the organization. Eventually, the CI operative is viewed as part of the team, just as many anthropologists have become honorary members of the cultures they have studied. Hofstede describes three types of artifacts that CI professionals can quickly recognize in order to come to an understanding of the CI client's culture.

2.1 *Symbols*

These are the most apparent manifestations of an organization's culture. The symbols discussed here include language, actions, or objects that have a particular meaning only understood by members who share the culture. Symbols such as terminology that is particular to the client can help the CI professionals find some common ground. Appearing more familiar with the client will help eliminate some of the difficulties that arise when uncertainty exists between the CI staff and client. Anthropologists analyze everything they find from the fieldwork; the CI professionals can do the same in order to gain useful insight into the client's culture. A simple example is the client's office space. Looking at the office layout, the CI professional may be able to develop a better understanding of how the client views business. Private offices for upper-level management, social areas such as break rooms, employee cubicles (or lack of cubicles) can all provide the CI professionals with important insights regarding the values and beliefs of the client. How they are signposted, where they are in relation to each other, and how people behave in them, are all significant in developing an understanding of the client. Together these artifacts will help the CI professionals to get a feel for how best to proceed with that particular client.

2.2 *Rituals*

In anthropology rituals are ceremonies that help to define a culture, the rituals discussed here refer to activities that are a common part of the way the client of CI professional operates on a day-to-day basis. Often, rituals are carried out for their own sake and are considered as a socially essential element of the client's organization. Examples of rituals include meetings, performance reviews, the ways in which members of an organization greet one another, and after work social events. The CI staff that is able to participate in these rituals and also understand what they mean to their clients will be able to become a more socially accepted part of their client's organization. This process is important if the CI operative is able to move from an outsider to an insider.

2.3 *Myths*

For business anthropologists, myths are the stories about the individuals that the client considers important as their heroes. They can be alive or dead, current or past. The characteristics of the heroes help the CI professionals to understand what the client values in its members. Organizational heroes such as Henry Ford, Bill Gates, or Motorola's Robert Galvin all have significant meanings to their companies. Recognizing the types of individuals that succeed within the client's culture tells CI managers how to best present themselves to the client. Again, this knowledge will help the CI professional seem more familiar to the client, and thereby become more accepted and trusted by the client (Simon et al., 2001).

2.4 *External and internal culture*

In order for CI to be effective, the corporate of internal culture defined by internal factors of influences must be supportive. These factors include management preferences, organizational structure, and resources that affect CI initiatives. The internal culture plays a large role in how CI is practiced, how companies view CI, how will they use it, and what aspects of CI they see as being important. It affects how information is collected and what information is collected. Equally important is an external culture that supports CI efforts. By exploring the external factors of influence such as political, social, and economic factors, a business firm can draw conclusions of a country's CI friendliness. Certain environments are more conductive to CI practices because these external factors allow practitioners to access information easily (Kahaner & Larry, 1996). Investigating a new culture requires a global mindset; understanding the social and political dynamics of countries is required for business success in overseas markets. For instance, an American corporate culture is known for its short-term outlook in business dealings, thereby influencing decisions accordingly. In contrast, Japanese firms focus on long-term initiatives, where patience is the key factor. Therefore, knowing the impact of culture on business practices will allow the CI professionals to make informed analyses. A historical background will assist in obtaining this knowledge, in addition to comparative and case study analysis. Specifically, global CI requires gathering information on sociological, educational, economic, political, and geographic criteria. These environmental factors can indicate competitive opportunities or threats (Toczydlowski & Betty, 2005).

2.5 *Ethics concerns*

Ethics has been a long-held issue of discussion amongst CI practitioners. Essentially, the questions revolve around what is and is not allowable in terms of CI practitioners' activity. A number of very excellent scholarly treatments have been generated on this topic, most prominently addressed through Society of Competitive Intelligence Professionals publications (Fleisher et al., 2003). There is considerable concern about such problems, but the majority of executives, when surveyed, state that they would be unwilling to engage in unethical intelligence gathering practices (Gordon & Ian, 1989). It is also unnecessary, especially in the context of Internet, which is above all a publicly available channel. It is emphasized that a key maxim of CI is that 90% of all information that you and your business need to make key decisions and to understand your market and your competitors is already public or can be collected from public sources. Professionals working in special libraries and information centers will be so busy exploring the riches of the Internet that they will not have time for or interest in unethical activities (McGonagle et al., 1993).

3 FROM COMPETITIVE INTELLIGENCE KNOWLEDGE MANAGEMENT

Competitive intelligence starts with managing and deploying knowledge systematically within one's own organization. Increasingly, knowledge is recognized as a key organizational asset, to be leveraged and exploited for competitive purposes. As such knowledge management becomes one of the hottest topics today in both the industry world and information research world (McLellan & Hilary, 2001). In our daily life, we deal with huge amount of data and information. Data and information is not knowledge until we know how to dig the value out of it. This is the reason we need Knowledge Management (KM). Unfortunately, there is no universal definition of knowledge management, just as there is no agreement as to what constitutes knowledge in the first place. We adopt the following definition for knowledge management for its simplicity and broad context for the purpose this book.

A KM system organizes the intellectual assets of a corporation. This includes recorded information, corporate experience, third-party information, and tacit knowledge of employees. Knowledge management is a newly emerging, interdisciplinary business model dealing with all aspects of knowledge within the context of the firm, including knowledge creation, codification, sharing, protection, and how these activities promote learning and innovation. The proper information architecture design is a key component in creating knowledge. The design incorporates web development, library science, cultural anthropology, and literary theory (Toczydlowski & Betty, 2005).

KM efforts have a long history, to include on-the-job discussions, formal apprenticeship, discussion forums, corporate libraries, professional training and mentoring programs. Many large companies and non-profit organizations have resources dedicated to internal KM efforts, often as a part of their business strategy, information technology, or human resource management departments (Addicott et al., 2006). More recently, with increased use of computers in the second half of the 20th century, specific adaptations of technologies such as knowledge bases, expert systems, knowledge repositories, group decision support systems, intranets and computer supported cooperative work have been introduced to further enhance such efforts.

In practice, KM encompasses both technological tools and organizational routines in overlapping parts. It efforts typically focus on organizational objectives such as improved performance, competitive advantage, innovation, the sharing of lessons learned, and continuous improvement of the organization. KM efforts overlap with competitive intelligence, and may be distinguished from that by a greater focus on the management of knowledge as a strategic asset and a focus on encouraging the sharing of knowledge. KM efforts can help individuals and groups to share valuable organizational insights, to reduce redundant work, to avoid reinventing the wheel per se, to reduce training time for new employees, to retain intellectual capital as employees' turnover in an organization, and to adapt to changing environments and markets (McAdam et al., 2000). These efforts can be better operated with the help of anthropologists.

What does anthropology have or do that is of value to Knowledge Management? The applied anthropologist and the founding director of Workspace International Patricia Burke developed an anthropological model to knowledge management through her longitude practice (Burke & Patricia, 1998). Burke indicates that anthropologists are interested in how different people constitute knowledge and how knowledge is managed in terms of how it is secured and deployed. Understanding KM in an organizational context brings anthropology together with a number of other disciplines such as psychology, business theory and information modeling. This can enhance our understanding while simultaneously creating new forms of organizational knowledge. The relevance of an anthropological approach to KM touches on a gamut of issues, including the popularity of the culture concept and its concomitant, ethnicity, issues of globalization and rapid change, issues of difference and sameness, and the deconstruction of how an organization organizes.

Anthropology's main distinguishing method is participant observation. This involves the anthropologist spending a protracted period doing fieldwork in an effort to gain an in-depth understanding of the society under study. By virtue of its eclecticism and experience of facilitating understanding of the processes of change across institutions and other social phenomena, anthropology can make a significant contribution to the implementation of KM.

According to Burke, observing employees going about their day-to-day tasks is an important way of externalizing tacit knowledge is just one of the ways in which anthropology can make a significant contribution to the implementation of KM.

4 SUMMARY AND CONCLUSION

Due to the rapidly changed contemporary business environments, CI and KM become two of most growing fields in the business world. Every business manager needs intelligence to find suppliers, mobilize capital, win customers and fend off rivals. The executive who has a conscious, systematic approach to acquiring intelligence will be better placed to carry away the most desirable prizes, while safeguarding their firm against the most menacing perils. As such, the top leaders of business firms must base on more reliable and valuable competitive intelligence for their decision-making. All these require CI staff must broaden their ways of collecting, storing, analyzing information to meet the need of business leaders for competitive intelligence. With their unique training and skills, business anthropologists are able to help CI staff for the best solution in their ways of conducting CI programs.

As CI evolve a better and through understanding of the various links of entities and knowledge sources becomes important. Consequently, the field of knowledge management requires more attention than ever before by the business leaders for the success. KM involves best leveraging knowledge internally and externally in an organization and creating a process for valuing the organization's intangible assets. KM staff has been applying social network analysis techniques to map the knowledge flows and detect knowledge gaps in the organization. Social net work analysis has grown out of the anthropology and sociology disciplines, but can be of great assistance to mapping knowledge flows within the organization for enhancing CI methods.

REFERENCES

Addicott, Rachael; Gerry McGivern & Ewan Ferlie. 2006. Networks, Organizational Learning and Knowledge Management: NHS Cancer Networks. Public Money & Management 26 (2): pp. 87–94.

Burke, Patricia. 1998. Anthropological Approach. Inside Knowledge, Vol. 2, Issue 1.

Fleisher, Craig S. and David Blenkhorn. 2003. Controversies in Competitive Intelligence: The Enduring Issues. Westport, CT: Praeger.

Fleisher, Craig S. and Babette E. Bensoussan. 2003. Strategic and Competitive Analysis: Methods and Techniques for Analyzing Business Competition. Upper Saddle River: Prentice Hall.

Galvin, Robert W. 2001. Competitive Intelligence at Motorola, in John E. Prescott and Stephen H. Miller (Eds.) Proven Strategies Competitive Intelligence: Lessons from Trenches. Wiley, John and Sons.

Gordon, Ian. 1989. Beat the Competition: How to Use Competitive Intelligence. Oxford: Basil Blackwell.

Hohhof, Bonnie Ed. 2010. Competitive Intelligence Anthology. Alexandria, VA: SCIP.

Hofstede, G. 1980. Cultures Consequences: International Differences in Work-related values. London: Beverly Hills.

Kahaner, Larry. 1996. Competitive Intelligence. New York: Simon Schuster; Fleisher, Craig S. and David Blenkhorn 2003. Controversies in Competitive Intelligence: The Enduring Issues. Westport, CT: Praeger.

McAdam, Rodney & Sandra McCreedy. 2000. A Critique Of Knowledge Management: Using A Social Constructionist Model. New Technology, Work and Employment 15 (2).

McGonagle Jr., John J., and Carolyn M. Vella. 1993. Outsmarting the Competition: Practical Approaches to Finding and Outsmarting the Competition. (New ed.). London: McGraw-Hill, 1993.

McLellan, Hilary. 2001. Introduction to Competitive Intelligence, retrieved in July, 2009 from http://www.saratogamedia.net/CI2004/cintel1.htm.

Simon, Neil and David Grzelak. 2001. Business anthropology: clues to culture, SCIP, Society of Competitive Intelligence Professionals, Vol. 04 No. 04, 7–8.

Toczydlowski, Betty. 2005. Knowledge, Skills, and Abilities of Domestic and International Competitive Intelligence Practitioners. In Blenkhorn, David and Craig S. Fleisher (Eds.) Competitive Inteeligence and Global Business. Greenwood Publishing Group.

Walle, Alf H. III. 2001. Qualitative Research in Intelligence and Marketing. Westport, CT: Quorum Books.

Modeling and Computation in Engineering II – Xie (ed)
© 2013 Taylor & Francis Group, London, ISBN 978-1-138-00058-2

Compensation mechanism and policy on pricing full-cost of demand in households in Liaoning

Yanli Song
Medicine of Dalian University, Dalian, Liaoning Province, China

Xieyong Wang
Economic Management of Dalian Univercity, Dalian, Liaoning Province, China

ABSTRACT: In China, the contradiction between water supply and water demand is becoming a critical problem because of deteriorating water shortage and increasing water demand. Uniform price and policy of residential water cannot promote sustainable utilization of water. In this paper, grade of water price bearing capacity for 14 cities in Liaoning Province is defined based on Fuzzy Comprehensive Evaluation. Taking Dalian for example, first the full-cost is given through calculating resources water price, engineering water price, environmental water price and marginal cost. Then the compensation mechanism and policy of Dalian is provided after comparing actual water price, full-cost and water price bearing capacity. The specific compensation mechanism and policy is given according to the different relations of full-cost, actual water price and water price bearing capacity which may occur in other 13 cities of Liaoning Province.

1 INTRODUCTION

China is short of water seriously, whose per-capita water resources stands at 2300 m^3, just less than a quarter of the world's average. Liaoning as an important province of China has a population of 40.67 millions. But its per-capita water resources is 34.2 billion cubic meters on average, about one third of the national average, and one fifth of the world average [Li Pei, 2001; Ma Pingsheng, 2009; Yang Lina, 2010]. It is reported that the full-cost price method can not only reflect water's commercial property, but promote use efficiency of water resources. "The notice of the State Council approving State Development and Reform Commission about Deepen the reform of economic system key work opinion in 2010" pointed that they would promote the reformation of the cost on water price into the public explicitly.

Pricing water reasonable is an effective way to realize the sustainable utilization of water resources. In this article, pricing the full-cost model was established by researching on water price bearing capacity of Liaoning Province, especially urban residents'. In the model, a reasonable compensation mechanism and guarantee system of policy are presented by combing the accounting of water price and national economy. This method may provide a theoretical basic for water price reformation, and sustainable development of Liaoning Province.

2 ANALYZING WATER PRICE BEARING CAPABILITY OF RESIDENTS IN LIAONING PROVINCE [SUN JING, 2008]

2.1 *Evaluating water price bearing capacity*

According to actual condition of Liaoning, grade of water price bearing capacity and affordable price interval of demand in households is shown in Table 1.

Table 1. The grade of water price bearing capacity and affordable price interval of urban household water.

Grade of water price bearing capacity	Water price bearing capacity	Affordable price interval (RMB/month)
I	Lowest affordable price	<4.0
II	Lower affordable price	4.0~5.5
III	Affordable price	5.5~7.0
IV	Higher affordable price	7.0~9.0
V	Highest affordable price	>9.0

Table 2. Results of water price bearing capacity of residents in Liaoning.

Grade of water price bearing capacity	Cities	Affordable capacity interval (RMB/month)
I	Dandong	<4.0
II	Anshan, Jinzhou, Shenyang	4.0~5.5
III	Shenyang, Yengkou, Chaoyang, Fuxin, Tieling	5.5~7.0
IV	Benxi, Fushun, Panjin	7.0~9.0
V	Dalian, Hulu Island	>9.0

2.2 *Analyzing water price bearing capacity of residents in Liaoning [Wang Huihui, 2012]*

According to "Liaoning province statistical yearbook 2010", the method Fuzzy Comprehensive Evaluation is used to get the affordable price interval of Liaoning. The results are shown in Table 2 [Liaoning province statistical yearbook 2010].

From Table 2, we know that water price bear ability of Dandong, Jinzhou, Liaoyang and Hulu Island is lower than others; Dalian and Fushun can accept a higher water price; other cities' is between the two kinds above.

3 ANALYZING RESULTS OF PRICING FULL-COST IN URBANS—TAKING DALIAN AS AN EXAMPLE [WANG YI-NING, 2010; WANG XIEYONG, 2011]

Full-cost is defined as the summation of all kinds of cost during exploiting water resources, which can be described as follows:

$$P = P_1 + P_2 + P_3 + P_4$$

Here, P is full-cost, which consists of resources water price P_1, engineering water price P_2, environmental water price P_3 and marginal cost P_4.

Dalian is seriously short of water resources with low per-capita water resources, misdistributions on area and mixed with sea water. What's more, its water price deviates from the law of value, which makes water-supply enterprises run in red deficit and the financial pressure of governments is heavy. Therefore, it is high time to make a reasonable scheme about pricing full-cost for Dalian.

3.1 *Pricing resources water price P_1 of Dalian*

Resources water price is the center of water pricing and the standard of pricing full-cost. Here, the revised Malthus and BP network algorithm are used to overcome weaknesses appeared in Fuzzy Comprehensive Evaluation, so that relations of resources water price and the value price are quantified as:

$$P_1 = f(p,Q) = \lambda \times p \times \left(\frac{Q_d}{Q}\right)^{E_0}$$

Here, p is the use value's price of water resources; Q is total water. We define supply elasticity coefficient is 0.28. So the resources water price is calculated as $P_1 = 0.99$ RMB/m^3.

3.2 *Pricing engineering water price P_2 of Dalian*

The commercialization of water includes engineering cost of prior period and supply cost of deliver period. The former changes water into product of labor, while the later commercialized it. Thus engineering water cost P_2 of Dalian is described as $P_2 = P_{21} + P_{22}$. P_{21} is the engineering water price of prior period and P_{22} is water supply price.

The engineering water price of prior period P_{21} and water supply price P_{22} is defined as:

$$\begin{cases} P_{21} == \dfrac{C}{W_1} \\ P_{22} = \dfrac{a+b+c+d}{W_2} \end{cases}$$

Here, C and W_1 are cost of all kinds of outlays in prior period and water supply each year, respectively. a, b, c, d are supply cost, water fees, taxes and profit, respectively. W_2 is water enterprises supplies. According to the year book, we get P_{21} is 0.21 RMB/m^3, and P_{22} is 1.12 RMB/m^3.

3.3 *Pricing environmental water price P_2 of Dalian*

Environmental water price is defined as the water pollution control and precaution costs when it is polluted during getting or using it. It's mainly decided by two aspects: the sewerage price; control or precaution cost.

According to "*The measures for the administration of the pollutant discharge fee charging*", environmental water price consists of loss during developing economic and cost during recovering polluted water., which is described as:

$$P_3 = P_{31} + P_{32} + \frac{C_{33}}{W_3}$$

Here, P_{31} the cost in draining off water is 0.6 RMB/m^3. We get P_3 is 1.24 RMB/m^3.

3.4 *Pricing marginal utilization cost P_4 of Dalian*

Full-cost includes not only value cost of water and human's labor, but marginal utilization cost. Here, marginal utilization cost P_4 is defined as the balance of actual marginal production cost of new and old products.

Marginal production cost P_4 is defined as:

$$P_4 = (P_{42} - P_{41})e^{qt}$$

Here P_{41}, P_{42} are the marginal production cost of new and old products. q is defined as $0 < q < 1$, while t is the time of substitutions appear.

We supposed that new products or technology appear in 5 years, which makes the marginal production cost fall by 0.5 RMB. Thus, $P_{42} = P_1 + P_2 + P_3 = 3.08$ RMB/m³. P_4 is 0.55 RMB/m³.

To conclude, we calculate that the full-cost in Dalian is described as:

$$P = f(p,Q) = \lambda p\left(\frac{Q_d}{Q}\right)^{E_0} + \frac{C}{W_1} + \frac{a+b+c+d}{W_2} + P_{31} + P_{32} + \frac{C_{33}}{W_3} + (P_{42} - P_{41})e^{qt}$$

According to the data we have got, $P = 0.99 + 1.12 + 1.24 + 0.55 = 3.9$ RMB/m³.

3.5 *Results*

In our report, it concluded that the residents' affordable price in Dalian is 10.43 RMB/m³, a relatively high grade compared to the other 13 cities of Liaoning. Also, the Neural Network method was used to show that full-cost in Dalian is 3.9 RMB/m³. In June 2012, it is reported that the price of demand in households is 2.9 RMB/m³. So the price of demand in households in Dalian is low, and still has a large space to adjust.

4 RECOMMENDATION AND POLICY ON PRICING WATER IN LIAONING PROVINCE

4.1 *The necessity of reforming price for demand in households*

Now, the proportion of per-capita water price to residents' disposable income is less than 0.5% in Dalian. Improving water price will accumulate fund, which contributes to the optimum circle for enterprises and the development of economy. What's more, it may arouse residents' awareness on saving water. From economic theories, we observe that residents can accept if pricing is on the basic of cost and keep open.

4.2 *The method of pricing demand in households in Liaoning Province*

In order to limit water consumption, we recommend Liaoning government conduct a water supply function that resembles a staircase (called price ladder). What's more, the price of water demand in households and extravagant must be distinguished. For example, the government can set 8 tons of water per month as the basic consumption, which be charged as normal. While the extravagant can be divided into 2~3 parts to charge. The more water residents have used, the more they will pay. Here, we suggest price of industrial, business and tourism water can be managed by price ladder, too.

Estimating cost of enterprises to make the highest price according to supply-demand method whatever tap water or water decontaminated from sewage, which is the best way of pricing water. However, a reasonable water price must consider government's macro-control, democratic consultation and market regulation.

4.3 *Constructing reasonable compensation mechanism and guarantee system of pricing water*

Water as a kind of renewable resource, has social attribute so that it's hard to commercialize completely. Governments must bear the social part of water. They often intervene into pricing water by making compensation mechanism (especially agricultural water supply). In this way, social stability and safeguard fairness can be guaranteed. What's more, that all reflect the spirit of the 18th National Congress of the Communist Party of China.

The compensation mechanism and guarantee system of water supply are ways of reallocation, which combine the accounting of water price and national economy. According to water price bearing capacity, full-cost and actual water price, four conditions and compensation policy are given (shown as Table 3).

Table 3. Compensation policy of pricing water.

Relations of three kinds of water price	Measures government may take
Actual water price ≥ water price bearing capacity > full-cost	Scheme 1: lower the price to water price bearing. Scheme 2: maintain actual water price and give the allowance produced by actual water price and water price bearing capacity.
Actual water price ≥ full-cost > water bearing capacity	Scheme 1: maintain actual water price or lower it to price full-cost, and government provide the allowance produced by actual water price and water price bearing capacity. Scheme 2: lower actual water price to water price bearing capacity, and government provide allowance caused by actual water price and full-cost.
Water price bearing capacity ≥ full-cos > actual water price	Scheme 1: raise actual water price to full-cost. Scheme 2: government gives the allowance to water enterprises.
Full-cost ≥ water price bearing actual water price	The worst condition, which actual water price capacity > can be raised to water price bearing capacity and government should be asked to compensate the actual water price to full-cost.

5 CONCLUSIONS

A reasonable water price is the key factor to effect sustainable utilization of water, which may raise our consciousness of saving water. Pricing water can make consumers ponder over their behavior again and put water environmental value into their economic behavior. However, Liaoning government should increase the proportion of eco-compensation during financial transfer payment and make reasonable water price. Putting the money which comes from water fees (such as "price ladder") into eco-compensation to improve the awareness of saving water and living condition of residents who is poor. Of course, it is not actual for Liaoning government to resolve all the problems in a short time for the integrated management of water just started.

All in all, the method of pricing full-cost provide a theoretical basis for water price reformation, which has great significance in reforming the price of resource products. Liaoning government should construct reasonable compensation mechanism and guarantee system for pricing water and protect residents' living whose water price bearing capacity is low. Only when water prices and costs are considered together, can the reasonable utilization of water resources, social harmony and sustainable development of economic be guaranteed in the process of making economic decision.

ACKNOWLEDGEMENT

This work was supported by National Training Programs of Innovation and Entrepreneurship for Undergraduates in 2012 (201211258016) and Economic and Social Development Project topic of Federation of Liaoning Province Social Science Circles in 2012 (2012 lslktzijjx-11).

Wang Xie-yong, the corresponding author, comes from Economic management of Dalian Univercity, Dalian, Liaoning Provinc, China.

REFERENCES

Li Pei, Jin Yiwan. Brief Discussion on Strategic measure of Sustaining Utilization for water Resources of Baicheng City. Jilin water resourcces. 2001, (1): 36–37.

Liaoning province statistical yearbook 2010.
Ma Pingsheng, Wan Yanhua etc. Discussion on Strategy of Sustainable Use of Urban Water Resources. Enviroment and sustainabke development. 2009, 34(4): 22–25.
Sun Jing, Shen Bi-feng. Forecast endure ability of water price of Beijing residents. BEIJING WATER, 2008,(6).44–46.
Wang Huihui Wang Xieyong etc. Application of Variable Fuzzy Evaluation Model in Water Price Bearing Capacity of Demand in households Use. Water Resources and Power, 2012, 30(7):140–143.
Wang Xieyong, Tan Xinxin etc. Study on Construction of Full Cost Water Pricing Model. WATER RESOURCES AND POWER. 2011, 29(5): 109–112.
Wang Yi-ning. Urban water price in China under market-oriented urban water services. Journal of Economics of Water Resources, 2010, (2):31–35.
Yang Lina, Ma Chuanbo, Wang Xin. Analysis of water price bearing capacity of urban residents in Huludao City. Technical supervision in water resources, 2010, 18(3): 20–22.

Modeling and Computation in Engineering II – Xie (ed)
 ISBN 978-1-138-00058-2

Analysis on esterase isozyme of *Auricularia polytricha*

Dan Zhang, Yu Zhang & Jianwei Gao
Chengdu Institution of Mountains Hazards and Environment, Chinese Academy of Sciences, Chengdu, Sichuan, China

Youliang Zheng
Sichuan Agricultural University, Sichuan, China

Bo Wang
Sichuan Agricultural Academy, Chengdu, Sichuan, China

ABSTRACT: Esterase isozyme zymogram polymorphisms of 56 strains in *Auricularia polytricha* were investigated by using vertical slab polyacrylamide gel electrophoresis. The results showed there is diversity among 56 strains. 10 zymogram bands were observed and 26 zymogram types existed in all tested strains. Each strain has 1~6 bands. Differences of strains were expressed in individual characteristic zymograms in this study. The results of cluster analysis at a 67% similarity level showed that all tested strains could be clustered into 8 groups, with the the 1st group including 34 strains, 2nd group including 21 strains, 3rd and 4th group only single strain of *A. auricla* and 915 respectively; 5th group including 4 strains, 6th group including 4 strains, 7th group including 2 strains, 8th group including 2 strains. The analytical technique of esterase isozyme is an effective method for identifying strains in species and under species.

1 INTRODUCTION

Isozymes are enzymes that differ in protein molecular patterns but catalyze the same chemical reaction. Amino acid substitutions in isozymes lead to changes in Electrophoretic Mobility. Specific staining for particular enzyme activity after electrophoresis could be used for distinguishing the differences among allele morphologies of a specific enzyme. In the process of hybridization, loci of isozyme usually appear in the Mendel pattern (Anderson IC et al. 1993). The amino acid alterations of isozymes are coded by corresponding DNA, so isozyme analysis could be treated as an effective method for studies of group genetic differentiation at the protein molecular level (Wanneng Hu & Xianguo Wan 1985, Shuting Zhang & Fangcan Lin 1997).

Structure similarities of isozymes reflect biological genetic relationships, so the analytical technique of isozyme is available for the classification of species and sub-species. Zymogram materials could be used as important indices in the study of species classification, evolution, heredity and variation. Isozyme could reveal genetic variations at a single site, and detect the existence of dominant and recessive alleles. This protein marker technique has been widely used in genetic studies in 1970s (Xun Li et al. 2003). Since 1980s, cross breeding workers found that the transfers of good traits that were difficult to be identified by morphological traits could be detected through analyzing isozymes in terms of both quality and quantity at molecular level. At present, esterase and peroxidase isozyme markers were widely used as isozyme markers in mushroom studies.

Until now, no systematic analysis of esterase isozyme zymgram of *Auricularia polytricha* has been reported. In this paper, esterase isozyme zymogram polymorphisms of 56 strains in *A. polytricha* were investigated so as to develop a theoretical framework for the precise application of esterase isozyme electrophoresis materials to identifications of *Auricularia polytricha* strains as well as to genetics and breeding research.

2 MATERIALS AND METHODS

2.1 *Stains tested*

There were 56 strains tested with one *Auricularia auricular*, one *Auricularia peltata Lloyd* and 54 *Auricularia polytricha*, which were listed on Table 1. *Auricularia auricular* is a kind of ear mushroom with the longest cultivation history in China, whose morphology is different greatly from *Auricularia polytricha*. *Auricularia peltata Lloyd* was distributed in the tropics holding a similar morphology with *Auricularia polytricha*.

2.2 *Reagents preparation and instruments*

① Separating gel buffer: 36.6 g trihydroxymethyl aminomethane(Tris), 48 ml, 1 mol/L HCl, with water added to 100 ml, pH 8.9;
② Separating gel storage solution: 29.2 g acrylamide(Acr), 0.8 g Bis, with water added to 100 ml, and filtration was conducted before using;
③ Ammonium persulfate solution: 0.56 g ammonium persulfate was dissolved in 100 ml water (prepared when using);
④ Stacking gel buffer: 5.98 g Tris, 48 ml 1 mol/L HCl, with water added to 100 ml;
⑤ Stacking gel storage solution: 10 g Acr, 2.5 g Bis, with water added to 100 ml, and filtration was conducted before using;
⑥ 40% (W/V) sucrose solution;
⑦ TEMED: 1% bromophenol blue water solution, acetone, 7% acetic acid;
⑧ Electrode buffer: 6 g Tris, 28.8 g glycine, with water added to 1 L, pH 8.3;
⑨ Esterase staining solution: 100 mg Luxol fast blue was dissolved in 150 ml 1 mol/L pH 6.2 phosphoric acid buffer and was filtrated and set aside; before using, 50 mg α-naphthyl acetate, 50 mg β-naphthyl acetate were dissolved in 3 ml acetone, then gradually added into the former solution with constant stirring at the same time for uniform mixing.

Preparation of separation gel: separation gel was prepared by mixing reagent ①,②,③ and water in the proportion of 1:2:4:1, with additional 24 ml EDTA.

Preparation of stacking gel: stacking gel was prepared by mixing reagents ④,⑤,⑥,⑧ in the proportion of 1:5:1:4, with additional 33 ml EDTA.

Electrophoresis apparatus type DDY-III2 and electrophoresis tank type DYC-24B were used.

2.3 *Extraction of enzyme solution*

The strains tested were inoculated into enrichment PDA media and were cultured at 25 °C for 15 d. The mycelia of different strains scraped from tubes were put in mortar. Mycelia, quartz sand and stacking gel storage solution were mixed with a proportion of 1:1:1 staying overnight at –20 °C. After removal and grinding to homogenates, the mixture was centrifuged at 4000 r/s for 10 min, and then supernatant was placed in the refrigerator in reserve.

2.4 *Gel production and spotting*

Gel production: The square and concave glasses were washed and dried in the air. Smear vaseline along the edges of glasses (3 edges) and then lay the PMMA bars on the glasses. Vase line was infused to the PMMA bars which were then covered by rectangular glass, pressed

Table 1. The strains tested and their sources.

Number	Name	Source
AP1	Huang er zh	Resources in Sichuan province
AP2	913	The strain factory of Edible Fungi in Sichuan province
AP3	AP6	The strain factory of Edible Fungi in Sichuan province
AP4	Huang10	The strain factory of Edible Fungi in Sichuan province
AP5	AP8	The strain factory of Edible Fungi in Sichuan province
AP6	Dashang3	The strain factory of Edible Fungi in Sichuan province
AP7	AP7	The wild poplar from Sichuan Academy of Agriculture Science
AP8	781	Institution of Fungi in Sanming, Fujian province
AP9	Hybridization one	Guangda edible fungi Center in Jining, Shandong province
AP10	Xiaoshang3	The strain factory of Edible Fungi in Sichuan province
AP11	3039	Yunong in Taiwan province
AP12	43	Institution of Light Industry in Fujian province
AP13	Guangzhen	Institution of Light Industry in Fujian province
AP14	AP5	The strain factory of Edible Fungi in Sichuan province
AP15	Chuan er one	Institution of Soil and Fertilizer, Sichuan Academy of Agriculture Sciences
AP16	915	Institution of Soil and Fertilizer, Sichuan Academy of Agriculture Sciences
AP17	951	The strain factory of Edible Fungi in Sichuan province
AP18	Shanghai one	The strain factory of Edible Fungi in Sichuan province
AP19	243	Institution of Light Industry in Fujian province
AP20	3043	The strain factory of Edible Fungi in Sichuan province
AP21	Jindi	Institution of Soil and Fertilizer, Sichuan Academy of Agriculture Sciences
AP22	Qingyou	The strain factory of Edible Fungi in Sichuan province
AP23	Sanyou	The strain factory of Edible Fungi in Sichuan province
AP24	Teda	The strain factory of Edible Fungi in Sichuan province
AP25	Daguang	Institution of Light Industry in Fujian province
AP26	Purple Auricularia	The strain factory of Edible Fungi in Sichuan province
AP27	Hupo	The strain factory of Edible Fungi in Sichuan province
AP28	50835	Institution of Soil and Fertilizer, Chinese Academy of Agriculture Sciences
AP29	50142	Institution of Soil and Fertilizer, Chinese Academy of Agriculture Sciences
AP30	50269	Institution of Soil and Fertilizer, Chinese Academy of Agriculture Sciences
AP31	50359	Institution of Soil and Fertilizer, Chinese Academy of Agriculture Sciences
AP32	Taimao one	Institution of Light Industry in Fujian province
AP33	50664	Institution of Soil and Fertilizer, Chinese Academy of Agriculture Sciences
AP34	99Feng	Tianda edible fungi institution in Jiangdu, Jangsu province
AP35	Teda A series	Tianda edible fungi institution in Jiangdu, Jangsu province
AP36	Taier319	Tianda edible fungi institution in Jiangdu, Jangsu province
AP37	719	Huanyu edible fungi institution in Jiayu, Hubei province
AP38	AU912	Institution of Fungi in Sanming, Fujian province
AP39	Hybridization 34	Institution of Fungi in Gaoyou, Jiangsu province
AP40	Hybridization Six	Institution of Fungi in Gaoyou, Jiangsu province
AP42	Baibei Auricularia	Huangzhong Agriculture University
AP43	99	Institution of Biology, Henan Academy of Sciences
AP44	Kangza Auricularia	Shifang, Sichuan province
AP45	Hongda er	Edible fungi institution, Shanghai Academy of Agriculture Sciences
AP46	583	Institution of Microbiology, Chinese Academy of Sciences
AP47	584	Institution of Microbiology, Chinese Academy of Sciences
AP48	m1	Institution of Microbiology, in Hebei province

(*Continued*)

Table 1. Continued.

Number	Name	Source
AP50	AP067	Cathaya argyrophylla in Wolong Nature Reserve(Wild)
AP52	8711	The strain factory of Edible Fungi in Sichuan province
AP53	143	The strain factory of Edible Fungi in Sichuan province
AP54	50385	Institution of Soil and Fertilizer, Chinese Academy of Agriculture Sciences
AP55	Huang er one	Institution of Biology, Henan Academy of Sciences
AP56	Jiyou one	Institution of Microbiology, in Hebei province
AP57	M1	Institution of Microbiology, in Hebei province
AP58	APy33	The strain factory of Edible Fungi in Sichuan province
AP63	Shan er one	The strain factory of Edible Fungi in Sichuan province
Chuan 7	Chuan er seven	Institution of Soil and Fertilizer, Sichuan Academy of Agriculture Sciences

hard, held together with stationery iron clamp and placed on the shelves. The separation gel was infused into two glasses and one layer of distilled water was covered afterward. When clear interface appeared between the polymerized gel and water layer, a syringe was used to remove the water layer. Stacking gel was injected 3 mm to the mouth of the glass, and a comb was inserted into the notch. When the gel presented milk white, polymerization was finished, and the comb was taken out carefully.

Spotting: microsyringe was used to draw in 20 μl sample and inserted into the bottom of sample cell. Sample was carefully infused, and then a layer of 10% sucrose was added carefully on the sample.

2.5 *Electrophoresis and staining*

Vertical slab polyacrylamide gel electrophoresis method was used[2]. After electrode buffer was added in the upper and lower tanks, one drop of 2% bromophenol blue was added in the tanks. The tanks were put in the refrigerator levelly and connected to power supply with the upper tank as positive polarity and the lower tank as the negative polarity. Electrophoresis was processed at 4 °C in refrigerator. When the indicator was at stacking gel, voltage was 180 V. After the indicator was in separation gel, voltage was increased to 210 V. A stable gear was adopted for electrophoresis. When indicator was 2 cm to the lower edge of the gel plate, electrophoresis was stopped.

The gel was immediately transferred into dye liquid, and color reaction was processed at 25 °C. After appear the brown esterase isozyme bands, the gel was rinsed several times in water and preserved in 7% acetic acid.

2.6 *Record and analysis of the data*

Drawing of schematic scheme; determination of relative mobility (Rf); photograph; scanning; storage of chart. Relative mobility (Rf): Migration rates of separated bands were demonstrated by relative mobility (Rf), a mark was made on the forward line of the position where the indicator moved to. After staining, the distances indicator and enzyme bands moved were determined. $Rf = X1/X2$ (X1 is migration distance of indicator in the gel before fixed staining; X2: migration distance of enzyme protein in the gel after fixed staining.)

Vernier caliper was used to determine the Rf value by the middle of the enzyme bands. Classification and drawing of the bands appeared on the gel was made. Values were assigned according to with and without bands, 1 for with band and 0 for without band. Jaccared Genetic Similarity (GS) between materials was calculated. Cluster was made with GS value using Unweighted Pair Group Method with Arithmetic cluster analysis (UPGMA) to build the polygenetic dendrogram. Statistical analysis was performed on NTSYS-pc.

3 RESULTS AND ANALYSIS

3.1 *Esterase isozyme zymograms of Auricularia polytricha strains*

3.1.1 *Zymogram types*

Esterase isozyme zymograms of 56 strains are shown in Figure 1. Migration rates of 10 different bands were detected between 0.262–0.655 in the tested strains. The number and migration rates of isozyme bands varied among strains with enzyme bands ranging from 1 to 6. Thus 56 strains were divided into 26 zymogram types which differ in both stains and species.

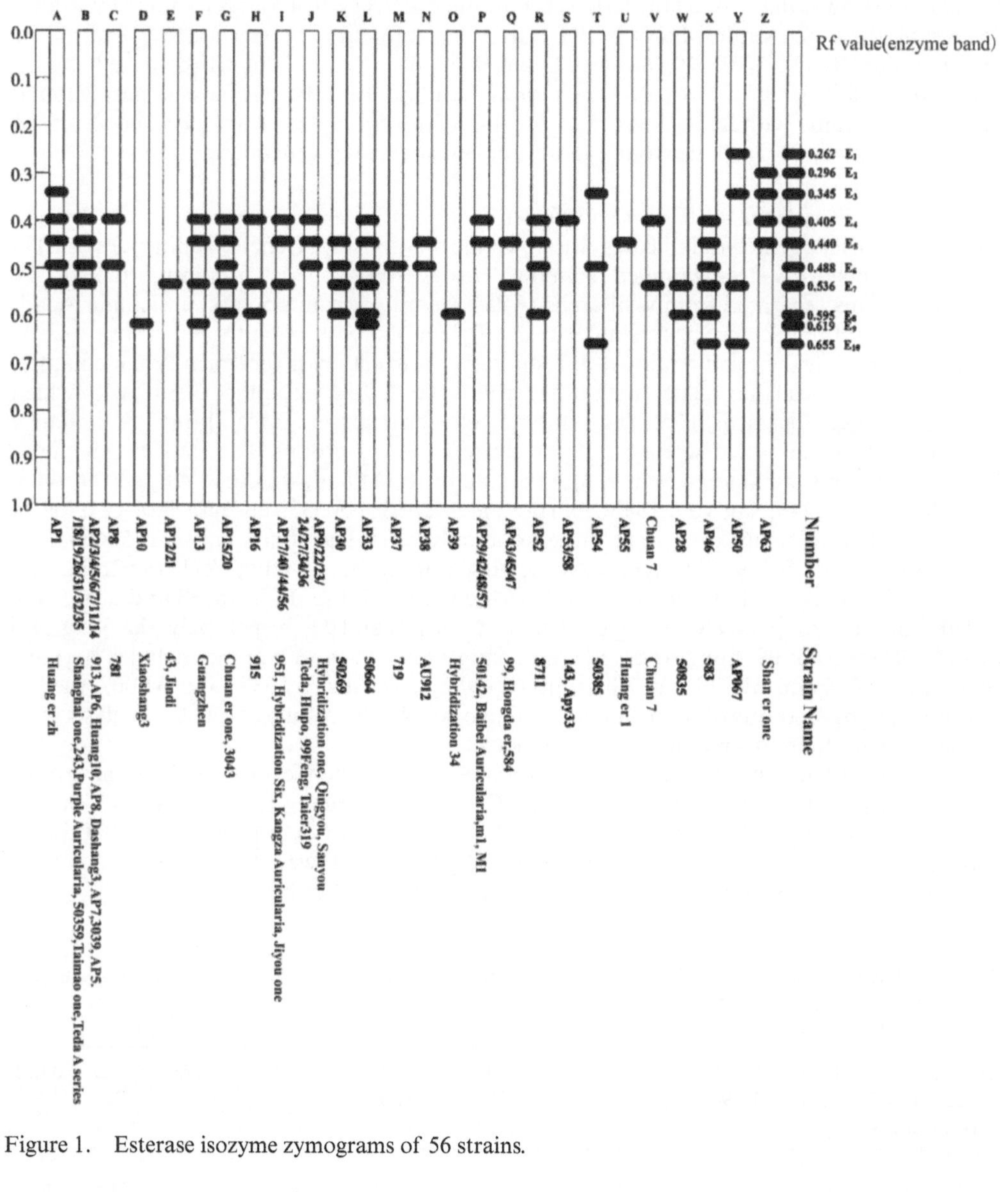

Figure 1. Esterase isozyme zymograms of 56 strains.

The 14 strains that include AP2, AP3, AP4, AP5 fell into zymogram type B, which was most abundant and was 25% of all the strains. Of the 14 strains, 9(64.3%) were from Edible Fungi Institution of Sichuan province and longer culturing history made the phenotypic characteristics of the strains show a larger similarity. Following type B was type J with 7 strains which was 12.5% of all the tested strains and 4(57%) of the 7 strains originated from Edible Fungi Institution of Sichuan province. The majority of the strains with zymograms of type B and J were artificially cultivated strains which were widely planted in Sichuan province. The wild strains separated from poplars in Sichuan Academy of Agricultural Science had the same zymogram with the cultivated strains. Other strains of various zymogram types are from different areas.

3.1.2 *Characteristics and distribution frequencies of the bands*

Distribution frequencies of the enzyme bands were calculated by the esterase isozyme zymogram patterns of *Auricularia polytricha* germplasm resources (Table 2). Band E(Rf = 0.440) was found in 43 strains, 76.8% of all the tested strains. Band E4(Rf = 0.405) was found in 42 strains, 75% of all the tested strains. Band E7(Rf = 0.536) was found in 34 strains, 60.7% of all the tested strains. Band E6(Rf = 0.488) was found in 32 strains, 57.1% of all the tested strains. These were stable bands of *Auricularia polytricha*. Band E8, E9, E10, E1, E2 were hardly found in the tested strains, with frequencies of 16.1, 7.14, 5.36, 3.57, and 1.79% respectively. None of the bands was shared by all the strains, indicating high degree of genetic variation.

3.1.3 *Genetic similarity coefficient and cluster analysis of esterase isozyme*

Higher similarity coefficient indicates closer genetic relationship. Genetic Similarities (GS) in any two strains of 56 strains ranged from 0 to 1, suggesting genetic variation in the strains. The two strains separated from wild areas, AP47 separated from poplars in Sichuan Academy of Agricultural Science has a similarity coefficient of 1 with 14 cultivated strains such as Ruhuang 10, while the other strain AP067 separated from cathay silver firs in Wolong natural reserve was genetically far related with other strains (GS < 0.24) except for 50385, with which the similarity coefficient is 0.75. The results of cluster analysis (Table 2) in 67% similarity level showed that all tested strains could be clustered into 8 groups. The 1st group includes 34 strains which were huang-er zh, 913, AP6, huang 10, AP8, da-shang 3, AP7, 3039, AP5, shanghai No. 1, 243, san-you, te-da, purple muer, 50359, 50385, tai-mao No. 1, te-da A strain, 951, hybrid No. 6, kang-za muer, ji-you No. 1, tai-er 319, chuan-er No. 1, 50664, 583, guang-zhen, 3043, 50269, 781, hybrid No. 1, qing-you, hu-po, 99 feng, 8711; the 2nd group includes 21 strains of white back moer, m1, M1, chuan 7, 50142, 143, Apy33 and 50835; the 3rd and the 4th group only have single strain of *A. auricla* and 915 respectively; the 5th group includes 4 strains of 99, hong-da-er, 584, and hang-er No. 1; the 6th group includes 4 strains of 43, jin-di, 719 and AU912; the 7th group including 2 strains of 50385 and AP067; the 8th group including 2 strains of xiao-shang 3 and hybrid 34. The results of cluster analysis were consistent with those from morphology analysis.

It was also indicated by dendrogram in Figure 2 that cluster relationships of the strains were not related to their geographic origins. For example, huang-er zh, 3039, 243, 50359, te-da A strain, ji-you No. 1 and hybrid No.1 in the 1st group were from Sichuan, Taiwan, Fujian, Beijing, Jiangsu, Hebei and Shandong respectively. However, strains such as jin-di, 915, 913, Apy33, 143, chuan 7 from Sichuan were divided into different groups.

Table 2. Distribution frequencies of the enzyme bands of *Auricularia polytricha* germplasm resources.

Enzyme band	E1	E2	E3	E4	E5	E6	E7	E8	E9	E10
Mobility	0.262	0.292	0.345	0.405	0.440	0.488	0.536	0.595	0.619	0.655
Frequency of occurences	1.79	1.79	7.14	75.0	76.8	57.1	60.7	16.1	5.36	3.57

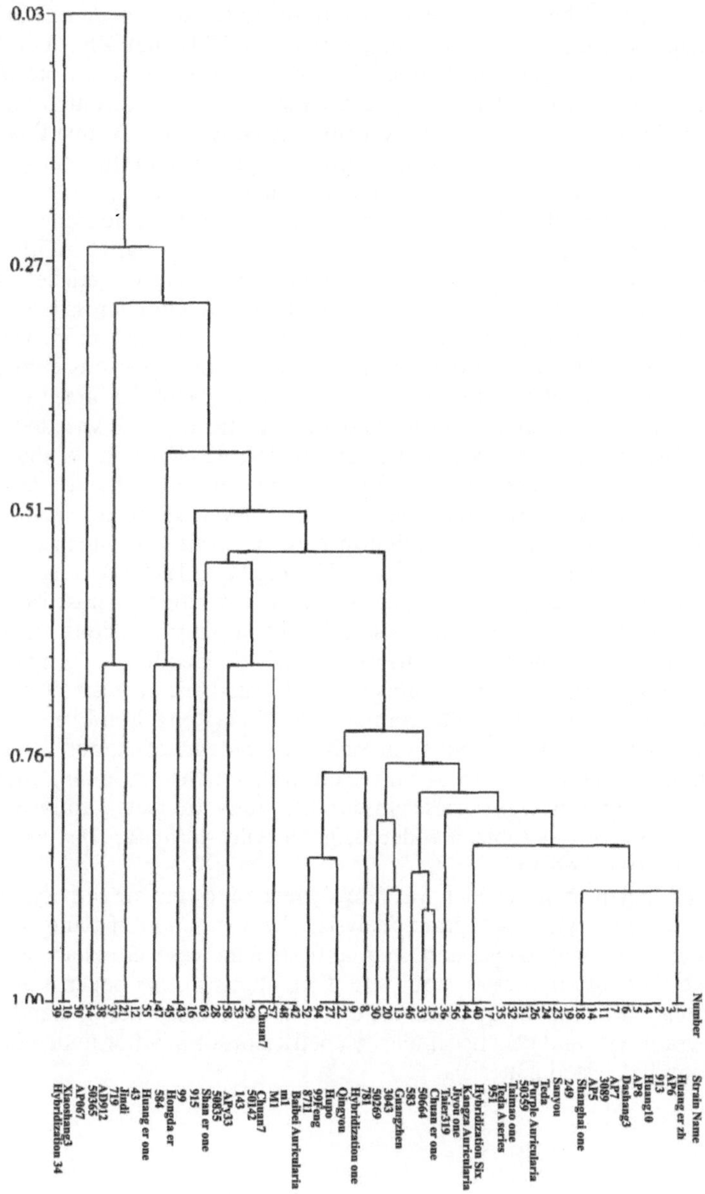

Figure 2. The dendrogram is concluded by cluster analysis on esterase isozyme of strains.

4 CONCLUSION

As the encoding product of genes, the amino sequences in the polypeptide chains of the isozyme reflect base sequences in DNA chains (via RNA), and express information at DNA molecular level. As one of the genetic marker techniques, the isozyme analysis improves traditional fungi classification which based on colony morphology, color, spore morphology, fruiting body morphology and so on, and becames an important method in the field of genetic differentiation at protein level and were widely used for the identification and genetic diversity researches of edible fungi (Suyue Zheng et al. 2003, Zesheng Wang 1991, Royse DJ & May B 1982,

Jian hang Jia et al. 1997). There have been some reports on the species identification of edible fungi by using esterase and peroxidase (Junling Wang et al. 2004, Jian Zhu 2000).

Record of *Auricularia auricular* cultivation of China could be dated back to 600 years ago (Shuting Zhang & Fangcan Lin 1997), and *Auricularia polytricha* cultivation and promotion came into being in Sichuan province since late 1970s and early 1980s. At that time yellow back muer introduced by Fujian Light Industry Institution from Taiwan was the main variety. As one of the leading organizations in agricultural soil microorganism research, promotion and application, Microorganism Centre of Soil and Fertilizer Institution, Sichuan Academy of Agricultural Science played the role of an initiator for the promotion of the new variety. With the extension of production and development of research work, more new strains or varieties were obtained through different means such as tissue isolation, variety introduction, cross breeding, induced mutation, cell fusion and so on. In the process of breeding and selection, more new varieties on genetic level were obtained because of the influence of cultivation conditions, cultivation history and different ways of strain introduction. The new strains may be of the same variety as the old ones in spite of their different names. During 20 years of cultivation and selection, some strains may have a little change with great genetic similarities. These may be one of the main reasons that cluster results in this test were not related to geographic origins of the strains.

Slight variations in enzyme structures may lead to differentiations of electrophoresis patterns, which make it very useful for identifying species and strains in spite of their similarities in morphology and physiology (Juxing Hetian & Shuang Fang 1989). As an effective method for identifying strains in species and sub-species, it was shown that esterase isozyme analysis was a practical and effective method in the solution of strain confusions. In this paper, by esterase isozyme analysis method, 56 strains were clearly divided into 8 groups.

Apparent differences between wild strains and cultivated strains were also confirmed in this paper, which agrees well with results from other studies (Rongchun Li et al. 2002). As for the fact that AP47 separated from poplars in Sichuan Academy of Agricultural Science had genetic similarity coefficient of 1 with 14 cultivated strains such as ru-huang 10, the explanation may be that the separation sites were close to the cultivation sites, and were affected by the cultivated strains. For this topic, a wider range of wild *Auricularia polytricha* collection will be done for a further research.

The strength of enzyme bands, which was directly related to enzyme activity, should also be considered in esterase isozyme zymogram analysis. The distinction of band strength largely depends too much on the naked eyes and no quantitative indices were established for precise measurement, so there are no descriptions and distinctions in this paper. Further research should pay more attention to the quantitative distinction of band strength and its combination with zymogram type and genetic similarity coefficient for a better understanding of the genetic background of tested materials.

REFERENCES

Anderson IC, Chambers SM, Cairney JWG, Mycol Res, 1993, 102:295–300.

Jianhang Jia, Guoxia Liu, Liyun Li. Journal of Agricultural University of Hebei, 1997, 20(1): 1–5.

Jian Zhu, Yushan Huang, Ping Jiang. Fujian Journal of Agricultural science, 2000, 15(3): 46–50.

Juxing Hetian(Author), Shuang Fang (Translator). Experimental Methods of Microbial Chemical Classification. Guizhou: Guizhou People Press. 1989, 266–276.

Junling Wang, Ming Li, Jinghua Tian. Journal of Agricultural University of Hebei, 2004, 27(3): 29–32.

Rongchun Li, Xuqing Liang, Zhilei Yang. Edible Fungi of China, 2002, 21(4): 34–36.

Royse DJ, May B. Mycologia, 1982, 74: 93–102.

Shuting Zhang, Fangcan Lin, Genetics and Breeding of Mushroom, Beijing: China Agriculture Press, 1997, 9.

Suyue Zheng, Jinxia Zhang, Chenyang Huang. Acta Edulis Fungi, 2003, 10(4): 1–6.

Wanneng Hu, Xianguo Wan, Isoenzyme Technique and Application. Changsha: Hunan Science and Technology Press, 1985, 10–13.

Xun Li, Yihuai Hu, Jianjun Pei. Edible Fungi of China, 2003, 22(4): 6–8.

Zesheng Wang, Fu Jian Edible Fungi, 1991, 1(2): 20–2.

Modeling and Computation in Engineering II – Xie (ed)
© 2013 Taylor & Francis Group, London, ISBN 978-1-138-00058-2

Investigation of diffusion of CO_2 in decane-saturated porous media

H.F. Zheng, Y.C. Song, Y. Liu, M. Hao, Y.C. Zhao, B. Su, Z.J. Shen & L.Y. Chen
Key Laboratory of Ocean Energy Utilization and Energy Conservation of Ministry of Education, Dalian University of Technology, Dalian, P.R. China

ABSTRACT: The diffusion coefficient of gas in oil-saturated porous media is a critical parameter in petroleum industry. Unfortunately, there are still not a standard method to measure the diffusion coefficient of gas in oil-saturated porous media. One difficulty is how to determine the tortuosity, which is essential for the calculation of diffusion. The main object of this work was to obtain the diffusion coefficient of gas in oil-saturated porous media by Dual-chamber Pressure-decay and CT technique. At last the relationship of pressure, tortuosity and diffusivity at T = 23.5 °C was discussed.

1 INTRODUCTION

CO_2 injection has been widely practiced to improve oil recovery since the early 1900s and proved to be an excellent contributor to the light oil EOR. The diffusion coefficient of the gas phase is critical for the prediction of the miscible process, and is essential to project design, risk assessment, economic evaluation for oil recovery. Therefore, relevant experimental study of gas-liquid diffusion in porous media is of much importance. Pomeroy et al. (1933) evaluated the diffusion coefficient and the solubility of lighter oil in quiescent liquids. They concluded that diffusion coefficient is not directly determined by the pressure or concentration of methane gas in solution while the pressure ranges under 300 psi (~2 MPa). Reamer et al. (1956) calculated the diffusion coefficient of methane in hydrocarbon mixtures through the method of assuming a resistance at the interface, discovered to be proportional to the mass transfer rate. Renner (1988) determined the diffusion coefficient by measuring the volume of dissolved gas in the solution with time at constant pressure and found the phenomenon of incubation period. Riazi (1996) developed a PVT experimental method and established a semi-analytical model to estimate the gas diffusion coefficient. In this method, the rate of change of pressure and the interface position as a function of time is considered to depend on the rate of diffusion in each phase, in other word, on the diffusion coefficients. Zhang et al. (2000) measured the rate of decrease in the pressure of a constant volume of gas in heavy oils. Their technique is improved on the base of the method by Riazi (1996), not necessary to measure the interface position with time. The diffusion equation was coupled with the gas material balance equation, and to match the gas absorption data using the diffusion coefficient as an adjustable parameter, while the gas compressibility factor was assumed to be constant the effect of bitumen swelling and the resistance of interface were neglected. Tharanivasan et al. (2004) used the approach of Zhang et al. (2000) and compared with the result of Riazi (1996) and Zhang et al. (2000). They found that the different interface boundary conditions resulted in different values of diffusion coefficient for the same data.

Recently, the PVT method (i.e. Pressure-decay method) has become more and more prevalent, and many researches are based on it, such as the dual-chamber technique and Zhang et al. (2000)'s method. However, both of them didn't measure the diffusion in porous media. The aim of this paper is using CO_2 and decane to get the diffusion coefficient of gas in oil-saturated porous media by Dual-chamber Pressure-decay technique. The relationship of pressure, tortuosity and diffusivity at T = 23.5 °C will also be discussed.

2 EXPERIMENTAL SECTION

2.1 *Measurement of diffusion coefficient*

Schematic diagram of the test apparatus is shown in Figure 1. It mainly consists of a diffusion cell, a high-pressure cylinder, an ISCO pump, a high-accuracy digital pressure gauge, and a data acquisition system.

The diffusion cell (with a volume of 40 ml) was used to hold the 1 cm diameter test tube which was filled with oil-saturated sand packed porous media, and placed in the vertical position. The gas cylinder used to supply pressurized CO_2 was with a volume of 140 ml. Both the diffusion cell and gas cylinder ware placed in water bath. A high-accuracy was used to measure the pressure of system during the experiment. The ISCO pump was an automatic gas injection pump, which connected to the bottom of the gas cylinder.

The procedure for measuring diffusion coefficients was as follows. First, the desired oil-saturated sand packed of the height of 2 cm was put in the test tube, place the test tube into the diffusion cell and connected the apparatuses as Figure 1 shows. After testing the system for leakage using nitrogen at about 6 MPa, a vacuum was applied for 24h. The temperature of system was controlled using water bath. Then CO_2 was injected the into the gas cylinder with desired pressure slowly using ISCO pump. The ISCO pump could record the total amount of gas which is extremely significant to the calculation of diffusivity. Next, the valve connected to the gas cylinder and diffusion cell was turned on and the gas was introduced to the diffusion cell The data acquisition system started to work until the equilibrium was reached. The CO_2-decane diffusion coefficients were eventually determined by the recorded data.

2.2 *Measurement of tortuosity*

The measurements of tortuosity were conducted on a laboratory micro-CT (InspeXio SMX-225CT, at Dalian University of Technology). The maximum object size that can be scanned in the InspeXio SMX-225CT instrument is 200 mm in diameter and 300 mm high. The detector is a cooled CCD camera with a dynamic range of 16 bit (65536 intensity levels). In this study, there are two kinds of samples, both of them are sand packed bed, one is BZ-3 glass beads (short for BZ-3) with a 2.600–3.500 mm diameter and the other is 20–40 mesh number refined quartz sand (short for 20–40 m).

The image of pore structure could be obtained using the micro-CT. By the calculation of Mathematic the effective area and pore network model was extracted. And then random walk technique was applied to acquire the tortuosity of the samples. Thus, Table 1 shows the experimental result and calculation result.

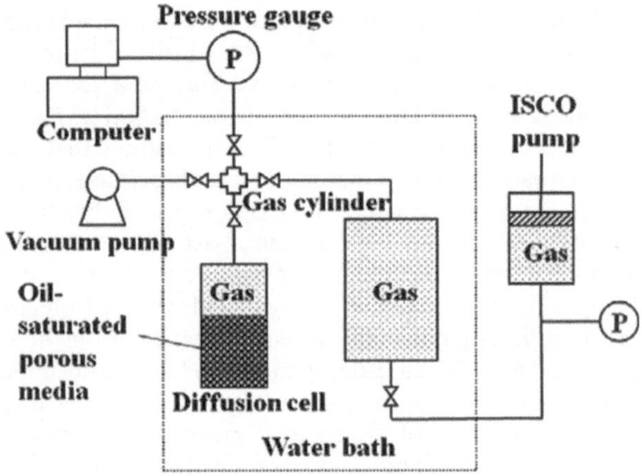

Figure 1. Schematic diagram of the experimental setup for diffusion coefficient measurement.

Figure 2. The image of pore structures of BZ-3 sand packed samples.

Figure 3. The image of pore structures of 20–40 m sand packed samples.

Table 1. Parameters of BZ-3 and 20–40 m sand packed samples.

Samples	BZ-3	20–40 m
Porosity	40.3	37.47
Tortuosity	1.65	2.06

According Yu and Li's approximate model (2004), the averaged tortuosity is given by

$$\tau = \frac{1}{2}\left[1 + \frac{1}{2\sqrt{1-\phi}} + \sqrt{1-\phi}\frac{\sqrt{\left(\frac{1}{\sqrt{1-\phi}}-1\right)^2 + \frac{1}{4}}}{1-\sqrt{1-\phi}}\right] \quad (1)$$

ϕ is the porosity of porous media. Use equation (1), the values of tortuosity are 1.8851 (20–40 m) and 1.8094 (BZ-3). It can be seen that the differences between these two methods are less than 10%.

3 DATA ANALYSIS

The following assumptions are made for the analysis of the diffusion process:

1. The CO_2 effective diffusion coefficient is constant during the pressure decay experiment.
2. There is no resistance to mass transfer at the gas-liquid interface, the concentration at the interface is the equilibrium.
3. Temperature remains constant during the experiment, and the swelling of the liquid phase is ignored.

4. The density-induced and swelling-induced natural convection is negligible.
5. The liquid phase is non-volatile.

Based on one-dimensional second Fick's law and continuity equation, Guo et al. (2004) had obtained the following equation:

$$D = \frac{k^2\pi}{4A^2C_0^2} = \frac{k^2\pi}{4(\phi Aseaction)^2\tau^2C_0^2} \tag{2}$$

Where D represents the CO_2 in porous media diffusion coefficient, and C_0 represents the mole concentration of the gas in liquid phase when the experiment began. ϕ and τ is porosity and tortuosity respectively. k is obtained by the relation of the amount of diffused CO_2 and quadratic root of time.

4 RESULTS AND DISCUSSION

The purities of CO_2 and decane are 99.999% and 98% respectively. The system temperature controlled by water bath is 23.5 °C. Six measurements totally were conducted for CO_2 diffusion in decane-saturated porous media as indicated in Table 2.

Figures 4 and 5 show that the diffusion coefficient exhibits a slowly increasing trend with increasing pressure. This trend reflects that the initial pressure have a great influence on diffusion coefficient, and as the pressure rises, the increasing ratio of diffusion coefficient slow down.

At the same time, it can be seen that the diffusion coefficient in the 20–40 m is lower than in the BZ-3 when the initial pressure is similar. It is reasonable because the tortuosity of the former is larger than the latter, and the less one is more close to bulk liquid.

Table 2. Summary of the measurements of CO_2 diffusion coefficients in decane-saturated porous media at 23.5 °C.

Samples	Initial pressure (kPa)	D (10^{-9} m²/s)
BZ-3	2614	2.89
	3552	3.12
	5003	3.43
20–40 m	2317	1.36
	3655	2.62
	4900	3.51

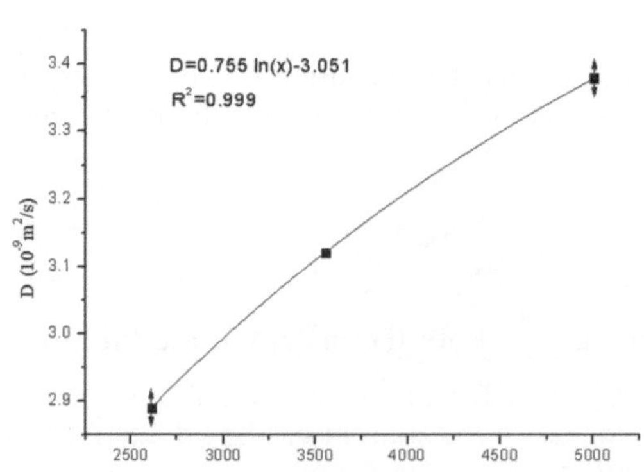

Figure 4. Diffusion coefficient of CO_2 in BZ3 under different pressure.

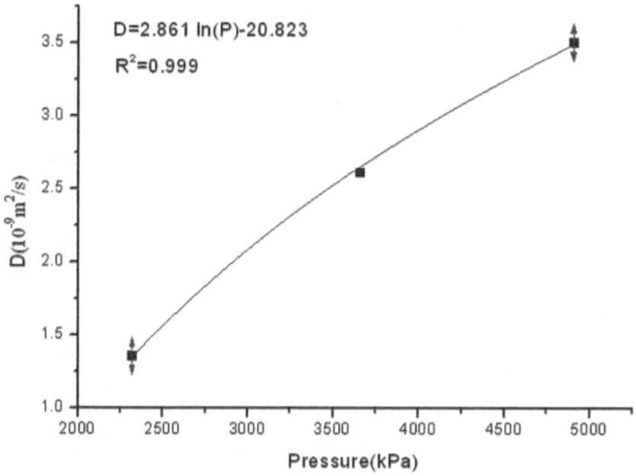

Figure 5. Diffusion coefficient of CO_2 in 20–40 m under different pressure.

5 CONCLUSIONS

Dual-chamber Pressure-decay was used to record the pressure-decay data. And CT technique, as a new technique, was used to determine the tortuosity of porous media, which is approximately with Yu and Li's model. Based on these results obtained, the diffusion coefficients of CO_2 in decane-saturated porous media have been calculated. This work discussed the relationship of initial pressure and diffusion coefficients at the same temperature. The effect of tortuosity to diffusion has been also indicated.

ACKNOWLEDGEMENT

This study has been supported by Key program of National Natural Science Foundation of China (Grant No. 51106019, No. 51206018), the National High Technology Research and Development of China (863) Program (Grant No. 2009AA63400), the National Basic Research Program of China (973) Program (Grant No. 2011CB707300), the Specialized Research Fund for the Doctoral Program of Higher Education (Grant No. 20120041120057), and the Ph. D. Startup Foundation of Liaoning Province, China (Grant No. 20121022).

REFERENCES

Guo B, Hou J.R. and Yu C.L. et al. 2009. Determination of diffusion coefficient of CO_2 in porous media. Journal *of Petrochemical Universities.* 22(4): 38–40 (in Chinese).

Pomeroy, R.D.; Lacey, W.N.; Scudder, N.F.; Stapp, F.P. 1933. Rate of Solution of Methane in Quiescent Liquid Hydrocarbons. *Ind. Eng. Chem.* 25, 1014–1019.

Reamer, H.H.; Opfell, J.B.; Sage, B.H. 1956. Diffusion Coefficients in Hydrocarbon Systems Methane-Decane-Methane in Liquid Phase. *Ind. Eng. Chem.* 48, 275–282.

Renner, T.A. 1988, Measurement and Correlation of Diffusion Coefficients for CO_2 and Rich Gas Applications. *SPE ReserVoir Eng.* (May), 517–523.

Riazi, M.R., 1996. A new method for experimental measurement of diffusion coefficients in reservoir fluids. *Journal of Petroleum Science and Engineering.* 14, 235–250.

Tharanivasan, A.K., Yang, C., Gu, Y., 2004, Comparison of three different interface mass transfer models *used* in the experimental measurement of solvent diffusivity in heavy oil, *Journal of Petroleum Science and Engineering,* vol. 44, no. 3–4: p. 269–282.

Yu B.M. and Li J.H. 2004. A Geometry Model for Tortuosity of Flow Path in Porous Media.*CHIN. PHYS.LETT.* 21(8).

Zhang, Y.P.; Hyndman, C.L.; Maini, B.B. 2000. Measurement of Gas Diffusivity in Heavy Oils. *J. Pet. Sci. Eng. 25*, 37–47.

Modeling and Computation in Engineering II – Xie (ed)
© 2013 Taylor & Francis Group, London, ISBN 978-1-138-00058-2

A numerical simulation study on horizontal well hydraulic fracturing

Guangming Zhang, Chunming Xiong, Jiandong Liu & Juan Jin
Research Institute of Petroleum Exploration and Development, PetroChina, Beijing, China
Key Laboratory of Oil and Gas Production, CNPC, Beijing, China

Yuguang Geng
PetroChina Huabei Oilfield Company, Renqiu, Hebei, China

ABSTRACT: Horizontal well hydraulic fracturing is a complex technology and there is no clear understanding about the mechanics of hydraulic fracturing. The Fluid-solid coupling elements were used to describe the behavior of rock, and the pore pressure cohesive elements were employed to simulate the process of fracture initiation and propagation. The fluid flow in the fractures and fracturing fluid leakoff effects were also taken into account. By adopting field data, the staged fracturing process of a horizontal well in Daqing Oilfield, China was simulated. The simulated results are fit well with the field treatment data.

1 INTRODUCTION

Hydraulic fracturing can be broadly defined as the process by which a fracture initiates and propagates due to hydraulic loading applied by a fluid inside the fracture. Even in the most basic form, hydraulic fracturing is a complicated process to model, as it involves the coupling of at least three processes (Adachi et al. 2007): (1) the mechanical deformation induced by the fluid pressure on the fracture surfaces; (2) the flow of fluid pressure on the fracture surfaces; (3) the fracture initiation and propagation.

Due to the complication of hydraulic fracturing, mathematical solutions are impossible. Usually, numerical simulation methods are used to study the behaviors of hydraulic fracturing, however, many assumptions and simplifications are made in many model to solve the problem of complicated hydraulic fracturing. Perkins and Kern (Perkins & Kern 1961) adapted the classic Sneddon plane strain crack solution to develop the so-called PK model. Later, Nordgren (Nordgren 1972) adapted the PK model to formulate the PKN model, which included the effects of fluid loss. Khristianovic and Zheltov, and Geertsma and de Klerk (Geertsma & de Klerk 1969) independently developed the so-called KGD (plane strain) model. Pseudo-3D (P3D) model (Weng 1992) and planar 3D (PL3D) model (Siebrits & Peirce 2002; Jeffery & Bunger 2007) are based on PKN model and KGD model, and also have a few of assumptions and simplifications. There have also been attempts to model fully 3D hydraulic fractures (Setaari & Cleary 1984) with limited success. The computational burden on such coupled systems is still excessive, even with today's powerful computational resources.

Varieties of numerical methods (Fischer et al. 1994; Siebrits et al. 2001; Smith et al. 2001; Miskimins & Barree 2003; Hustedt et al. 2006) have been developed for hydraulic fracture simulations. With the fast development of computer technology over the past decades, the finite element analysis method is more and more used in geotechnical engineering. In this paper, a non-linear full fluid-solid coupling finite element model was proposed with the finite element software ABAQUS, the fluid-solid coupling theory is used to capture the behavior of rock, the damage mechanics criterion is adopted to simulate the fracture initiation and propagation. The fluid flow in the fractures and fracturing fluid leakoff effect are also taken into account. Based on the field data, the finite model was established and the staged fracturing process of a horizontal well in Daqing Oilfield, China was simulated with the model.

2 THE BASIC EQUATIONS

2.1 *Fluid-solid coupling equation of rock*

The equilibrium equation of rock is expressed by writing the principle of virtual work (Xue et al. 2006)

$$\int_{\Omega} \delta\boldsymbol{\varepsilon}^{\mathrm{T}} \boldsymbol{\sigma} \mathrm{d}\Omega - \int_{\Gamma} \delta\mathbf{u}^{\mathrm{T}} \mathbf{t} \mathrm{d}\Gamma - \int_{\Omega} \delta\mathbf{u}^{\mathrm{T}} \mathbf{b} \mathrm{d}\Omega = 0 \tag{1}$$

where $\delta\boldsymbol{\varepsilon} = \mathrm{sym}(\partial\delta\mathbf{u}/\partial\mathbf{x})$ is virtual rate of deformation, **t** and **b** are the surface traction per unit area and body force per unit volume respectively.

2.2 *Continuity equation of liquid flow in porous medium*

ABAQUS has capabilities for the treatment of single phase flow through porous media. A continuity equation is required for the fluid, equating the rate of increase in fluid volume stored at a point to the rate of volume of fluid flow into the point within the time increment.

$$\frac{\mathrm{d}}{\mathrm{d}t}\left(\int_{\Omega} \rho n \mathrm{d}\Omega\right) = -\int_{\Gamma} \rho n \mathbf{n} \cdot \mathbf{v} \mathrm{d}\Gamma \tag{2}$$

where ρ and n are the density of the liquid and the porosity of the medium. **v** is the seepage velocity.

2.3 *Pore pressure cohesive element in ABAQUS*

Pore pressure cohesive element is used to simulate hydraulic fracture in software ABAQUS. The element has two functions, one is to catch the behavior of fracture initiation and propagation, the other is to simulate the fluid tangential flow within fracture and fluid normal leakoff.

Fracture initiation and propagation is simulated by a traction-separation law, as depicted in Figure 1. The initial response of cohesive element is assumed to be linear (Camanho & Davila 2002). Once the traction of cohesive element reaches the damage initiation value, material damage occurs. The traction monotonically degrades as the separation of cohesive element increases after damage initiation (Turon et al. 2006).

For the quadratic stress mode, damage initiation is expressed by assuming a quadratic interaction function of involving the ratios of stresses (tractions) reaches the value of one.

$$\left\{\frac{\langle t_n \rangle}{t_n^o}\right\}^2 + \left\{\frac{t_s}{t_s^o}\right\}^2 + \left\{\frac{t_t}{t_t^o}\right\}^2 = 1 \tag{3}$$

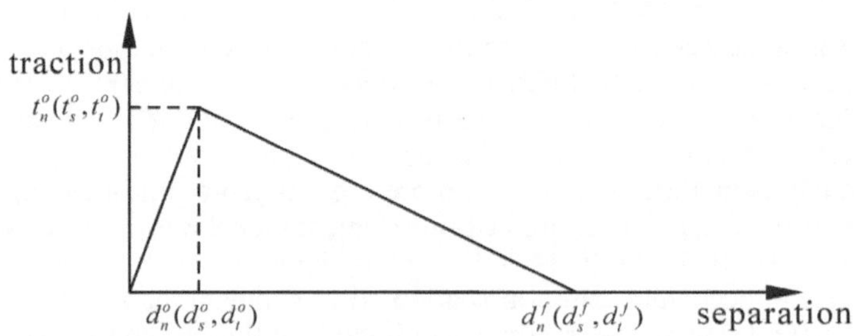

Figure 1. Typical traction-separation response.

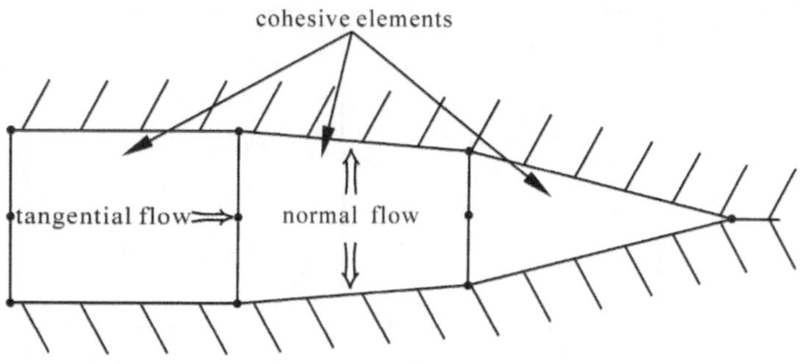

Figure 2. The flow patterns of pore fluid in cohesive elements.

where t_n^o, t_s^o, and t_t^o are the tensile strength and shear strengths in the corresponding directions, respectively.

For pore pressure cohesive element, there are fluid tangential flow and normal flow, as depicted in Figure 2.

The fluid tangential flow is described as

$$\boldsymbol{q}d = -\frac{d^3}{12\mu}\nabla p \tag{4}$$

where $\boldsymbol{q}$ is the volume flow rate density vector, d is the opening displacement of cohesive element.

The fluid normal flow is described as

$$u_l = c_l(p_i - p_o) \tag{5}$$

where u_l is the fluid leakoff velocity, c_l is the leakoff coefficient, p_i and p_o are the internal pressure and outer surfaces pore pressures, respectively.

3 SIMULATION MODEL

A staged fracturing process of a horizontal well in Daqing Oilfield, China is simulated with the software ABAQUS. The field data of Daqing Oilfield are adopted in computation. The true vertical depth of the well is 1592.05 m, the drilled well depth is 2345 m. Provided that the shape of hydraulic fracture is axial symmetry about the wellbore center line, it will be needed to establish a finite element model only containing half of the meridian ellipse, as depicted in Figure 3. The dimension of the model is 300 m and 200 m in the X, Y directions, respectively. Perforation, wellbore, cement sheath, reservoir, micro-annulus and transverse fracture are included in the model. The diameter of perforation is 8.8 mm, the outer diameter and the thickness of wellbore is 139.7 mm and 7.72 mm, respectively. The outer diameter of cement casing is 200 mm. In-situ stresses of reservoir in the X, Y directions, are −11.8 MPa, −21.8 MPa, respectively. The saturation and the porosity of formation is 1 and 0.2, respectively. The initial pore pressure of formation is 14 MPa. All the normal direction displacements of outer boundary surfaces of the model are restricted and the outer boundary keeps 14 MPa pore pressure during the process of simulation. The elastic modulus and poisson's ratio of wellbore are 210 GPa and 0.3, respectively. The elastic modulus and poisson's ratio of cement sheath are 30 GPa and 0.25, respectively. The reservoir geologic parameters and the material properties of cohesive element in transverse fracture and micro-annulus are listed in Tables 1 and 2, respectively. Number of nodes and elements in the numerical model are 79 622 and 77 653, respectively.

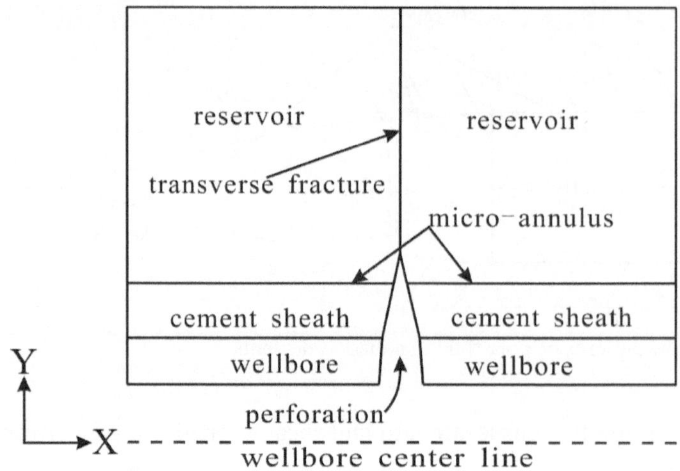

Figure 3. Schematic plot of the model.

Table 1. Reservoir geologic parameters.

	Elastic modulus (GPa)	Poisson's ratio	Permeability (mD)	Fluid specific gravity ($N \cdot m^{-3}$)
Reservoir	35	0.22	2	8624

Table 2. Material properties of cohesive element in transverse fracture and micro-annulus.

	Elastic modulus (GPa)	t_n^o (MPa)	t_s^o (MPa)	t_t^o (MPa)	d_n^f (mm)
Transverse fracture	35	4	1.5	1.5	5
Micro-annulus	35	6	2	2	5

Cohesive elements are embedded in reservoir to describe the process of transverse fracture initiation and propagation during treatment history. Cohesive elements are embedded between cement casing and pay zone to catch the behavior of micro-annulus fracture. Transverse fracture will initiate and propagate according to the principle of least principal stress. Transverse fracture and micro-annulus fracture both connect to the perforation, so they could initiate and propagate simultaneously under the injected hydraulic loading. All the cohesive elements are undamaged and the opening displacements are zero initially. The fractures volume increase as cohesive elements damage and fail according to the damage initiation criterion and corresponding damage evolution law and a typical T-shaped fracture is likely to occur.

The treatment design also considers a pumping schedule for both fluid and proppants. As the proppant-laden fluid is injected, there will be an interaction of solid particles and fluid. Consideration of these effects in detail is challenging, it is a common practice to "lump" all these effects into a modified viscosity of the slurry, which is usually expressed as (Barree & Conway 1994)

$$\mu = 0.1 \times (1 - c/0.65)^{-1.7} \tag{6}$$

where μ is the viscosity of proppant-laden fluid, c is proppant concentration.

ABAQUS field variable technology and user subroutine UFIELD are used to implement equation (6).

4 RESULTS

The radial fracture geometry is obtained because the finite element model is axis symmetry about wellbore center line. Figure 4 presents pore pressure distribution and the fracture configuration at the final moment of treatment history. The deformation magnification factor is taken 400 and the axis symmetry model is revolved 180 around the wellbore center line for clear observation of the simulation results. At the beginning of simulation, cohesive elements both in micro-annulus and in transverse direction damage simultaneously, a T-shaped fracture generates. At the later time, micro-annulus disappears due to the large stress concentration factor near wellbore. Then, only the transverse fracture remains and propagates. The half length and width of the transverse fracture are 98.1 m and 10.34 mm, respectively.

Fracturing fluid is pumped into the wellbore, part of fracturing fluid leaks into formation and the other leaves in the fracture. Fracturing fluid leaks into formation, which results in increasing formation pore pressure, formation effective stress increases according to the effective stress principle of porous medium. When the formation effective stress rises up to formation tensile strength, hydraulic fracture initiates and propagates forward a small distance, which produces a new fracture tip, in which the formation effective stress is lower than the formation tensile strength. Hydraulic fracture will not propagate until the effective stress at the new fracture tip reaches the formation tensile strength again. As more fracturing fluid is injected into the fracture, the width of fracture enlarges and the effective stress at the new fracture tip increases. Hydraulic fracture will propagate once more when the effective stress at the new fracture tip reaches the formation tensile strength. When the normal stress of a cohesive element rises up to the tensile strength, the cohesive element initiates damage and with further loading the stiffness of cohesive element decreases monotonically as depicted in Figure 1.

Fracturing fluid flow rate keeps 3.46 m^3/min and lasts for about 30 min. The results of the simulated and field measured treatment history are plotted in Figure 5. The proppant concentration and fracturing fluid flow rate in simulation are taken as the same as in the

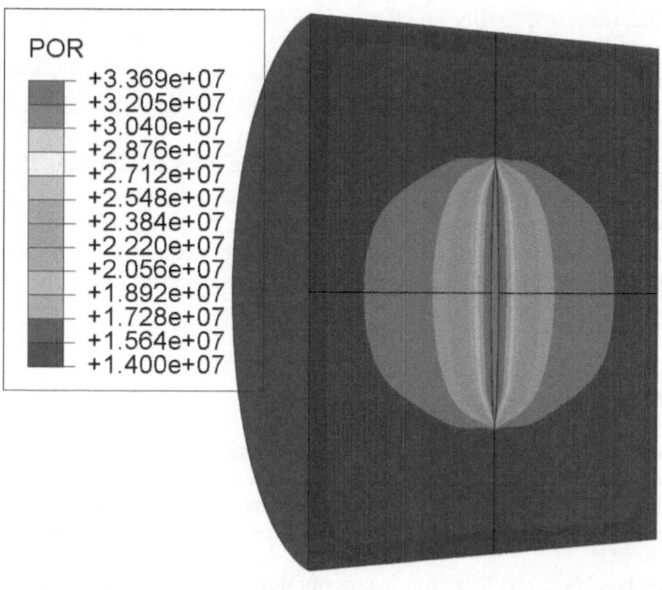

Figure 4. Pore pressure distribution in the model at the final moment of treatment history.

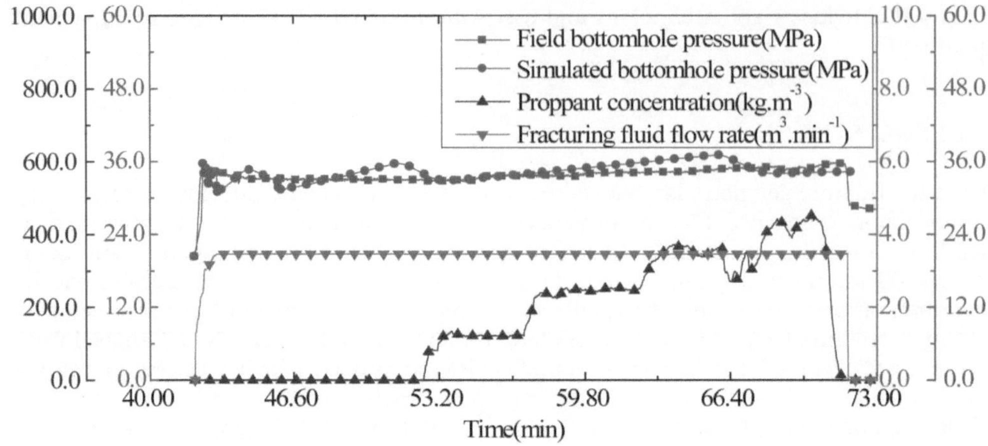

Figure 5. Curves of field treatment history and simulated bottomhole pressure history.

field implementation process (presented also in Fig. 5). The evolution of the bottomhole pressure obtained from the simulation is consistent very well with the corresponding field data, which approves the validation of the proposed finite element model. The evolution of bottomhole pressure is the most important judgment of success or failure of a hydraulic fracturing treatment. The obtained results demonstrate that the proposed model is applicable to hydraulic fracturing designs and treatments for reservoirs lithology similar to Daqing Oilfield, China.

5 CONCLUSIONS

Hydraulic fracturing analysis is inherently a complicated problem, in which fluid flow, deformation of porous medium and fracture initiation and propagation are fully coupled, mathematical solutions are impossible. Usually, Numerical simulation method is employed to catch and study the characters in the process of hydraulic fracturing. Finite Element Analysis (FEA) has been approving an effective method to simulate and forecast the process of hydraulic fracturing.

Multiple transverse fractures can be generated in the process of horizontal well hydraulic fracturing, the stimulation effect is more considerable than vertical well. It is significant to study the mechanics of horizontal well hydraulic fracturing.

A non-linear fluid-solid coupling finite element model was established with the finite element software ABAQUS. A staged fracturing process of a horizontal well in Daqing Oilfield, China is simulated with the model. The initiation and propagation of hydraulic fractures are simulated by using the cohesive element based on damage mechanics. A good match between simulation results and field measurement data is obtained. Validation of the numerical model is approved.

REFERENCES

Adachi, J., Siebrits, E., Peirce, A., Desroches, J. 2007. Computer simulation of hydraulic fractures. *International Journal of Rock Mechanics & Mining Sciences* 44(5): 739–757.

Barree, R.D., Conway, M.W. 1994. Experimental and numerical modeling of convective proppant transport. SPE Paper 28564, 1994 SPE Annual Technical Conference and Exhibition, New Orleans, USA. Sep. 25–28.

Camanho, P.P., Davila, C.G. 2002. Mixed-mode decohesion finite elements for the simulation of delamination in composite materials. NASA/TM-2002-211737, 1–42.

Fischer, M.P., Gross, M.R., Engelder, T., Greenfield, R.J. 1994. Finite-element analysis of the stress distribution around a pressurized crack in a layered elastic medium: implications for the spacing of fluid-driven joints in bedded sedimentary rock. *Tectonophysics* 247(1): 49–64.

Geertsma, J., de Klerk, F. 1969. A rapid method of predicting width and extent of hydraulically induced fractures. *Journal Petroleum Technology* 21: 1571–1581.

Hustedt, B., Zwarts, D., Bjoerndal, H.P., Masfry, R., van den Hoek, P.J. 2006. Induced fracturing in reservoir simulations: application of a new coupled simulator to a waterflooding field example. SPE Paper 102467, 2006 SPE Annual Technical Conference and Exhibition, San Antonio, Texas, USA. Sep. 24–27.

Jeffrey, R.G., Bunger, A.P. 2007. A detailed comparison of experimental and numerical data on hydraulic fracture height growth though stress contrasts. SPE Paper 106030, 2007 SPE Hydraulic Fracturing Technology Conference, College Station, Texas, U.S.A. Jan. 29–31.

Miskimins, J.L., Barree, R.D. 2003. Modeling of hydraulic fracture height containment in laminated sand and shale sequences. SPE Paper 80935, 2003 SPE Production and Operations Symposium, Oklahoma, USA. Mar. 22–25.

Nordren, R.P. 1972. Propagation of a vertical hydraulic fracture. *SPE J* 12(8): 306–314.

Perkins, T.K., Kern, L.R. 1961. Widths of hydraulic fractures. *Journal of Petroleum Technology* 13(9): 937–949.

Setaari, A., Cleary, M.P. 1984. Three-dimensional simulation of hydraulic fracturing. *Journal of Petroleum Technology* 36(7): 1177–1190.

Siebrits, E., Gu, H.R., Desroches, J. 2001. An improved pseudo-3D hydraulic fracturing simulator for multiple layered materials. Proceeding of 10th International Conference on Computer Methods and Advances in Geomechanics. Tucson, USA. Jan. 7–12.

Siebrits, E., Peirce, A.P. 2002. An efficient multi-layer planar 3D fracture growth algorithm using a fixed mesh approach. *Internal Journal for Numerical Methods in Engineering* 53:691–717.

Smith, M.B., Bale, A.B., Britt, L.K., Klein, H.H., Dang, X. 2001. Layered modulus effects on fracture propagation, proppant placement and fracture modeling. SPE Paper 71654, 2001 SPE Annual Technical Conference and Exhibition, New Orleans, Louisiana, USA. Sep. 30–Oct. 3.

Turon, A., Camanho, P.P., Costa, J., Davila, C.G. 2006. A damage model for the simulation of delamination in advanced composites under variable-model loading. *Mechanics of Materials* 38(11): 1072–1089.

Weng, X.W. 1992. Incorporation of 2D fluid into a pseudo-3D hydraulic fracturing simulator. *SPE Production Engineering* 7(4): 331–337.

Xue, B., Wu, H.A., Wang, X.X., Lian, Z.L., Zhang, J., Zhang, S.C. 2006. A Three-Dimensional Finite Element Model of Hydraulic Progressive Damage, *Key Engineering Material* 324–325: 375–378.

Modeling and Computation in Engineering II – Xie (ed)

Nitrogen runoff loss from paddy-pond system based on water cycling

L.Q. Zhu, Z.W. Zhang & X.M. Bian
College of Agriculture, Nanjing Agriculture University, Nanjing, Jiangsu, China

L.Q. Bian
Lishui Plant Science Test Base, Jiangsu Academy of Agricultural sciences, Nanjing, Jiangsu, China

R.F. Jiao
College of Resources and Environmental Science, Nanjing Agriculture University, Nanjing, Jiangsu, China

ABSTRACT: In order to reduce the nitrogen runoff loss from paddy fields, a paddy-pond system with water cycling was constructed based on the existing drainage-irrigation system in rice cropping region in plain polder of China. According to the characteristic of the paddy-pond system, we established a paddy-pond-external environment waters nitrogen runoff loss simulation model (PPE). The model was developed to estimate nitrogen runoff loss from paddy-pond system in rice cropping region. According to the conventional planting technique of rice in Nanjing, we analyzed the characteristics of the runoff water and runoff nitrogen from pond to external environment, irrigation water and irrigation nitrogen from external environment to pond during the whole rice-growth period based on the daily rainfall data of Nanjing from 1951 to 2010. The simulation results indicated that the irrigation and drainage unit of the paddy-pond system would improve nitrogen cycling rate, reduce runoff nitrogen loss from paddy fields.

1 INTRODUCTION

Agriculture modernization and intensive farming have being resulted in serious water environmental issues in the countryside, such as water eutrophication. The main trait of water eutrophication in countryside is high nutrient concentration in water such as nitrogen and phosphorus. Nitrogen loss from cropland has been one of the most important causes of water eutrophication in dense waterway-net region in south China, and nitrogen runoff loss from paddy fields accounts for a large proportion of total nitrogen losses.

Most studies indicated that nitrogen concentration and load of surface water runoff from paddy fields were related to rainfall intensities, runoff volumes, fertilization levels, fertilizer types, irrigation technologies and so on. Informed research mostly focused on field experiments or simulation experiments at a certain time or on a certain site. Because rainfall and runoff volumes show great spatial-temporal variability, it is difficult to evaluate the general status and characteristics of runoff nitrogen in paddy fields macroscopically at field-scale experiments. It is known that mathematical model is a valuable tool for such evaluation. Several models have been developed to estimate the nitrogen runoff loss from paddy fields to external rivers, such as PRM model, NPS model, FPRNP model and PRNSM model.

Aiming at the characteristics of water and nutrient management of rice cultivation, a Precipitation-Runoff-N loss Simulation model (PRNSM) was established to estimate the nitrogen runoff loss from paddy fields. Due to lack of validation experiments, they validated the model only based on the comparison of the simulation values in Nanjing area and the measured values in Shanghai area. Nowadays, rice cropping regions in plain polder in China are well equipped with drainage-irrigation systems, these systems consist of different sized drainage-irrigation units, which are independent of external environment waters. They offer an existing framework of water system to control the pollution caused by agriculture production on external environment waters. Based on the existing drainage-irrigation system, we built a *paddy-pond system* with water cycling function, and established a paddy-pond-external environment waters nitrogen runoff loss simulation model (PPE) based on the PRNSM model.

According to the PPE model, we simulated the nitrogen runoff loss from paddy fields and the nitrogen balance between the *paddy-pond system* and external environment waters in Nanjing during different years through rainfall data (1951–2010), so as to reveal the traits and the properties of *paddy-pond system*, and provide engineering design method and theoretical basis to construct this system which could take full advantage of the existing water system resources to control nitrogen runoff loss to external environment waters.

2 METHODS

2.1 *Construction of paddy-pond system based on water cycling*

The structure of the *paddy-pond system* based on water cycling is shown in Figure 1. One subunit of the system is paddy fields, the other is pond. Water for irrigation is supplied into paddy fields from pond by pumps, runoff water from paddy fields flow into pond through ditches. If the pond water were not enough for irrigation, we would draw external environment waters into pond by opening the valve or by pumps. If the pond water exceeded the highest level, the water would overflow to external environment waters from the pond.

2.2 *Conceptualization of the PPE model*

The PPE model was developed, which integrates rainfall, evaporation, surface water management, dynamic changes of nitrogen in surface water of paddy fields, runoff water and runoff nitrogen from paddy fields, irrigation water from pond to paddy fields, irrigation water from external environment waters to pond, and runoff water and runoff nitrogen loss from pond to external environment waters as a systematic procedure of paddy-pond system. Assuming that all variables of the model are the functions of time in days, which means that the system status of the previous day and the factors of rainfall, evaporation and water management of the present day affects runoff volume and nitrogen concentration of surface water in

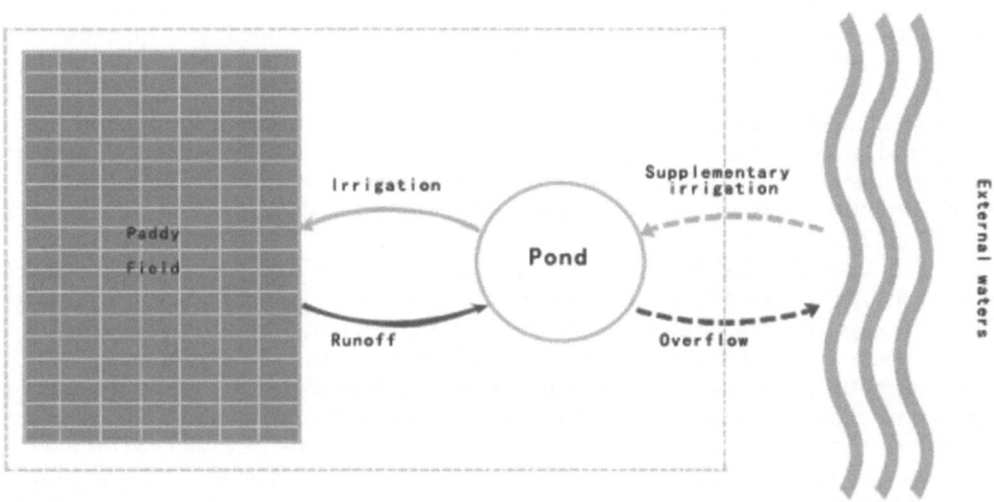

Figure 1. Caption of a typical figure. Photographs will be scanned by the printer. Always supply original photographs.

paddy fields of the present day; runoff volume and nitrogen concentration of surface water in paddy fields influence runoff nitrogen load from paddy fields to pond; runoff volume of surface water in paddy fields, rainfall, irrigation water from external environment waters to pond and from pond to paddy fields affect water storage in pond, nitrogen concentration of water in pond, runoff volume and runoff nitrogen loss from pond to external environment waters. The summation of runoff nitrogen losses from *paddy-pond system* per day during the whole rice-growth period is the total load of nitrogen runoff from the system during the rice cropping season (Fig. 2).

2.3 *State variables and parameters of the PPE model*

2.3.1 *State variables of paddy fields per day in rice growth period*

$H_T^{(t)}$: total surface water in paddy fields on the *t*th day (mm); $H_{RF}^{(t)}$: depth of water runoff from paddy fields to pond on the *t*th day (mm); $H_I^{(t)}$: depth of irrigation water from pond to paddy fields on the *t*th day (mm); $H_R^{(t)}$: depth of surface water retained in paddy fields on the *t*th day (mm); $N_I^{(t)}$: nitrogen load in irrigation water from pond to paddy fields on the *t*th day (kg hm^{-2}); $C^{(t)}$: nitrogen concentration of surface water in paddy fields on the *t*th day (mg L^{-1}); $N^{(t)}$: nitrogen load of surface water in paddy fields on the *t*th day (kg hm^{-2}); $N_{RF}^{(t)}$: nitrogen load runoff from paddy fields to pond on the *t*th day (kg hm^{-2}).

2.3.2 *State variables of pond per day in rice growth period*

$PH_T^{(t)}$: total water in pond on the *t*th day (mm); $PH_{RF}^{(t)}$: depth of runoff water from pond to external environment on the *t*th day (mm); $PH_I^{(t)}$: depth of irrigation water from external environment to pond on the *t*th day (mm); $PH_R^{(t)}$: depth of water retained in pond on the *t*th day (mm); $PN_I^{(t)}$: nitrogen load in irrigation water from external environment to pond on the *t*th day (kg hm^{-2}); $PC^{(t)}$: nitrogen concentration of water in pond on the *t*th day (mg L^{-1}); $PN^{(t)}$: nitrogen load of water in pond on the *t*th day (kg hm^{-2}); $PN_{RF}^{(t)}$: runoff nitrogen loss from pond to external environment on the *t*th day (kg hm^{-2}).

2.3.3 *Meteorological and technical parameters of paddy-pond system*

$H_P^{(t)}$: rainfall on the *t*th day (offered by China Meteorological Administration or obtained from the rainfall gauge, mm); $H_E^{(t)}$: depth of surface water evaporated from paddy fields on the *t*th day (based on irrigation volume in Southern Jiangsu, set as 7.5 mm in sunny day, and 2.5 mm in rainy day); $H_{\max}^{(t)}$: Max proper depth of surface water in paddy fields (according to experiences of production practice, set as 80 mm); $H_{\min}^{(t)}$: Min proper depth of surface water in paddy fields (according to experiences of production practice, set as 10 mm); $PH_{\text{full}}^{(t)}$: Maximum depth of water in pond (according to actual situation in Southern Jiangsu, set as 2.5 m); $PH_{\max}^{(t)}$: Maximum proper depth of water in pond (according to actual situation in Southern Jiangsu, set as 1.5 m); $H_{\min}^{(t)}$: Min proper depth of water in pond (according to actual situation in Southern Jiangsu, set as 1.0 m); *S*: paddy fields area (obtained according to actual situation, m^2); *PS*: pond area (obtained according to actual situation, m^2); $PC_I^{(t)}$: nitrogen concentration of irrigation water from external environment to pond (set as 2 mg L^{-1} according to our experiment result); $C_D^{(t)}$: natural decay rate of nitrogen concentration of surface water in paddy fields, that is the ratio of nitrogen concentration after 24 hours to the original concentration, in condition of surface water retains constant (set as $e^{-0.335}$according to our experiment result); $PC_D^{(t)}$: natural decay rate of nitrogen concentration of surface water in pond (same as $C_D^{(t)}$).

2.4 *Progressive functions of state variables of the PPE model*

2.4.1 *Progressive functions of state variables in paddy fields of the model*

1. Total surface water in paddy fields

$$H_T^{(t)} = H_R^{(t-1)} + H_P^{(t)} - H_E^{(t)} \tag{1}$$

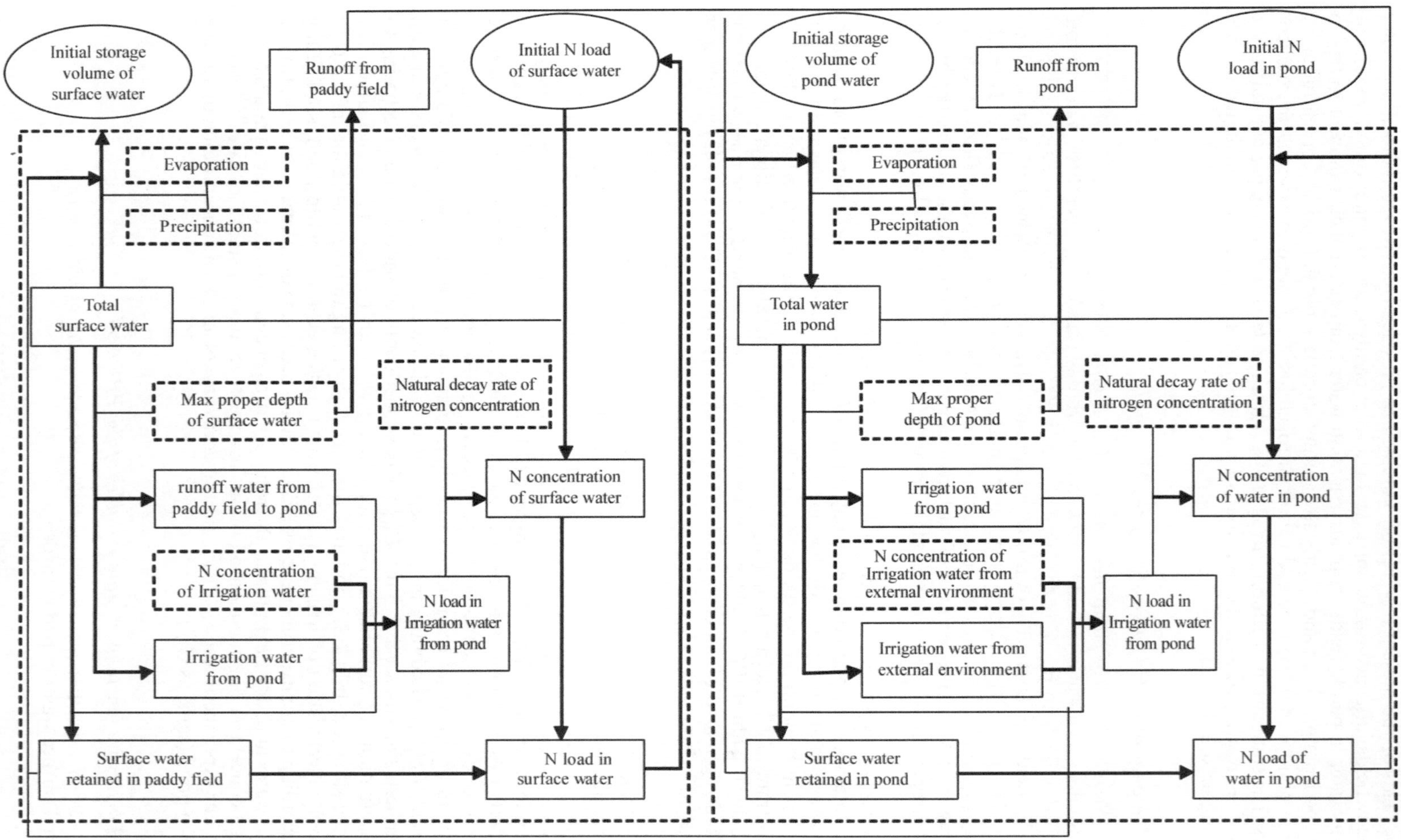

Figure 2. Concept map of *PPE* model.

2. Depth of daily runoff water from paddy fields to pond

$$H_{RF}^{(t)} = H_T^{(t)} - H_{\max}^{(t)} \qquad |_{H_T^{(t)} \geq H_{\max}^{(t)}} \tag{2}$$

3. Depth of daily irrigation water from pond to paddy fields

$$H_I^{(t)} = H_{\max}^{(t)} - H_T^{(t)} \qquad |_{H_T^{(t)} < H_{\min}} \tag{3}$$

4. Depth of surface water retained in paddy fields

$$H_R^{(t)} = H_T^{(t)} - H_{RF}^{(t)} + H_I^{(t)} \tag{4}$$

5. Nitrogen load in daily irrigation water from pond to paddy fields

$$N_I^{(t)} = PC^{(t)} \times H_I^{(t)} \times S \times 10^{-6} \qquad H_I^{(t)}>0 \tag{5}$$

6. Nitrogen concentration of surface water in paddy fields

$$C^{(t)} = \frac{\left(N^{(t-1)} + N_I^{(t)}\right)}{H_T^{(t)} \times S} \times C_D^{(t)} \times 10^6 \tag{6}$$

7. Nitrogen load of surface water in paddy fields

$$N^{(t)} = C^{(t)} \times H_R^{(t)} \times S \times 10^{-6} \qquad N^{(0)} = \textit{nitrogen load from fertilizer application} \tag{7}$$

8. Daily nitrogen load runoff from paddy fields to pond

$$N_{RF}^{(t)} = N^{(t-1)} \times \left(H_{RF}^{(t)}/H_T^{(t)} \right) \tag{8}$$

2.4.2 *Progressive functions of state variables in pond of the model*

1. Total water in pond

$$PH_T^{(t)} = PH_R^{(t-1)} + PH_P^{(t)} - PH_E^{(t)} \tag{9}$$

2. Depth of daily runoff water from pond to external environment

$$PH_{RF}^{(t)} = PH_T^{(t)} - PH_{\text{full}}^{(t)} \qquad |_{PH_T^{(t)} \geq PH_{\text{full}}^{(t)}} \tag{10}$$

3. Depth of daily irrigation water from external environment to pond

$$PH_I^{(t)} = PH_{\max}^{(t)} - PH_T^{(t)} \qquad |_{PH_T^{(t)} < PH_{\min}} \tag{11}$$

4. Depth of surface water retained in pond model

$$PH_R^{(t)} = PH_T^{(t)} - PH_{RF}^{(t)} + PH_I^{(t)} \tag{12}$$

5. Nitrogen load in daily irrigation water from external environment to pond

$$PN_I^{(t)} = PC_I^{(t)} \times PH_I^{(t)} \times PS \times 10^{-6} \qquad |_{PH_I^{(t)} > 0} \tag{13}$$

6. Nitrogen concentration of water in pond

$$PC^{(t)} = \frac{\left(PN^{(t-1)} + PN_I^{(t)}\right)}{PH_T^{(t)} \times PS} \times PC_D^{(t)} \times 10^6 \tag{14}$$

7. Nitrogen load of water in pond

$$PN^{(t)} = PC^{(t)} \times PH_R^{(t)} \times PS \times 10^{-6} \qquad PN^{(0)}\text{=nitrogen load of pond on the initial day} \tag{15}$$

8. Daily runoff nitrogen loss from pond to external environment

$$PN_{RF}^{(t)} = PN^{(t-1)} \times (PH_{RF}^{(t)}/PH_T^{(t)}) \tag{16}$$

3 MODEL VALIDATION

We carried out two validation experiments in different sites. One was done in Lishui Plant Science Test Base of Jiangsu Academy of Agricultural Sciences from early June to late October in 2009, the other was done in Maoshan, Jiangsu from mid-June to early November in 2010.

In Lishui experiment, the rice transplanting time was June 7th, the complete rice growth period was 140 days, nitrogen quantity (measured by pure N) in base, tilling and panicle fertilizers were 120 kg hm^{-2}, 60 kg hm^{-2} and 60 kg hm^{-2}, respectively. Base, tilling and panicle fertilizer were applied on the 1st, 8th and 38th day of transplanting, the period of sunning the field was from the 28th to 34th day after transplanting, and irrigation was stopped 7th days before harvest.

In Maoshan experiment, the rice transplanting time was June 19th, the complete rice growth period was 143 days, nitrogen quantity (measured by pure N) in base, tilling and panicle fertilizers were 150 kg hm^{-2},75 kg hm^{-2} and 75 kg hm^{-2}, respectively. Base, tilling and panicle fertilizer were applied on the 1st, 11th and 45th day of transplanting, the period of sunning the field was from the 30th to 36th day after transplanting, and irrigation was stopped 8th days before harvest. The rainfall data were obtained from the rainfall gauge installed beside the experimental field in the two experiments.

Table 1. Simulation and measured values of nitrogen runoff loss from paddy fields.

Site	Date	Rainfall (mm)	Simulation values (kg hm^{-2})	Measured values (kg hm^{-2})
Lishui	June 20, 2009–June 21, 2009	17.8	0.02	0.03
	June 28, 2009–June 30, 2009	117.9	3.85	4.13
	July 6, 2009–July 7, 2009	106.8	1.61	1.97
	July 21, 2009–July 23, 2009	143.2	2.98	2.55
	July 27, 2009	71.1	1.17	1.08
	July 29, 2009–August 1, 2009	84.9	1.15	0.96
	August 9, 2009–August 12, 2009	75.3	0.87	0.97
Maoshan	June 22, 2010–June 23, 2010	82.7	4.67	4.21
	July 12, 2010–July 14, 2010	97.6	1.63	1.58
	August 23, 2009–August 24, 2010	41.0	0.86	0.81
	August 27, 2010–August 28, 2010	26.9	0.27	0.25
	September 6, 2010	14.7	0.12	0.13
	Sept 13, 2010–Sept 14, 2010	57.5	0.27	0.24
	Sept 23, 2010–Sept 24, 2010	40.3	0.10	0.12
	October 2, 2010	18.8	0.03	0.02

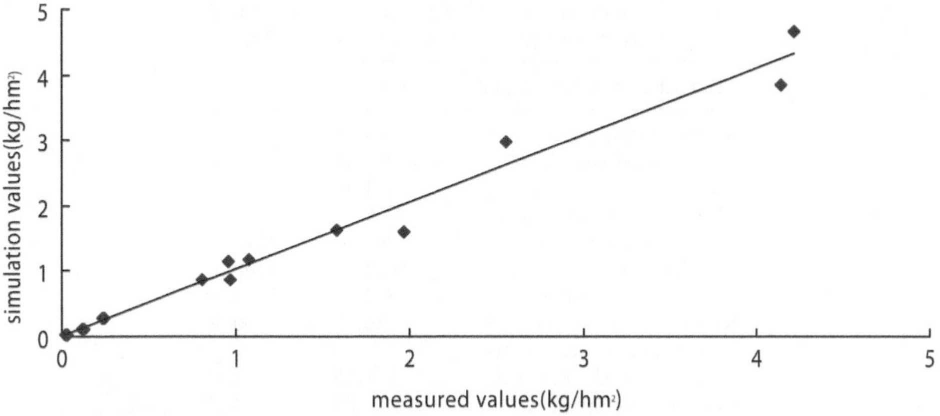

Figure 3. Validation between simulation values and measured values.

We obtained the simulation values of runoff nitrogen load from paddy fields while inputting the parameters and rainfall data from the validation experiments to the PPE model, and got the measured values of nitrogen load runoff from paddy fields through the validation experiments (Table 1).

It could induce a linear regression between the simulated values and measured values (Fig. 3), the regression equation is $y = 0.0004 + 1.0285x$. The results of t-test on intercept and slope of the regression equation showed no significant difference, which indicated that the PPE model is credible.

4 RESULTS

4.1 *Parameters setup and data acquisition*

According to the rice production practice in Nanjing, we set the rice transplanting time as June 15th, the complete rice growth period as 135 days, nitrogen quantity (measured by pure N) in base, tilling and panicle fertilizers as 135 kg $hm^{-2,}$ 27 kg hm^{-2} and 108 kg hm^{-2} respectively, phosphorus quantity (measured by P_2O_5) in base fertilizer as 68 kg hm^{-2}. Base, tilling and panicle fertilizer were applied on the 1st, 7th and 35th day of transplanting. The period of sunning the field was from the 25th to 31th days after transplanting, and irrigation was stopped 7th days before harvest.

The rainfall data were obtained from daily meteorological observation data of Nanjing Observing Station from 1951 to 2010 offered by China Meteorological Administration.

4.2 *Runoff water and runoff nitrogen from paddy-pond system*

According to the actual area ratio of controlled ponds to paddy fields in Southern Jiangsu with 10%, we set the area ratio of paddy to pond as 10:1 and the water storage ratio of paddy to pond as 800:1500. The simulation results indicated that it could be decreased by 30.04% of average annual irrigation water from external environment to paddy fields in the 60 years with controlled water cycling pond, by 64.65% of the runoff water discharging into external environment and by 93.40% of the nitrogen runoff losses, compared to the paddy fields without controlled water cycling ponds.

Nitrogen balance between production system and external water environment turned from discharging nitrogen outwards (polluting environment) to absorbing external nitrogen (purifying environment). There were more significant improvements in minimum runoff volume, minimum irrigation, maximum nitrogen loss, extreme environment pollution and so on during the 60 years (Table 2).

Table 2. Simulation results of runoff water and runoff nitrogen losses from paddy-pond system (1951–2010).

Area ratio of paddy to pond	10:1	10:0
Mean rainfall (mm)	588.35	598.35
Mean runoff water (m^3)	1201.54	3398.65
Max runoff water (m^3)	6031.36	7605.21
Min runoff water (m^3)	0	508.75
Mean irrigation water (m^3)	3296.94	4712.58
Max irrigation water (m^3)	6414.85	7160.27
Min irrigation water (m^3)	812.54	2760.02
Rate of runoff to external environment waters (%)	80.13	100.00
Mean N loss (kg hm^{-2})	1.25	18.94
Max N loss (kg hm^{-2})	11.84	74.72
Min N loss (kg hm^{-2})	0	0.92
Mean N balance (kg hm^{-2})	5.39	−9.41
Max N balance (kg hm^{-2})	12.83	11.06
Min N balance (kg hm^{-2})	−5.93	−64.73

Table 3. Simulation results of runoff water and runoff nitrogen losses from *paddy-pond system* under different ratio of paddy to pond (1951–2010).

Area ratio of paddy to pond	10:0	20:1	10:1	6.5:1	2:1	1.1:1
Rate of runoff to external environment waters (%)	100.00	93.31	80.13	50.11	30.03	0
Mean runoff water (m^3)	3398.65	1845.31	1201.54	799.43	143.32	0
Max runoff water (m^3)	7605.21	6722.73	6031.36	5001.92	2183.03	0
Min runoff water (m^3)	508.75	0	0	0	0	0
Mean N loss (kg hm^{-2})	18.94	3.28	1.25	0.64	0.08	0
Max N loss (kg hm^{-2})	74.72	25.82	11.84	8.61	2.67	0
Min N loss (kg hm^{-2})	0.92	0	0	0	0	0
Mean irrigation water (m^3)	4712.58	3769.51	3296.94	2987.43	2598.51	2209.43
Max irrigation water (m^3)	7160.27	6480.23	6414.85	6457.94	6980.83	6220.31
Min irrigation water (m^3)	2760.02	1739.84	812.54	0	0	0
Mean N balance (kg hm^{-2})	−9.41	4.26	5.39	5.51	5.65	6.51
Max N balance (kg hm^{-2})	11.06	12.49	12.83	12.92	13.96	22.22
Min N balance (kg hm^{-2})	−64.73	−18.43	−5.93	−3.36	−2.67	0

4.3 *Effects of different area ratio of paddy to pond on runoff water and runoff nitrogen loss*

Simulation results based on the 60-year rainfall data indicated that the proportion of years with runoff occurring in rice growth period could be risen from 80.13% to 93.31% if the area ratio of paddy to pond was adjusted from 10:1 to 20:1, the proportion could be decreased to 50.11% if adjusted to 6.5:1, the proportion could be decreased to 30.03% if adjusted to 2:1, the runoff occurring in rice growth period could be eliminated completely if adjusted to 1.1:1 (Table 3).

In order to reduce 5/6 of the nitrogen load released into external environment waters which runoff from *paddy-pond system*, we could design a paddy-pond unit offering recycling water with an area ratio of 20:1. As a result, reutilization rate (nitrogen recycled by irrigation/total nitrogen fertilization) could reach 5.84%, and could achieve 7.44% if its environment purification function is considered.

In order to reduce to less than 10% of the nitrogen load released into external environment waters through *paddy-pond system*, we could design a paddy-pond unit offering recycling water with an area ratio of 10:1. As a result, recovery utilization rate (nitrogen recycled by irrigation/total fertilizer nitrogen) could reach 6.53%, and could arrive at 8.63% when its environment purification function was considered.

5 DISCUSSION

Nitrogen runoff loss from paddy fields is one of the most important causes of water eutrophication in south China. How to control nitrogen runoff from cropland has become an urgent problem. The control technology which has been formed currently concentrated in situ emission reduction technology, artificial wetland technology, buffer zone technology and artificial reservoirs. Ding et al. indicated that after spring tillage, in comparison with the conventional water management whose height of overflow weir was 3 cm, the appropriate height of overflow weir should be 6 cm, it could reduce the emission of nitrogen by 35.76%~72.13%. Liu showed that artificial wetland technology could remove 60% of nitrogen from runoff water. Zhang et al. claimed that there would be more than 70% of nitrogen removal from pollution water if strengthen purification pretank programs. In the effect of the riparian buffer zone on the nitrogen removal from agricultural runoff water, Lowance et al. found that the remaining amount of nitrogen is only the original value of 1/7 after runoff water flow through a riparian buffer zone. Aiming at the severity of farmland non-point pollution in Taihu Lake basin, Yang et al. set forward a new ecological engineering solution—ecological ditch system with interception function, and the removal rate of runoff nitrogen by the ditch system was 48.36%. The *paddy-pond system* based on water cycling constructed in this paper under the area ratio of paddy to pond of 10:1 could reduce the nitrogen runoff loss from paddy fields by 93.40%, which is the highest among all the technologies.

The PPE model is strict in logic, its structure is simple and its parameters are easy to obtain. It also well reflects the characteristics of runoff and runoff nitrogen of *paddy-pond system*. Nitrogen runoff not only followed with the loss of runoff water, but also with the loss of soil erosion. Without considering the nitrogen loss followed with the soil erosion is the shortcomings of the model. The PPE model should be improved by adding parameters of soil erosion, so as to provide instructions for improving the available *paddy-pond system*.

6 CONCLUSIONS

The *paddy-pond system* based on water cycling constructed in this paper could not only reduce nitrogen runoff loss, but also purify the environment by absorbing nitrogen from outward to the paddy-pond unit. If we set the area ratio of paddy to pond as 10:1 in this system, it could be decreased by 30.04% of average annual irrigation water from external environment to paddy fields in the 60 years, by 64.65% of the runoff water discharging into external environment and by 93.40% of the nitrogen runoff losses, compared to the paddy fields without controlled water cycling ponds. Recovery utilization rate (nitrogen recycled by irrigation/total fertilizer nitrogen) could reach 6.53%, and could arrive at 8.63% when its environment purifying function is considered.

REFERENCES

Bian L.P, Jiao J, Li H, Zhang W.J. 2008. Modeling of nitrogen loss with runoff from paddy field during rice growth period in Nanjing area. Ecol. Rural Environ 24(3): 89–93.

Carpenter S.R, Caraco N.F, Correll D.L, Howarth R.W, Sharpley A.N, Smith V.H. 1998. Nonpoint pollution of surface with phosphorus and nitrogen. Ecol. Appl 8(3): 559–568.

Ding M, Yang R.B, Feng G.L, Cai K.B. 2010. Research on loss characteristics of nitrogen and phosphorus in surface water of paddy field under controlled drainage condition and its performance of emission reduction. Acta Ag. Jiangxi 22(5): 122–124.

Donigian A.S, Carsel R.F. 1987. Modeling the impact of conservation tillage practices on pesticide concentrations in ground and surface waters. Environ. Toxicol. Chem 6: 241.

Evan R, Konrad M. 2000. Nitrogen losses from entire plants of spring wheat (Triticum aestivum) from tillering to maturation. Eur. J. Agron 13: 101–110.

Haith D.A. 1980. A mathematical model for estimating losses in runoff. Environ. Qual 9(3): 428–433.

Han J.G, Li Z.B, Li P, Tian J.L. 2001. Nitrogen and phosphorous concentrations in runoff from a purple soil in an agricultural watershed. Agr. Water Manage 97: 757–762.

He B.G, Zhou N.S, Hu X.F, Gao X.J, Wang S.P. 2001. On farmland precipitation-runoff nonpoint pollution model-farmland nitrogen pollution model in Shanghai. Res. Environ. Yangtze Basin 10(2): 159–165.

Huang Y.F, Zhang L.P, Hong H.S, Chen Q.W. 2006. Characteristics and influencing factors of nitrogen loss of subwatershed. Hydraul. Eng 37(7): 801–806.

Huett D.O, Morris S.G., Smith G, Hunt N. 2005. Nitrogen and phosphorus removal from plant nursery runoff in vegetated and unvegetated subsurface flow wetlands. Water Res 39: 3259–3272.

Ikuo Y, Asa M, Tadayoshi H, Koji H, Eisaku S. 2007. Runoff nitrogen from a large sized paddy field during a crop period. Ag. Water manag 87: 217–222.

Ji X.H, Zheng S.X, Lu Y.H, Liao Y.L. 2007. Study of dynamics of floodwater nitrogen and regulation of its runoff loss in paddy field-based two-cropping rice with urea and controlled release nitrogen fertilizer application. Agr. Sci.China 6(2):189–199.

Kothyari B.P, Verma P.K, Joshi B.K, Kothyarim U.C. 2004. Rainfall-run off-soil and nutrient loss relationships for plot size areas of bhetagad watershed in Central Himalaya India. Hydrol 293(1–4): 137–150.

Lang X.Q, Tian G.M, Li H, Chen Y.X, Zhu S. 2005. Study on characteristic of nitrogen and phosphorus loss from rice field by natural rainfall runoff. Soil Water Conserv 19(1): 59–63.

Liu P.B, Zhang Y.F. 1999. Research of field test and mathematical simulation on nitrogen losses in paddy rice fields. Agro-environ. Prot 18(6): 241–245.

Liu W.X. 1997. Study on the application of artificial wet-land in agriculture non-point source pollution. Res.Environ. Sci 10(4): 15–19.

Lowrance R, Mclntyre S, Lance C. 1988. Erosion and deposition in a field/forest system estimated using cesium-137 activity. Soil Water Conserv 43: 195–199.

Moldan F, Kjønaas O.J, Stuanes A.O, Wright R.F. 2006. Increased nitrogen in runoff and soil following 13 years of experimentally increased nitrogen deposition to a coniferous-forested catchment at Gårdsjön, Sweden. Environ. Poll 144: 610–620.

Ng Kee, Kwong K.F, Ng K, Bholah A, Volcy L, Pynee K. 2002. Nitrogen and phosphorus transport by surface runoff from a silty clay loam soil under sugarcane in the humid tropical environment of mauritius. Agr. Ecosyst. Environ 91: 147–157.

Ojeda G, Tarrasón D, Ortiz O, Alcañiz J.M. 2006. Nitrogen losses in runoff waters from a loamy soil treated with sewage sludge. Agr. Ecosyst. Environ. 117: 49–56.

Qiu W.G, Tang H, Wang C. Rule of lose of nitrogen in the surface water of rice fields and the control technology. 2000. Agro-environ. Sci 23(4): 740–744.

Ranjith P.U, Peter P.M, Harold E.G, John K.J. 2006. Nitrogen losses in runoff from three adjacent agricultural watersheds with claypan soils. Agr. Ecosyst. Environ 117: 39–48.

Schilling K.E, Libra R.D. 2000. The relationship of nitrate concentration in streams to row crop land use in Iowa. Environ. Qual 29: 1846–1851.

Si Y.B, Wang S.Q, Chen H.M. 2000. The lose of N and phosphorus in the farmland and eutrophication. Soil 32(4):188–193.

Smith D.R, Owens P.R, Leytem A.B, Warnemuende E.A. 2007. Nutrient losses from manure and fertilizer applications as impacted by time to first run off event. Environ. Poll 147(1): 131–137.

Tian Y.H, Yin B, Yang L.Z, Yin S.X, Zhu Z.L. 2007. Nitrogen runoff and leaching losses during rice-wheat rotations in taihu lake region, China. Pedosphere 17(4): 445–456.

Timmons D.R, Holtr F, Latterell J. 1970. Leaching of crop residues as a source of nutrients in surface runoff water. Water Resour. Res 6(5):1367–1375.

Ventura M, Scandellari F, Ventura F, Guzzon B, Rossi Pisab P, Tagliavini M. 2008. Nitrogen balance and losses through drainage waters in an agricultural watershed of the Po Valley (Italy). Eur. J. Agron 29:108–115.

Wang Q, Yang J.P, Shen J.G, Zheng H.F, Yu Y.Y. 2003. Study on dynamic change of three kinds of nitrogen in surface water of paddy field. Soil Water Conserv 17(3): 51–54.

Yang H.W, Zhang J.Q, Tang J.L, Xu C.J, Ding D.R. 2008. Soil, water and nutrient loss under different cropping systems in purple-soil slope-lands. Chinese. J. Eco-Agr 16(3): 615–619.

Yang L.Z, Zhou X.P, Wang J.G, Wang D.J, Shi W.M, Shi N.X. 2005. Ecological ditch system with interception function and its effects on controlling farmland non-point pollution. Chinese J. Ecol 24(11): 1371–1374.
Zhang M.K, Wang L.P, He Z.L. 2007. Spatial and temporal variation of nitrogen exported by runoff from sandy agricultural soils. Environ. Sci 19: 1086–1092.
Zhang Y.M, Zhang Y.C, Zuo Y.H. 2003. Discussion on application of pre-dam in the nonpoint pollution control of Lake Tai basin. Envrion. Poll. Control 25(6): 341–344.

Modeling and Computation in Engineering II – Xie (ed)
© 2013 Taylor & Francis Group, London, ISBN 978-1-138-00058-2

Author index